Dynamical Aspects of Nuclear Fission

Dynamical Aspects of Nuclear Fission

Proceedings of the 6th International Conference
Smolenice Castle, Slovak Republic 2 – 6 October 2006

editors

J. Kliman
Joint Institute for Nuclear Research, Russia &
Slovak Academy of Sciences, Slovakia

M. G. Itkis
Joint Institute for Nuclear Research, Russia

Š. Gmuca
Slovak Academy of Sciences, Slovakia

World Scientific

NEW JERSEY • LONDON • SINGAPORE • BEIJING • SHANGHAI • HONG KONG • TAIPEI • CHENNAI

Published by

World Scientific Publishing Co. Pte. Ltd.
5 Toh Tuck Link, Singapore 596224
USA office: 27 Warren Street, Suite 401-402, Hackensack, NJ 07601
UK office: 57 Shelton Street, Covent Garden, London WC2H 9HE

British Library Cataloguing-in-Publication Data
A catalogue record for this book is available from the British Library.

DYNAMICAL ASPECTS OF NUCLEAR FISSION
Proceedings of the 6th International Conference

ISBN-13 978-981-283-752-3
ISBN-10 981-283-752-3

Printed in Singapore by World Scientific Printers

Organized by:

Institute of Physics, Slovak Academy of Sciences, Bratislava
Flerov Laboratory of Nuclear Reactions, JINR, Dubna

International advisory committee:

N. Carjean (Bordeaux)
H. Faust (Grenoble)
W. Greiner (Frankfurt)
M.G. Itkis (Dubna)
M. Mutterer (Darmstadt)

Organizing committee:

Š. Gmuca (Bratislava)
M.G. Itkis (Dubna)
J. Kliman (Bratislava)

Local organizing committee:

M. Berešová
Š. Gmuca
J. Kliman
L. Krupa
M. Kubica
V. Matoušek

PREFACE

The 6th International Conference on Dynamical Aspects of Nuclear Fission was held from 2 to 6 October, 2006 at Smolenice Castle, Slovakia. It was organized by Institute of Physics, Slovak Academy of Science (Slovakia) and Flerov Laboratory of Nuclear Reactions, Joint Institute for Nuclear Research (Russia).

The scientific programme of the Conference covered a wide range of problems in the field of nuclear fission dynamics. The main discussed topics were: dynamics of fission, fusion-fission, superheavy elements, nuclear fragmentation, exotic modes of fission, structure of fission fragments and neutron rich nuclei and development in experimental techniques.

Around 40 scientists from 13 countries took part in the conference. We are especially pleased that the conference attracted young participants and speakers.

We wish to thank all the participants for their contributions to the conference and for lively and fruitful discussions. We would like to express our sincere gratitude to the International Advisory Committee for their excellent recommendations on speakers, invaluable remarks and suggestions. We are also grateful to the members of the Organizing Committee and to everyone who contributed to organizing the Conference.

Editors

CONTENTS

SUPERHEAVY ELEMENTS

FRAGMENTATION

EXOTIC MODES

STRUCTURE OF FISSION FRAGMENTS AND NEURTON RICH NUCLEI

DEVELOPMENTS IN EXPERIMENTAL TECHNIQUES

DEPENDENCE OF SCISSION-NEUTRON YIELD ON LIGHT-FRAGMENT MASS FOR $\Omega = 1/2$

N. CARJAN

Centre d'Etudes Nucleaires de Bordeaux - Gradignan,
UMR 5797, CNRS/IN2P3 - Universite Bordeaux 1,
BP 120, 33175 Gradignan Cedex, France
E-mail: carjan@cenbg.in2p3.fr

M. RIZEA

National Institute of Physics and Nuclear Engineering,
"Horia Hulubei",
PO Box MG-6, Bucharest, Romania
E-mail: rizea@theory.nipne.ro

The dependence of the scission-neutron multiplicity on the mass ratio of the fragments in asymmetric fission of ^{236}U was investigated in the frame of the sudden approximation. Only emission from neutron states characterized by the projection of the total angular momentum on the symmetry axis $\Omega = 1/2$ was considered. This dependence was found to be different and less pronounced than for prompt neutrons.

Keywords: Scission-neutrons; sudden-approximation; asymmetric fission; bound and unbound states; neutron multiplicity; fragment mass.

1. Introduction

In a previous study[1] the multiplicity of scission-neutrons emitted during the low-energy symmetric fission of ^{236}U was estimated. For this it was assumed a sudden transition between two nucleon configurations: one just before scission (two fragments connected by a neck characterized by r_{min}) and one immediately after scission (two newly separated fragments characterized by the distance between the inner tips d_{min}). Each initially occupied single-neutron state is thus transformed into a wave packet that is a linear combination of single-neutron states in the final potential well. Some of these states are unbound and the probability to populate such states gives the emission probability.

Since fission into equal fragments represents only 0.01% of the total yield in the thermal-neutron induced fission of ^{236}U, this previous calculation concerns a rare process. Moreover the most striking aspect of neutron emission during fission is the variation of the average neutron multiplicity with the fragment mass. It is therefore necessary to extend our approach to asymmetric fission.

Numerical calculations for each fragment mass ratio and all bound states requires however a considerable amount of CPU time. As a first step towards our goal we report here the results obtained only for a subset of neutron states, defined by a given value of the projection of the total angular momentum along the symmetry axis, namely for $\Omega = 1/2$. It was shown[1] that, during symmetric fission, more than 55% of the scission neutrons are emitted from 1/2 states. Since this precentage is expected to approximately hold for any mass asymmetry, the present results will give a good idea of the variation of the total number of scission neutrons with the fission-fragments mass ratio.

2. Sudden-approximation formula for the multiplicity of scission neutrons

The probability for a neutron, that just-before-scission had occupied a given state $|\Psi^i>$, to be emitted is

$$P^i_{em} = \sum_f |a_{if}|^2 \tag{1}$$

where $a_{if} =<\Psi^f|\Psi^i>$ and $|\Psi^f>$ are the eigenstates in the continuum of the immediately-after-scission single-particle hamiltonian.

To gain precision we replace Eq. (1) by

$$P^i_{em} = 1 - \sum_f |a_{if}|^2 \tag{2}$$

where the sum is now over all final bound states.

Summing these partial emission probabilities P^i_{em} for all initially occupied states one obtains the total number of scission neutrons per fission event:

$$\nu_{sc} = \sum_i v_i^2 P^i_{em} \tag{3}$$

where v_i^2 is the ground state occupation probability of $|\Psi^i>$. For independent neutrons it is a step function: it is 1 for all states below the Fermi level and 0 above.

3. The eigenvalue problem of the single-particle hamiltonian for arbitrary-shape nuclei solved on a grid of cylindrical coordinates

In the previous section we have seen that the main ingredients in our formalism are the single-particle wave functions $|\Psi^i(\epsilon_i)>$ and $|\Psi^f(\epsilon_f)>$ with negative energies $e^i(\epsilon_i)$ and $e^f(\epsilon_f)$ corresponding to the two nuclear configurations, ϵ_i and ϵ_f, between which the sudden transition is supposed to occur.

To describe the nuclear shapes just-before and immediately-after scission we have used as zeroth-order approximations Cassini ovals[2] with only one deformation parameter: $\epsilon_i = 0.985$ (i.e., $r_{min} = 1.5$ fm) and $\epsilon_f = 1.001$ (i.e. $d_{min} = 0.3$ fm) respectively. Note that $\epsilon = 1.0$ describes a zero neck scission shape. It is known that these ovals are very close to the conditional equilibrium shapes, obtained by minimization of the deformation energy at fixed value of the distance between the centers of mass of the future fragments.[3,4] To include asymmetric fission it is necessary to introduce a deviation from these ovals defined by a second parameter α_1 - see.[5]

We have recently developed a new numerical method to find the eigenstates (wave functions and energies) of the single-particle hamiltonian for an axially symmetric (otherwise arbitrary shape) nucleus. Rather than diagonalizing in a deformed oscillator basis (as in Nilsson model or in the deformed Woods-Saxon generalizations that followed) we solved the two-dimensional stationary Schrödinger equation on a grid in cylindrical coordinates (ρ, z). The numerical method consists in calculating the eigensolutions of the matrix resulting from the discretization by central finite differences of our two-dimensional hamiltonian with zero Dirichlet boundary conditions.

The wave functions have two components, corresponding to spin "up" and spin "down" as follows

$$\Psi = f(\rho, z)e^{i\Lambda_1\phi}|\uparrow> + g(\rho, z)e^{i\Lambda_2\phi}|\downarrow> . \tag{4}$$

The values $\Lambda_1,\ \Lambda_2$ are defined by:

$$\Lambda_1 = \Omega - \frac{1}{2},\quad \Lambda_2 = \Omega + \frac{1}{2}.$$

Ω is the projection of the total angular momentum along the symmetry axis and it is a good quantum number.

Taking into account the spin-orbit coupling, the hamiltonian has also two components: H_1 and H_2 (see[6]). Considering in addition the axial symmetry, we have

$$H_1\Psi = O_1 f - 2K(S_a g + S_c f), \tag{5}$$

$$H_2\Psi = O_2 g - 2K(S_b f + S_d g) \tag{6}$$

where K is a constant and

$$O_1 = -\frac{\hbar^2}{2\mu}(\Delta - \frac{\Lambda_1^2}{\rho^2}) + V, \quad O_2 = -\frac{\hbar^2}{2\mu}(\Delta - \frac{\Lambda_2^2}{\rho^2}) + V$$

with

$$\Delta = \frac{1}{\rho}\frac{\partial}{\partial\rho} + \frac{\partial^2}{\partial\rho^2} + \frac{\partial^2}{\partial z^2}$$

$$S_a = \frac{\partial V}{\partial\rho}\frac{\partial}{\partial z} - \frac{\partial V}{\partial z}\left(\frac{\partial}{\partial\rho} + \frac{\Lambda_2}{\rho}\right), \quad S_b = -\frac{\partial V}{\partial\rho}\frac{\partial}{\partial z} + \frac{\partial V}{\partial z}\left(\frac{\partial}{\partial\rho} - \frac{\Lambda_1}{\rho}\right),$$

$$S_c = \frac{\partial V}{\partial\rho}\frac{\Lambda_1}{\rho}, \quad S_d = -\frac{\partial V}{\partial\rho}\frac{\Lambda_2}{\rho}.$$

The approximation by finite differences leads to

$$H_1\psi_{i,j} = -\frac{\hbar^2}{2\mu}(\frac{1}{\rho_i}\frac{f_{i+1,j} - f_{i-1,j}}{2\Delta\rho} + \frac{f_{i+1,j} - 2f_{i,j} + f_{i-1,j}}{\Delta\rho^2} +$$

$$+\frac{f_{i,j+1} - 2f_{i,j} + f_{i,j-1}}{\Delta z^2} - \frac{\Lambda_1^2}{\rho_i^2} f_{i,j}) + V_{i,j} f_{i,j} - 2K \tag{7}$$

$$\left[\frac{\partial V_{i,j}}{\partial\rho}\frac{g_{i,j+1} - g_{i,j-1}}{2\Delta z} - -\frac{\partial V_{i,j}}{\partial z}(\frac{g_{i+1,j} - g_{i-1,j}}{2\Delta\rho} + \frac{\Lambda_2}{\rho_i} g_{i,j}) + \frac{\partial V_{i,j}}{\partial\rho}\frac{\Lambda_1}{\rho_i} f_{i,j}\right]$$

$$H_2\psi_{i,j} = -\frac{\hbar^2}{2\mu}(\frac{1}{\rho_i}\frac{g_{i+1,j} - g_{i-1,j}}{2\Delta\rho} + \frac{g_{i+1,j} - 2g_{i,j} + g_{i-1,j}}{\Delta\rho^2} +$$

$$+\frac{g_{i,j+1} - 2g_{i,j} + g_{i,j-1}}{\Delta z^2} - \frac{\Lambda_2^2}{\rho_i^2} g_{i,j}) + V_{i,j} g_{i,j} - 2K \tag{8}$$

$$\left[-\frac{\partial V_{i,j}}{\partial\rho}\frac{f_{i,j+1} - f_{i,j-1}}{2\Delta z} + \frac{\partial V_{i,j}}{\partial z}(\frac{f_{i+1,j} - f_{i-1,j}}{2\Delta\rho} - \frac{\Lambda_1}{\rho_i} f_{i,j}) - \frac{\partial V_{i,j}}{\partial\rho}\frac{\Lambda_2}{\rho_i} g_{i,j}\right]$$

where the subscript (i, j) corresponds to the grid point (ρ_i, z_j).

The deformed Coulomb plus nuclear potential $V(\rho, z)$ is defined in terms of the above mentioned Cassini ovals.

To obtain the eigenstates we are using the software package **ARPACK**, which solves large algebraic eigenvalue problems based on the implicitly restarted Arnoldi method.[7]

Since the above hamiltonian depends on Ω, the computation has, in principle, to be repeated for all possible values: $1/2, 3/2, 5/2, \ldots$. However bound states in ^{236}U have $\Omega < 11/2$.

The main advantages of our new approach are:

1. Reflexion asymmetric nuclear shapes are calculated with the same program as reflection symmetric ones, without additional numerical effort. The Nilsson-type models require another basis that doesn't conserve parity, i.e. another program.

2. Generalization to non-axiality can be simply done by keeping the 3rd cylindrical coordinate ϕ in the Schrödinger equation.

3. The tails of the wave functions are properly described and not inevitably cut by the finite dimension of the basis.

This last advantage is important at least in three situations:

a) When calculating properties of single-neutron or single-proton states near the drip-line or in hallo nuclei.

b) When preparing initial quasi-stationary states for the time-dependent approach to deep quantum tunnelling. Due to the extremely high precision necessary to calculate extremely small tunnelling probabilities, only high purity (essentially one component) initial wave packets can be numerically handled.

c) When calculating stripping or pick-up reaction cross sections that are extremely sensitive to the tail of the nucleonic wave functions.

4. Results and conclusions

For each light-fragment mass A_L we have first calculated the value of the parameter α_1 that defines a perturbed Cassini ovaloid that is asymmetric under reflection at a plane perpendicular to the axis of symmetry and has the required ratio A_L/A_H. For a given A_L, α_1 depends on the deformation parameter ϵ and we have thus obtained two different values α_1^i and α_1^f. Then we have calculated the two sets of bound states $|\Psi_i(\epsilon_i, \alpha_1^i) >$ and $|\Psi_f(\epsilon_f, \alpha_1^f) >$ as described in the previous section. We have finally used them in Eqs.(2) and (3) to estimate $\nu_{sc}(A_L)$. So far we have done this only for neutron states characterized by $\Omega = 1/2$.

The results are presented in Fig.1 where the approximate variation of all neutrons is also sketched. We notice the large difference between the two behaviours. This is due to the fact that the scission neutrons reflect the properties of the extremely elongated fissioning nucleus while the prompt neutrons reflect the properties of the primary fragments.[8] We have considered a step-function for v_i^2 in Eq. (3) (i.e., independent neutrons) that makes the results sensitive to the quantum numbers of the last occupied state. For pairing correlated neutrons the solid curve in Fig.1 is expected to be smoother.

In conclusion, the variation of the scission-neutron yield with the fragment mass ratio is predicted to be less pronounced but more complicated than for the rest of the prompt fission neutrons.

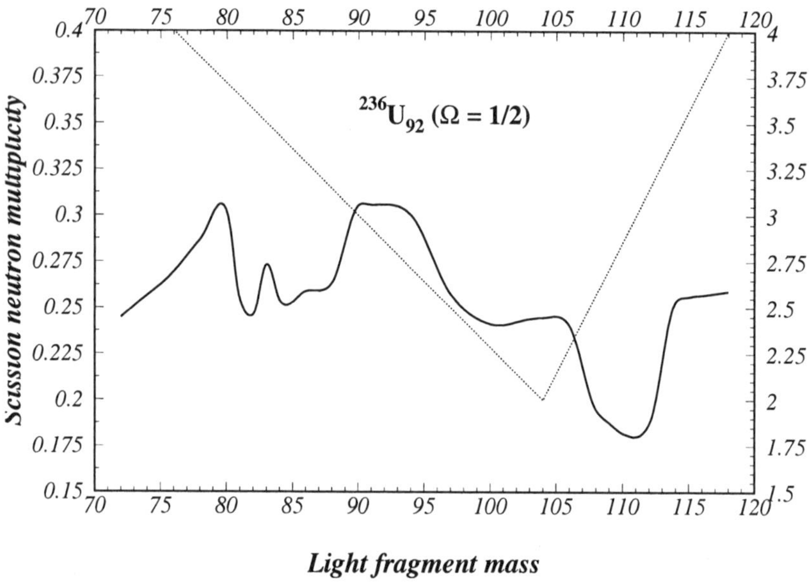

Fig. 1. Estimated scission-neutron yields (solid curve and left-hand axis) compared with experimental total-neutron yields (shown schematically by the dotted lines and right-hand axis).

References

1. N. Carjan, P. Talou, O. Serot, submitted for publication.
2. V. S. Stavinsky, N. S. Rabotnov, A. A. Seregin, *Yad. Fiz.* **7** (1968) 1051.
3. V. M. Strutinsky, N. Ya. Lyashchenko, N. A. Popov, *Nucl. Phys.* **46** (1963) 639.
4. A. A. Seregin, *Yad. Fiz.* **55** (1992) 2639.
5. V. V. Pashkevich, *Nucl. Phys. A* **169** (1971) 275.
6. J. Damgaard, H. C. Pauli, V. V. Pashkevich, V. M. Strutinsky, *Nucl. Phys. A* **135** (1969) 432.
7. D. Sorensen, R. Lehoucq, Chao Yang, *ARPACK - An implementation of the implicitly restarted Arnoldi method for computing a few selected eigenvalues and corresponding eigenvectors of a large sparse matrix*, **www.netlib.org** (1996).
8. N. Carjan, H. Goutte, to be published in *Yad. Fiz.*

NEW CLUES ON FISSION DYNAMICS FROM SYSTEMS OF INTERMEDIATE FISSILITY

E. Vardaci, A. Brondi, A. Di Nitto, D. Guadagnuolo, V. Fiorillo,
G. La Rana, R. Moro, M. Trotta, A. Ordine, A. Boiano

Dipartimento di Scienze Fisiche, Universitá di Napoli "Federico II", 80126 Napoli, Italy
**E-mail: Emanuele.Vardaci@unina.it*
www.unina.it

M. Cinausero, E. Fioretto, G. Prete, V. Rizzi

Laboratori Nazionali di Legnaro, Viale dell'Universitá, Legnaro, Padova, Italy

N. Gelli, F. Lucarelli

Dipartimento di Fisica, Universitá di Firenze, and INFN,, Firenze, Italy

P.N. Nadtochy

Department of Theoretical Physics, Omsk State University, Omsk, Russia

V.A. Rubchenya

Department of Physics, University of Jyväskylä, Finland,
Khlopin Radium Institute, St.Petersburg, Russia

Systems of intermediate fissility are characterized by an evaporation residues (ER) cross section comparable or larger than the fission cross section, and by a relatively higher probability for charged particle emission in the pre-scission channel. In a theoretical framework in which time scale estimates of the fission process rely on statistical model calculations, the analysis of particle emission in the ER channel is the source of additional constraints on the statistical and dynamical models. This contribution will offer an overall view of the systems studied so far and will provide new clues for interesting and contradictory physical cases. In particular, the analysis of the systems ^{32}S + ^{100}Mo at E_{Lab} = 200 MeV and ^{18}O + ^{150}Sm at E_{Lab} = 122 MeV will be presented. Recent advancement will be briefly discussed on the interplay between the fission and the evaporation residues channels as implied by a dynamical (Langevin) approach to the description of the fission process.

Keywords: Fission dynamics, fusion-fission, fusion-evaporation

1. Introduction

Fission dynamics has been the subject of a large variety of studies in the last two decades.[1,2] Most of the studies have shown that the pre-scission multiplicities of neutrons and charged particles increase monotonically with the bombarding energy in contrast with the calculations of the standard statistical model (SM). Since the fission process, as well as the fission time scale, is thought to be affected by nuclear dissipation,[3] this result is considered as the evidence that fission is a slow process with respect to the lifetime for the emission of light particles. With increasing excitation energy, the particle decay lifetime decreases and becomes smaller than the time necessary for the build up of the collective motion of the nuclear matter toward the saddle point. Consequently, fission does not compete as effectively as predicted by the SM in the early stages of the decay, and particles and GDR $\gamma - ray$ emissions can occur more favorably. The overall cause of the establishment of these transient effects is believed to be associated with the nuclear matter viscosity which slows down the collective flow of mass from equilibrium to scission and does not allow the fission decay lifetime to be reduced with increasing excitation energy as in the case of light particles. This is equivalent to consider that fission is delayed with respect to the picture of the SM in which the probability for fission has its full Bohr-Wheeler value already at the beginning of the decay where the compound nucleus is pictured fully equilibrated. An energy domain has further been identified[4] above which the SM predictions begin to deviate from the data.

A strong dissipation due to nuclear viscosity can indeed trigger a variety of effects of dynamical origin, between which the possibility that a compound nucleus committed to fission (already at the saddle point configuration) can still became an evaporation residue if enough particles are evaporated and the fissility reduced. This correlation between the enhanced yield of prescission particles and the survival of evaporation residues might be an important channel for the feeding of evaporation residues having large deformations in the mass region of A≈ 150-160.[5]

Several variants of the statistical model have been proposed in the literature to take explicitly into account time scales as well as viscosity.[1,2,6,8,12,17,18] Following the initial idea of the " neutron clock",[1,2,6] the common trend is to split the path from the equilibrium to the scission point configuration into two regions, the pre- and the post-saddle. The total fission time is defined as $\tau_f = \tau_d + \tau_{ssc}$, where τ_d is the pre-saddle delay, namely, the characteristic time which the composite system spends inside the barrier, and τ_{ssc} is the time necessary to travel the path from saddle to scission. The relevant observables are computed using τ_d and τ_{ssc} as free parameters, along with the other input parameters relative to the specific ingredients of the model, and fit to the experimental data. In spite of the extensive work, estimates of the fission time scales are however quite controversial, ranging from 0 to 400x10^-21s, depending on the system and on the experimental probe. Furthermore, such estimates are weakened by the fact that different sets of input parameters can result in equally good fits within the same model.[10,12,13]

Modified statistical models as well as dynamical models based on the Euler Lagrange, Fokker-Planck and Langevin equations [19-27] have been used in order to gain insight on the nature of dissipation. Following Kramer's work,[3] the inclusion of dissipative effects results in an effective fission decay width Γ_{eff} which is smaller than the standard Bohr-Wheeler decay width by a hindrance factor,

$$\Gamma_{eff} = \Gamma_{BW}\left(\sqrt{1+\gamma^2}+\gamma\right) \tag{1}$$

where γ is the nuclear friction coefficient and Γ_{BW} is the Bohr-Wheeler fission decay width. γ is usually associated to the fission delay time.[12,16,17] The estimates of γ from the fits to the particle multiplicities, both from statistical and dynamical models, provides a contradictory picture on the values of γ, which range over an order of magnitude, and on the one-body or two-body origin of the nuclear dissipation. More dramatic is the situation from the pure theoretical point of view, where the predicted values of γ, on the basis of microscopic models, are spread over three orders of magnitude.[28]

It must be pointed out that only neutron multiplicities have been measured in most of the studies and mostly for heavy systems (A$\geq$200), and the lack of a sufficient number of constraints to the models could, in several cases, be the source of discrepancies. In order to withdraw a more consistent picture of nuclear dissipation it seems crucial to take into account a larger number of observables which can be expected to be sensitive to the nuclear dissipation and to try to reproduce the variety of observables with a unique set of input parameters.

2. Dissipation in systems of intermediate fissility

Systems of intermediate fissility (χ = 0.5-0.6) are very little studied although they offer several advantages. They are characterized by an evaporation residue (ER) cross-section comparable or larger than the fission cross-section, and by a shorter path in the deformation space from the saddle-to-scission point.[29] Consequently: 1) the input parameters of the models can be further constrained by the energy spectra and multiplicities of the light particles in the ER channel; 2) the effect of the fission delay over the fission and ER cross section is much more pronounced with respect to heavier systems because the emission of a charged particle in the pre-saddle region strongly enhances the probability of producing an evaporation residue as consequence of both a reduction of the fissility and the large value of the angular momentum necessary to ignite fission.

The fact that the potential energy surface is characterized by a shorter path from the saddle to scission means that the role of the pre-saddle dynamics relative to the saddle to scission dynamics is enhanced and, therefore, some of the ambiguities on the not-well identified separation and interplay between pre- and post-saddle might be reduced in the interpretation of the data.

We expect that the measurements of neutron and charged particle multiplicities and energy spectra in the two channels as well as the measurements of the cross sections of the channels themselves will allow more severe constraints onto the models. This should provide more reliable values of fission delay and of the friction parameter, and contribute to a better comprehension of the nuclear viscosity. In this framework, the 8πLP collaboration has started a research program at the Laboratori Nazionali di Legnaro (Padova, Italy) aimed at studying the fission dynamics in systems of intermediate fissility.

This presentation will review some of the measurements performed recently with the 8πLP apparatus in Legnaro and will provide some clues as the outcome of recent preliminary results based on the SM code PACE2 and a dynamical model approach which is used in the attempt to overcome the limits of the picture of the SM. The paper is thus organized as follows. First, a short review of the 8πLP apparatus is given. Second, we present the study of the fission time scale in the system ^{32}S + ^{100}Mo in which the fission and the ER channels are both included in the SM analysis. Third, we show some preliminary results on the system ^{18}O + ^{150}Sm which clearly identify one of the major weakness of the SM. Finally, a brief account will be given about the application of a dynamical model to a more complete set of data in the system $^{32}S + ^{100}Mo$.

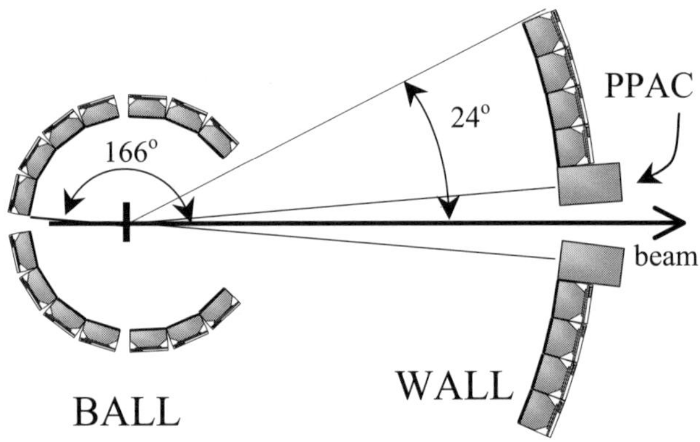

Fig. 1. Schematic layout of the 8πLP apparatus

3. The 8πLP apparatus

The 8πLP apparatus,[30] shown schematically in Fig.1, is a light charged particle detector assembly which covers about 80% of 4π . It consists of two detector subsystems, each made out of two-stage telescopes: the WALL and the BALL. The WALL contains 116 telescopes and is placed at 60 cm from the target. Each of the WALL telescopes consists of a 300μm Si detector backed by a 15 mm CsI(Tl) crystal and has an active area of 25 cm^2 which corresponds to an angular opening of about 4°. The WALL covers the angular range from 2° to 24°. The BALL has a diameter of 30 cm and consists of 7 rings placed coaxially around the beam axis. Each ring contains 18 telescopes and covers an angular opening of about 17°. The telescopes of the BALL are made out of a 300μm Si detector mounted in the flipped configuration (particle entering from the ohmic side) backed by a 5 mm CsI(Tl) crystal. The BALL has a total of 126 telescopes and covers the angular range from 34° to 166°. The rings are labeled from A to G going from backward to forward angles. The experimental method used to investigate on the fission dynamics consists in measuring light charged particles in coincidence with both fission fragments (triple coincidence) and with the evaporation residues (double coincidences). Particle identification is carried out by the ΔE-E method for the ions that are not stopped in the E stage. The particles stopping in the first stage are identified by the TOF method in the case of the WALL telescopes, and by the Pulse Shape Discrimination (PSD) technique in the case of the BALL telescopes. In this configuration we are able to measure energies up to 64 AMeV in the WALL and 34 AMeV in the BALL with energy thresholds of 0.5 MeV for protons and 2 MeV for alpha particles.

Heavy fragments can be detected in the telescopes of the BALL. The PSD technique allows the separation between heavy fragments and light particles stopping in the same detector but does not provide any information about the mass or charge of the fragments. The selection between symmetric and asymmetric binary mass splitting can nevertheless be achieved on a kinematics ground as shown later. A more selective trigger detector based on a time-of-flight spectrometer is presently under study.

In the 8πLP setup it is also possible to detect Evaporation Residues (ER). The WALL detectors between 2.5° and 7.5° around the beam axis are indeed replaced by 4 Parallel Plate Avalanche Counter (PPAC) modules, each one subtending a solid angle of about 0.3 msr. Each module consists of two coaxial PPACs mounted and operating in the same gas volume at a distance of 15 cm from each other. By adjusting the gas pressure, it is possible to stop the ER between the two PPACs, and let the lighter ions to impinge on the second PPAC. Consequently, ERs are sorted out from the first PPAC signals using the signals from the second PPAC as a veto.

4. Dynamical effects in the system $^{32}S + ^{100}Mo$ at E_{Lab} = 200 MeV

4.1. *Fission Channel*

This study follows our precedent report on the same system but at E_{Lab} = 240 MeV.[15] The main conclusion of that study was that there was no evidence of significant dynamical effects in the alpha particle pre-scission emission, regardless of the fact that a fission delay was expected on the basis of a systematic study on the threshold excitation energy for the onset of a non-statistical behavior of the fission decay,[4] from which we would expect a sizable deviation from the statistical description in the pre-scission particle multiplicities. It has then been argued[38] that the presence of fast fission might have been responsible for the overall lowering of the deduced fission time. To probe this hypothesis we have performed an experiment on the same system but at 200 MeV, where fast fission is expected to be negligible. Preliminary results on the data analysis and the fission time scale will discussed here in conjunction with the results on the evaporation residues channel.

Heavy fragments were detected in the telescopes of the ring F and G of the BALL. The selection between symmetric and asymmetric binary mass splitting has been achieved on a kinematical ground. An example of such a selection for the reaction at 200 MeV is in Fig.2 where we show the fragment-fragment energy correlation corresponding to two fragments detected by two different telescopes in the same ring F (a), in the same ring G (b) and in the rings F-G (c), respectively. The plots show clearly the transition between the symmetric and asymmetric mass splitting by a proper choice of the detecting geometry, which means, a variable coverage of the folding angle. The angular correlations are such that when the two fragments are detected in two different telescopes of the ring F, which corresponds to a folding angle $\Delta\theta$ between 105° and 138°, only the symmetric component of the fragment mass distribution is detected; in the case of ring G instead, only the most asymmetric component is selected. The case of rings F and G corresponds to an intermediate condition, namely to an angular range centered on the most probable folding angle for symmetric fission.

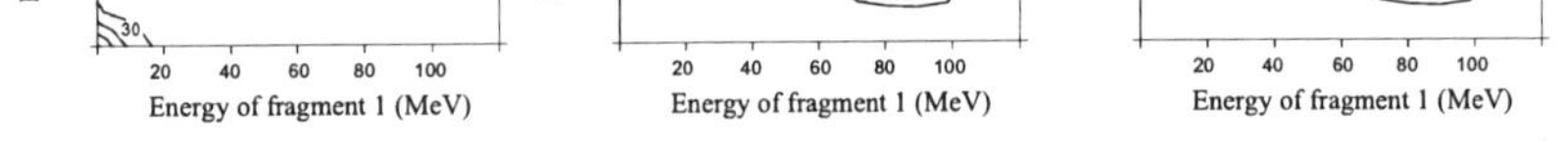

Fig. 2. Energy-energy correlation matrix of the measured fragments: a) both fragments in ring F, b) both fragments in ring G, c) one fragment in ring F and the other in ring G.

Laboratory energy spectra of protons and alpha particles in triple coincidence (ring F- ring G - particle) were extracted for all the correlation angles allowed by the geometry of the BALL (12 in plane and 56 out of plane). Some of the multiplicity spectra are shown as histograms in Fig.3 and 4.

The position of each particle detector with respect to the beam has been translated to a position relative to a trigger plane (defined by the position

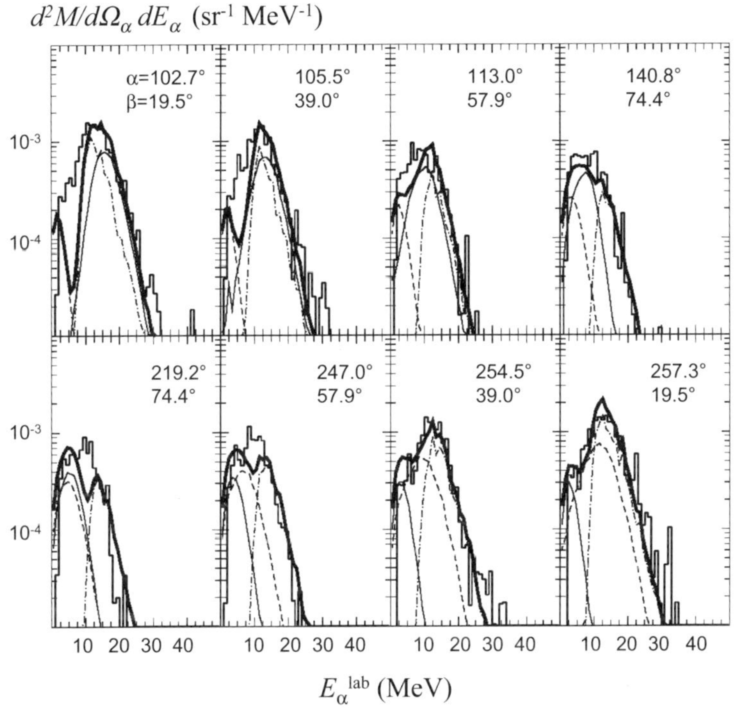

Fig. 3. Out-of-plane multiplicity spectra of alpha particles in ring D in coincidence with fission fragments detected in the rings F-G.

of the two fired fragment detectors) using the in-plane ($0° < \alpha < 360°$) and the out-of-plane ($-90° < \beta < 90°$) angles. The values of α and β are shown in Fig.3 and 4. Each particle spectrum has been obtained as the sum of the alpha particle spectra corresponding to the same in-plane and out-of-plane angles, and normalized to the number of its corresponding trigger fragment-fragment events.

To extract the pre- and post-scission integrated multiplicities, particle spectra have been analyzed considering three evaporative sources: the composite nucleus prior to scission (CE, Composite Emission) and the two fully accelerated fission fragments F1 and F2 (FE, Fragment Emission). We have used a well established procedure which employs the Monte Carlo Statistical code GANES.[31–33] Particle evaporative spectra are computed separately for each source of emission in the trigger configuration defined in the experiment, taking into account the detection geometry. Afterwards,

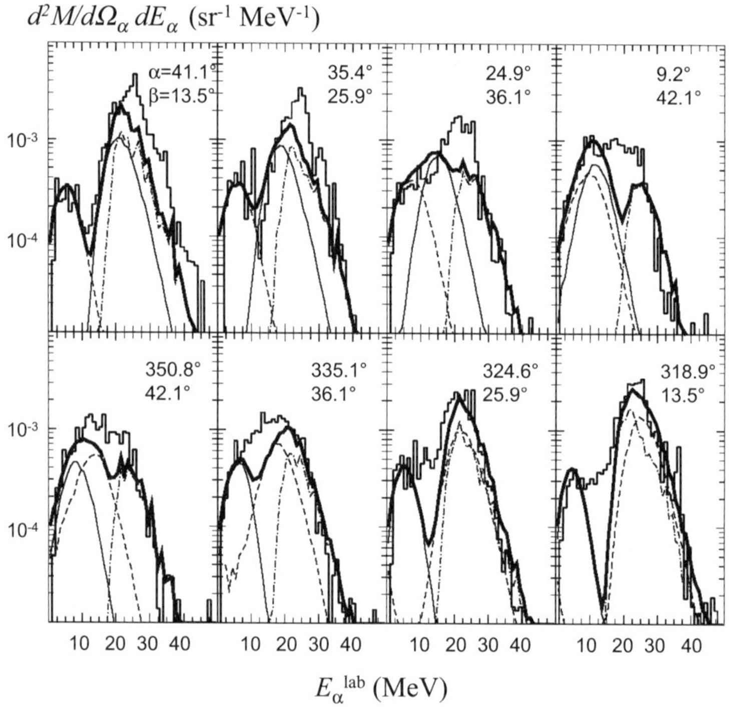

Fig. 4. Out-of-plane multiplicity spectra of alpha particles in ring G in coincidence with fission fragments detected in the ring F-G.

the calculated spectra are normalized to the experimental ones, and the integrated multiplicities are evaluated for each emitting source.

The curves superimposed on the histograms in the Fig.3 and 4 represent calculated multiplicity spectra for CE (dot-dashed line), F1 (thin solid line) and F2 (dashed line) components, along with their sum (thick solid line). A large deformation of the composite system prior to scission was necessary to fit simultaneously the energy spectra and the angular distributions. The deformation of the emitter affects both the mean energy of the evaporated charged particles (because of the change in the evaporation barriers) and the out-of-plane angular distribution (because of the increase in the moment of inertia). In our data the best fit to the energy spectra provides for the CE component a prolate shape with axis ratio $b/a = 3$, both at 200 and 240 MeV. This emitter deformation results into mean energies of the alpha particles which are ≈ 2 MeV lower than those expected for the case of a spherical emitter.

Although the bulk of the experimental spectra is very well reproduced at both energies, also considering the wide angular coverage of the detecting array, there are some important deviations which indicate contributions not accounted for by the CE and FE components which are mainly of two kinds: an excess of high energy alpha particles at most forward angles, and a surplus of particles with energies intermediate between those corresponding to CE and FE components. The same is observed for the protons. These two types of contributions have already been observed in other experiments of the same kind as presented here[31] and have been ascribed to pre-equilibrium and near-scission emission,[34–36] respectively.

From the fit to experimental spectra, we have extracted the pre and post scission particle multiplicities which we have compared with the prediction of a modified version of the code PACE2.[37] In particular, a fission delay parameter τ_d is included into the code so that the fission width is zero up to a time τ_d and has the full statistical value subsequently. Since we expect that the particle multiplicities are sensitive to the value of the ratio $a_d/a\nu$ and to the delay time τ_d, we performed a grid of calculations for $1.00 < a_d/a\nu < 1.10$ and $0 < \tau_d < 50\text{x}10^{-21}s$. The result relevant for this presentation is that the SM is able to reproduce the particle prescission multiplicities in the reaction at 200 MeV with a delay of about $25\text{x}10^{-21}s$ whereas at 240 MeV no fission delay is necessary. This result is at variance with what is expected with increasing excitation energy and at variance with the prediction of the systematics,[4] where a sizable deviation from the statistical description is expected at both energies. This point deserves more investigation, and in fact a new experiment has been performed for this purpose.

4.2. *Evaporation Residues Channel*

In Fig.5 we show, as solid points, the multiplicity angular distribution of protons and alpha particles, detected in coincidence with one of the PPAC versus the identification number of the BALL detectors. The predictions of the statistical model, as implemented in the code PACE2 (solid lines), are superimposed to the data. The detailed geometry of the detecting system has been properly included in PACE2. For each ring, we observe a strong dependence of the intensity on the detector position resulting from the different correlation angles with respect to the trigger detector, both for protons and alpha-particles. The same pattern of Fig.5 is observed for the case of the other trigger positions.

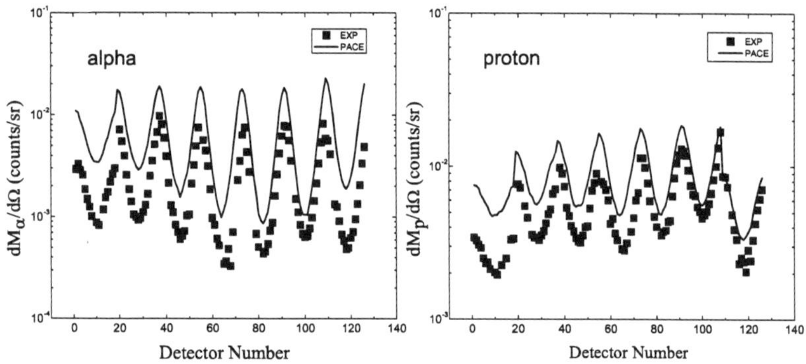

Fig. 5. Comparison between experimental and theoretical multiplicity angular distribution for alpha particles (left) and protons (right).

The code PACE2 reproduces well the observed pattern with the same input parameters that reproduce the prescission charged particle multiplicities, but overestimates the integral multplicities by a factor 1.8 for the protons and a factor 3.1 for the alpha particles. The code LILITA-N97,[39] that is another implementation of the SM without fission, also overestimates these multiplicities, but this time by a factor 2 for both charged particles. It interesting to notice that other few sets of data exists where this same behaviour of the SM is found.[40]

5. Discussion and Conclusions

The results obtained with the application of the SM in the ER channel open a series of questions that inevitably affect the estimates of the fission time scales, at least for the use so far done of the SM model to predict dynamical effects. If the SM overestimstes the charged particles multiplicities in the ER channel, we expect that neutron multiplicities should be underestimated. Unfortunately no data on neutrons are available at present for the system $^{32}S + ^{100}Mo$.

Yet, if the same behaviour is applied in the fission channel this means that the delay time may be overestimated if only neutrons are measured in the fission channel and no cross check is done on the charged particles. Besides this, if the delay time is measured from an excess of neutron multiplicities, the question arises about what is the baseline number from which the excess is to be determine. According to what we find in the ER channel, this number might not be reliable. At the same time, charged particles should behave exactly in the opposite way. The time delay extracted by using only the charged particles might be, in turn, underestimated or not necessary at all.

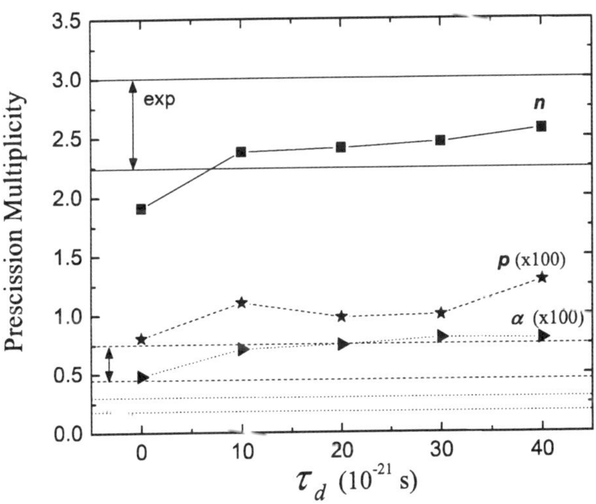

Fig. 6. Measured prescission particle multiplicities in the fission channels of the system ^{18}O + ^{150}Sm at E_{Lab} = 122 MeV compared to the prediction of the SM for different values of the delat time τ_d

This effect is clearly seen in Fig.6 for the system ^{18}O + ^{150}Sm at E_{Lab} = 122 MeV which shows the experimental prescission multiplicities of neutrons, protons and alpha particles compared to the predictions of the SM (PACE2) for increasing value of the fission delay. The neutrons have been measured in Ref.[41] while protons and alpha particle multiplicities have been recently measured with 8πLP setup. At zero delay, the SM underestimates the neutron multiplicities and overestimates the charged particles multiplicities. The effect of the delay is to increase all the particles multiplicities because of the suppression of the fission cross section. Although a value of τ_d = 10 would reproduce the neutron multiplicity, there is no possibility to reproduce the charged particles prescission multiplicities.

These contradictory results outlines the necessity of considering dynamical models. Recently we have coupled the LILITA-N97 code with a dynamical model[23,26] which describes the fission process by using a three-dimensional Langevin stochastic approach. This coupling was necessary in order to allow the evaporation of light particles from the composite system during the evolution along trajectories in the phase space. At the moment we have performed several sets of calculations for the system $^{32}S + {}^{100}Mo$ at E_{Lab} = 200 MeV assuming different prescriptions of transmission coefficients and level densities for particle evaporation, and by modulating the values of the strenght of the one-body dissipation. The model is able to reproduce most of the measured quantities, including the ones in the ER channel, assuming a reduction coefficient k_s=0.5. This value, which is consistent with systematics, implies sizable transient times for fission, ranging from 15 to 20 $\times 10^{-21}$s at high angular momenta of the composite system, where fission is relevant. This result supports the conclusion that the dynamical approach to fission decay is very promising in describing both fission and evaporation residues channel within the same model.

References

1. A. Gavron et al., *Phys. Rev.* **C35** 579 (1987).
2. D. J. Hinde et al., *Phys. Rev.* **C39** 2268 (1989); D.J. Hinde, D. Hilscher, H. Rossner, et al., *Phys. Rev.* **C45**, 1229 (1992)
3. A. Kramer, *Physica (Amsterdam)* **7** 284 (1940).
4. M. Thoennessen and G.F. Bertsch, *Phys. Rev. Lett.* **71** 71 (1993).
5. L. Fiore, G. D'Erasmo, D. Di Santo, et al., *Nucl. Phys.* **A620** 71 (1997).
6. J.P. Lestone et al., *Nucl. Phys.* **A559** 277 (1993).
7. H. Ikezoe, Y. Nagame, I. Nishinaka, et al., *Phys. Rev.* **C49** 968 (1994).
8. A. Saxena, A. Chatterjee, R. Choudhury et al., *Phys. Rev.* **C49** 932 (1994).
9. D.J. Hofman, B.B. Back, and P. Paul, *Phys. Rev.* **C51** 2597 (1995).
10. A. Chatterjee, A. Navin, S. Kailas et al., *Phys. Rev.* **C52** 3167 (1995).
11. V.A. Rubchenya et al., *Phys. Rev.* **C58** (1998) 1587.
12. I.Diószegi, N.P. Shaw, I. Mazumdar et al., *Phys. Rev.* **C61** 24613 (2000).
13. N.P. Shaw, I.Diószegi, I. Mazumdar et al., *Phys. Rev.* **C61** 44612 (2000).
14. A. Saxena, D. Fabris, G. Prete, et al., *Phys. Rev.* **C65** 64601 (2002).
15. G. La Rana et al., *Eur. Phys. J.* **A16** 199 (2003).
16. P. Grang et H. A. Weidenmuller, *Phys. Lett.* **B96** 26 (1980).
17. V.A. Rubchenya, *Proceedings of Int. Conf. Large-Scale Collective Motion of Atomic Nuclei*, eds. G. Giardina, G. Fazio and M. Lattuada, Brolo (Messina), Italy, 1996, World Sci., Singapore, p.534 (1997).
18. G. Giardina, *Proceedings of 6th Int. School-Seminar on Heavy Ion Physics*, eds.Yu.Ts. Oganessian and R. Kalpakchieva, Dubna, Russia, 1997, World Sci., Singapore, 1998) p.628.

19. T. Wada et al., *Phys. Rev. Lett.* **70** 3538 (1993).
20. C. Bhattacharya et al., *Phys. Rev.* **C53** 1012 (1996).
21. P. Frobrich and I.I. Gontchar, *Phys. Rep.* **292** 131 (1998).
22. A.K. Dhara et al., *Phys. Rev.* **C57** 2453 (1998).
23. A.V. Karpov et al., *Phys. Rev.* **C63** 54610 (2001).
24. G. Chaudhuri and S. Pal, *Phys. Rev.* **C65** 54612 (2002).
25. P. Frobrich and I.I. Gontchar, *Europhys. Lett.* **57** 355 (2002).
26. P.N. Nadtochy et al., *Phys. Rev.* **C65** 64615 (2002).
27. W. Ye, *Progr. Theor. Phys.* **109** 933 (2003).
28. C. Badimon, *Ph.D. Thesis.*- CENBG, Bordeaux, (2001).
29. M.G. Itkis, A. Ya. Rusanov, *Phys. Part. Nucl.* **29** 160 (1998).
30. E. Fioretto et al., *IEEE Trans. Nucl. Scie.* **44** 1017 (1997).
31. R. Lacey et al., *Phys. Rev.* **C37** 2540 (1988).
32. N.N. Ajitanand et al., *Nucl.Instr. Meth. Phys. Res.* **A243** 111 (1986).
33. N.N. Ajitanand, G. La Rana, R. Lacey, et al., *Phys. Rev.* **C34** 877 (1986).
34. E. Duek et al., *Phys. Lett.* **B131** 297 (1983).
35. L. Schad et al., *Z. Phys.* **A318** 179 (1984).
36. E. Vardaci, M. Kaplan et al., *Phys. Lett.* **B480** 239 (2000).
37. M.A. DiMeo, *Ph.D. Thesis.* - Universita' di Napoli "Federico II" (2002).
38. G. La Rana et al., *Proc. Int. Conf. on Nuclear Reaction Mechanisms*, June 5-9, Varenna, Italy, ed. E. Gadioli (Ricerca Scientifica ed Educazione Permanente, Grafiche Vadacca, Vignate (MI), 2000).
39. LILITA-N97 is an extensively modified version of the original LILITA program made by J. Gomez del Campo and R. G. Stockstad, Oak Ridge National Laboratory, (Rep. No TM7295, 1981 unpublished).
40. J. Alexander et al., *Proc. of the Symposium on Nuclear Dynamics and Nuclear Disassembly*, ed. J.B. Natowitz, Dallas, Texas, USA, April 10-14, 1989, World Sci., Singapore, p.211 (1989).
41. J.O. Newton et al., *Nucl. Phys.* **A483** 126 (1988).

DYNAMICS OF CAPTURE QUASIFISSION AND FUSION-FISSION COMPETITION

L. STUTTGÉ, C. SCHMITT, O. DORVAUX, N. ROWLEY
Institut Pluridisciplinaire Hubert Curien-Département de Recherches Subatomiques, IN2P3/CNRS-Université Louis Pasteur
F67037 Strasbourg, France

T. MATERNA, F. HANAPPE, V. BOUCHAT
Université Libre de Bruxelles, CP229
B1050 Brussels, Belgium

Y. ARITOMO, A. BOGATCHEV, I. ITKIS, M. ITKIS, M. JANDEL, G. KNYAJEVA, J. KLIMAN, E. KOZULIN, N. KONDRATIEV, L. KRUPA, Y. OGANESSIAN, I. POKROVSKI, E. PROKHOROVA, V. VOSKRESENSKI
Flerov Laboratory of Nuclear Reactions, Joint Institute for Nuclear Research, Dubna, Russia

N. AMAR, S. GRÉVY, J. PÉTER
Laboratoire de Physique Corpusculaire
F14050 Caen Cedex, France

G. GIARDINA
Istituto Nazionale di Fisica Nucleare, Sezione di Catania and Dipartimento di Fisica dell'Universita di Messina,
Messina, Italy

An overview of the different experimental approaches to disentangle the quasi-fission and the fusion-fission processes in the heavy and superheavy mass region is presented. Indeed the separation of these two processes is essential in order to get a correct and complete insight of the mechanisms leading to the synthesis of superheavy elements. The importance of the neutron information through a new analysis protocol is detailed. Future perspectives are presented.

1. Introduction

In the heavy mass region when two partners interact, a part of the cross section goes away by the deep inelastic channel. If the two nuclei undergo capture a big competition takes place with quasi-fission, where the two nuclei reseparate before an equilibrated compound nucleus has been formed. Again at this point, if fusion occurs the main part of the cross section goes into fusion-fission. And a tiny part only survives as an evaporation residue. The synthesis of superheavy elements is thus hindered at several stages of the process: quasi-fission hinders the fusion and fusion-fission the survival of the compound nucleus. A good understanding of the process leading to the synthesis of superheavy elements goes through a good knowledge of these intermediate processes.

A huge work is being done on the theoretical side in this domain. Many models succeed in reproducing the total capture cross section as well as the residue formation one. As an example, figure 1 shows the cross section as a function of the incident energy calculated by different models for ^{48}Ca + ^{244}Cm system leading to Z=114. The lines correspond to the capture cross section, the dots to the evaporation residue one. All the models agree quite well about the capture cross section as well as the evaporation residue formation and reproduce well the experimental data for the capture (squares) as well. But they disagree completely to reproduce the intermediate processes, especially at low energies.

A better inside in these mechanisms is essential in order to bring strong constraints to the theory and hopefully decrease the number of parameters that all the models contain.

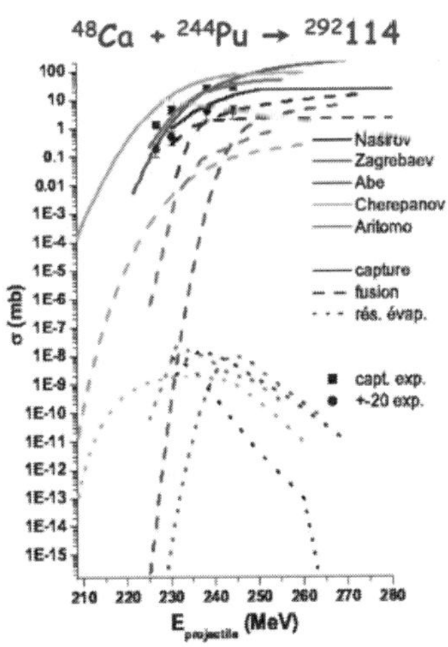

Figure 1. Various model calculations of the cross sections of the different processes as a function of the incident energy for ^{48}Ca +^{244}Pu leading to Z=114. The full lines and the squares correspond respectively to the theoretical and experimental total captures, the dashed lines to the theoretical fusion cross sections, the dots to the experimental symmetric scission fragments with A=A_{CN}±20 considered as fusion-fission fragments.

In particular the dissipation is still not well understood. Figure 2 adapted from reference [1] shows the dissipation coefficients deduced from the calculations of K. Wilczynska et al [2] as a function of the temperature for different systems measured by D. Hinde et al [3]. One can see that for the lighter systems induced by ^{16}O, a dissipation coefficient around 5 corresponding to one-body dissipation reproduces well the experimental data whereas the huge values obtained for the heavier systems induced by ^{40}Ar and ^{64}Ni are rather in favour of two-body dissipation.

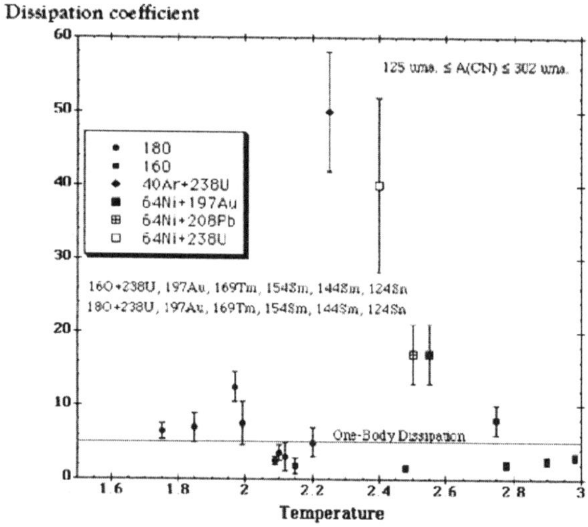

Figure 2. Dissipation coefficients for different systems deduced by the Siwek-Wilczynska et al model [2] from the Hinde et al data [3].

2. How can we disentangle experimentally the quasi-fission from the fusion-fission?

2.1. The scission fragments

On the experimental side, the difficulty lies in the fact that the separation between the quasi-fission and the true fusion-fission is almost impossible as the experimental observables of the two mechanisms present almost the same characteristics. The crucial point is in the very symmetric fragment mass region as shown in the examples of figure 3 [4]. On the left side are shown reactions induced by a ^{48}Ca beam. On the top figure, the TKE-fragment distribution for ^{48}Ca + ^{208}Pb leading to ^{256}No exhibits a well-known symmetric central component which corresponds to the fission of the compound nucleus. For the heavier targets ^{238}U, ^{244}Pu and ^{248}Cm leading to Z=112, 114 and 116 respectively it seams also clear that the asymmetric components are quasi-fission events. But for these heavier systems central symmetric fragments appear which origin is not clear. Do they correspond to true fusion-fission, to quasi-fission or to a mixture of both processes? The same features appear for reactions with a ^{208}Pb target, on the right hand figures. The case of ^{56}Fe + ^{208}Pb leading to ^{266}Hs, where the central part is quite well populated looks very different from the ^{86}Kr + ^{208}Pb case leading to Z=118 where only quasi-fission seems to occur. Thus the main problem lies in this very central region where the scission fragment characteristics don't allow the distinction.

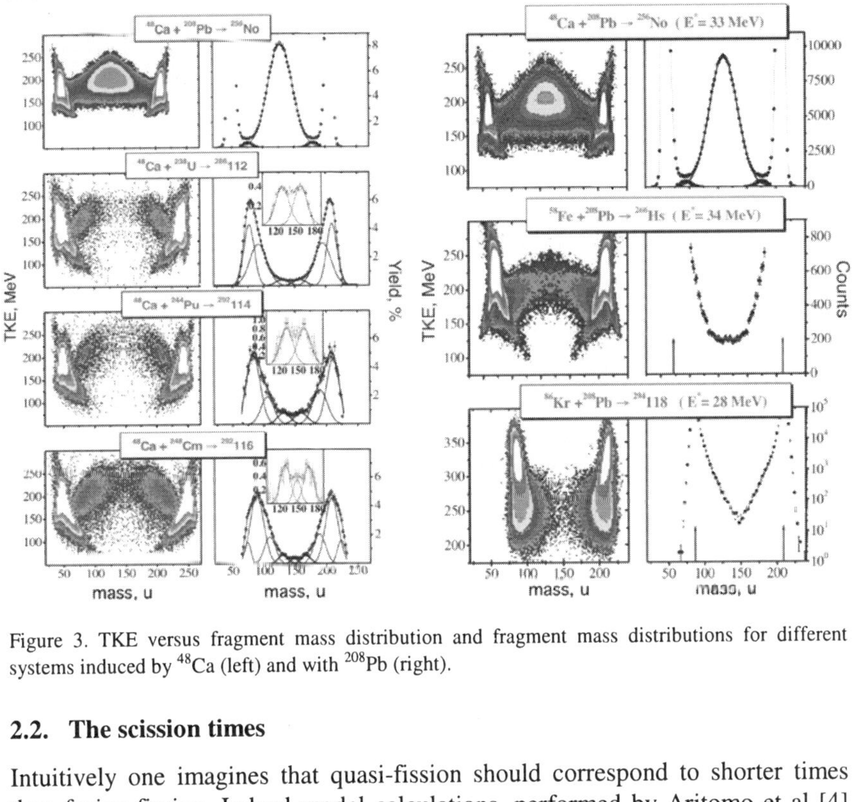

Figure 3. TKE versus fragment mass distribution and fragment mass distributions for different systems induced by ^{48}Ca (left) and with ^{208}Pb (right).

2.2. The scission times

Intuitively one imagines that quasi-fission should correspond to shorter times than fusion-fission. Indeed model calculations, performed by Aritomo et al [4] for the Ca + Pu system and shown in figure 4, give values of 10^{-20} s and 10^{-19} s for the symmetric quasi-fission and fusion-fission respectively.

However the scission times are not easy to access experimentally. If one considers for example the crystal blocking method which is a very powerful tool, there is no clear separation between different processes as one can observe in figure 5 which shows fission lifetimes of Uranium-like nuclei studied by Morjean et al [6].

Thus one has to rely on theoretical models to decide where to make the separation.

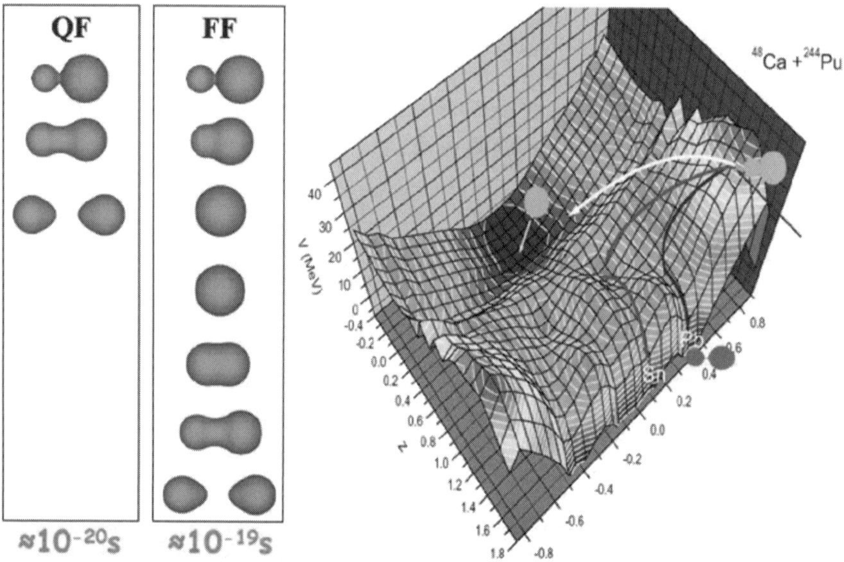

Figure 4. Multidimensional Langevin calculations performed by Aritomo et al on the ^{48}Ca + ^{244}Pu system showing the different exit channels (right): the asymmetric one around Pb and the symmetric one around Sn and the scission times deduced from the model (left) for the symmetric quasi-fission and fusion-fission.

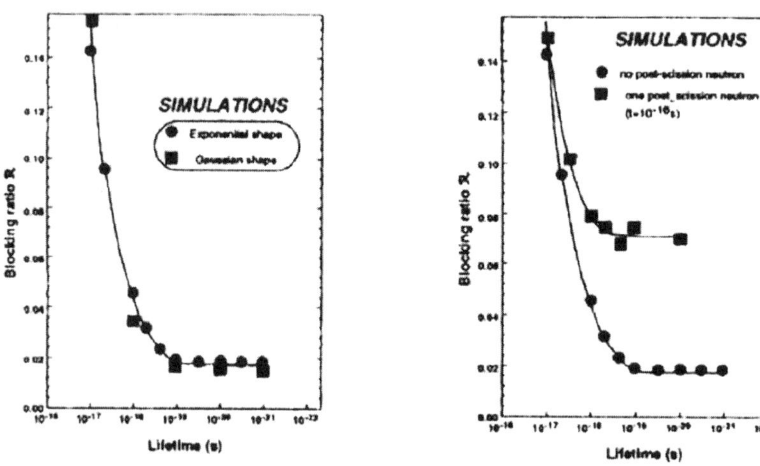

Figure 5. Blocking effect magnitudes as a function of the lifetime in the fission of U-like nuclei [6].

2.3. The neutron information

A new analysis procedure has been developed by the collaboration, the backtracing [7], which gives access not only to the mean value but to the distribution and the correlation of the pre- and post-scission neutron multiplicities. This procedure consists in a mathematical matrix inversion and is almost model independent.
To validate the method, a simple case where only fusion-fission occurs, the ^{28}Si + ^{98}Mo system leading to ^{126}Ba has been investigated through this method. The backtracing results [8] are shown on figure 6 and compared to the Pomorski et al model calculations [9], based on the resolution of the one-dimensional Langevin equation in which one-body dissipation is assumed. The two distributions are in a very good agreement as well as the mean values of the pre-scission multiplicity: 2.54 and 2.29 for the backtracing and the calculations respectively. One has to note that the backtracing as well as the model are in this case in perfect agreement with the mean value obtained by a conventional χ^2 minimization: 2.52. This is of course due to the fact that in this case only one process, the fusion-fission, occurs.

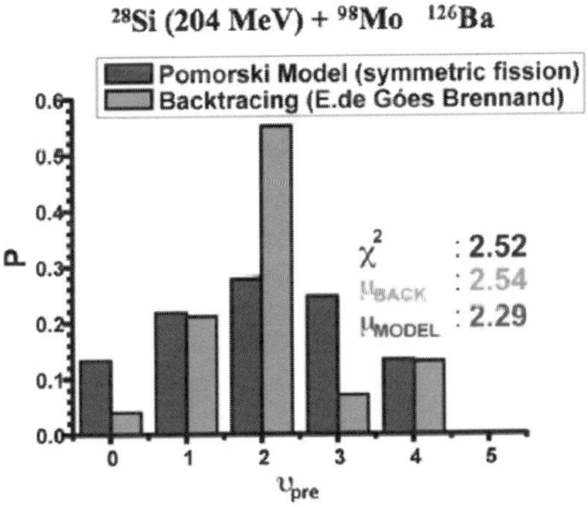

Figure 6. Pre-scission neutron multiplicity for the ^{28}Si + ^{98}Mo system leading to ^{126}Ba at 204 MeV incident energy: in green, the backtracing results, in red, the model calculations.

The procedure has been used in more complex systems as Z=110 obtained through two entrance channels: ^{58}Ni + ^{208}Pb [10] and ^{40}Ca+^{232}Th [11] measured at E*=186 and 166 MeV respectively. Figure 7 shows the correlation between the pre- and post-scission neutron multiplicities obtained by the backtracing. Two components appear clearly for both systems. Intuitively, one can attribute these two separated components to the two capture processes: the low pre-scission multiplicity, around 4, to quasi-fission which is a faster mechanism and the larger one around 7 to fusion-fission which is a slower process.

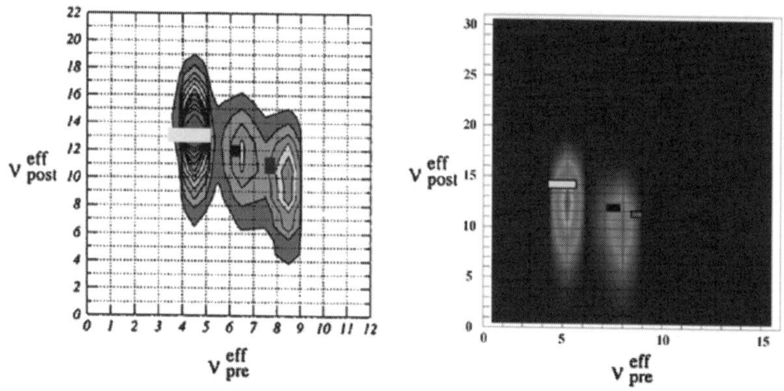

Figure 7. Pre- and post-scission neutron multiplicity correlations for the ^{58}Ni + ^{208}Pb (left) and ^{40}Ca + ^{232}Th (right) systems at 186 MeV and 166 MeV excitation energy, respectively. Results of calculations using HICOL + DYNSEQ from Siwek-Wilczynska *et al* are also shown. The rectangle to the left represents quasi-fission (30<l<120, where l is the angular momentum) with only one-body dissipation. The two squares stand for fusion-fission (0<l<30) with OBD (left) and two times OBD (right).

Calculations using HICOL and DYNSEQ have been performed by K. Wilczynska et al. The multiplicities obtained from the deduced dissipation coefficients are presented in figure 7. The left rectangles correspond to quasi-fission with one-body dissipation (γ=5). The two right squares are fusion-fission events assuming one time and two times one-body dissipation (γ=5-11). The two processes correspond to different angular momentum ranges. They were calculated in HICOL by comparison with the experimental mass distribution. The good agreement between the calculations and the backtracing support the conclusion that also in the heavy mass region the friction occurs through one-body dissipation. One has to note that the same code had been used to analyze the Hinde et al data presented in figure 2. In these data only mean values of the neutron multiplicities were accessible and led thus to the huge dissipation coefficients obtained for the heavy systems as ^{64}Ni + ^{208}Pb which is the system presented on the left of figure 7. This means that if the separation between quasi-fission and fusion-fission can be obtained, no discrepancy remains between the different mass regions and that one-body is large enough to reproduce the experimental data over the whole mass range.

This spectacular first-time agreement with one-body dissipation led us to investigate heavier systems. Nevertheless the backtracing procedure presents some weak points, the main one being that it requires huge statistics for the neutron experimental observables. Thus it cannot be used in the superheavy mass region where the cross sections are very small. Moreover the neutron multiplicities in this domain will drop.

A new protocol, T$_H$OMATE, able to handle low statistics data [12] has been developed and applied to ^{48}Ca + ^{208}Pb, ^{244}Pu leading to ^{256}No and ^{228}114 respectively, at 40 MeV excitation energy.

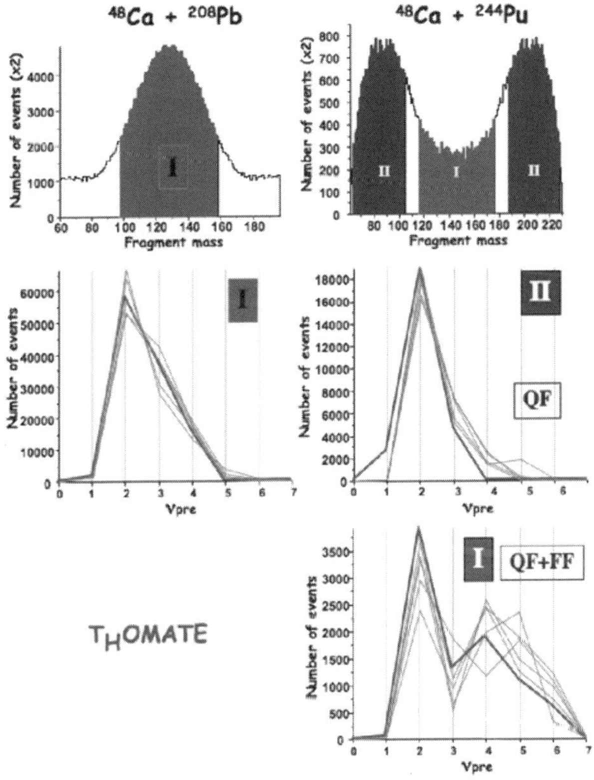

Figure 8. Fragment mass distributions (top) for ^{48}Ca + ^{208}Pb (left) and ^{48}Ca + ^{244}Pu (right) and the neutron pre-scission multiplicities obtained with the T$_H$OMATE protocol for a symmetric fragment mass selection (I) and an asymmetric one (II) in the second case.

Figure 8 shows the results obtained in both cases. In the ^{48}Ca + ^{208}Pb case, the symmetric fragment mass selection shown on the left top of the figure leads to a pre-scission neutron multiplicity (left) presenting only one component: fusion-fission. For ^{48}Ca + ^{244}Pu, two fragment mass selections have been used. The asymmetric one (II), leads to a neutron pre-scission multiplicity with one component with a smaller mean value: quasi-fission, whereas the symmetric selection (I) shows two components as for Z = 110 of figure 7: quasi-fission presenting the same distribution as for the asymmetric selection (II) centred around 2 and fusion-fission around 4. These results have been confronted to calculations of the Aritomo et al model [5] as shown in figure 9.

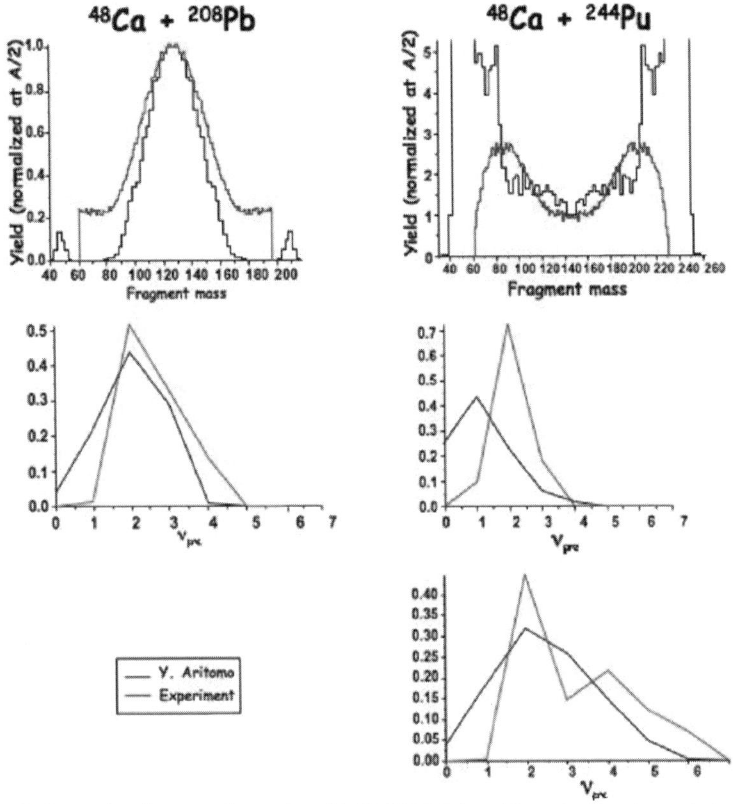

Figure 9. Comparison between the results of T_HOMATE deduced from the experimental observables and the calculations of Aritomo et al for ^{48}Ca + ^{208}Pb (left) and ^{48}Ca + ^{244}Pu (right). The top figures show the fragment mass distributions and the lower figures the neutron pre-scission distributions for the same fragment mass selections as in figure 8.

The theoretical calculations reproduce well the ^{48}Ca + ^{208}Pb data where only one process occurs: fusion-fission. For ^{48}Ca + ^{244}Pu, in which there is a mixing of quasi-fission and fusion-fission, the agreement is less good. In particular the model of Aritomo et al underestimates the fusion-fission which is almost absent.

The T_HOMATE protocol has then been applied to ^{58}Fe + ^{244}Pu leading to Z=120 [13]. Figure 10 shows the TKE-fragment mass distribution for this system (left). When different mass selections are applied, the pre- and post-scission multiplicity correlation (right) presents a similar behaviour as for the Z = 114 case. The asymmetric fragment masses correspond to the quasi-fission process. The symmetric mass fragments correspond to a mixing of the two processes: quasi-fission and fusion-fission. When the mass selection is taken more central, the fusion-fission component increases compared to that of the quasi-fission.

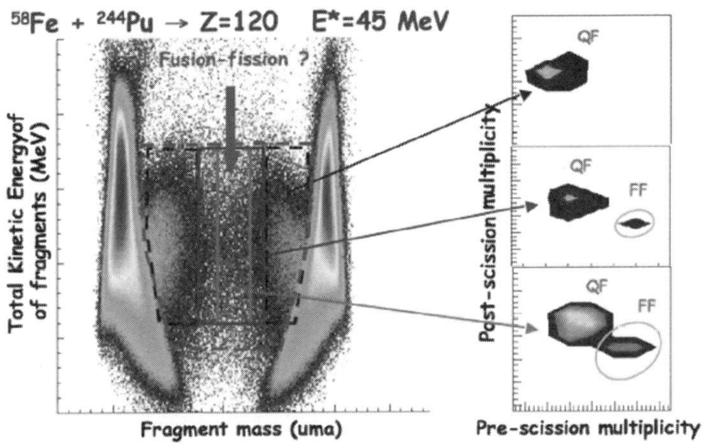

Figure 10. TKE-fragment mass distribution (left) for ^{58}Fe + ^{244}Pu leading to Z = 120 and the correlation between pre- and post-scission neutron multiplicities (right) obtained with T$_H$OMATE for different fragment mass selections: asymmetric fragment corresponding to quasi-fission (top), symmetric fragment masses corresponding to a mixture of quasi-fission and fusion-fission (middle and bottom).

Cross sections have been deduced from these results. They are shown on figure 11. It appears that the fusion-fission process represents 50% of the symmetric capture cross section in the case of ^{48}Ca + ^{244}Pu. This relative proportion decreases when the mass of the compound system increases and represents only 27% in the ^{58}Fe + ^{244}Pu case.

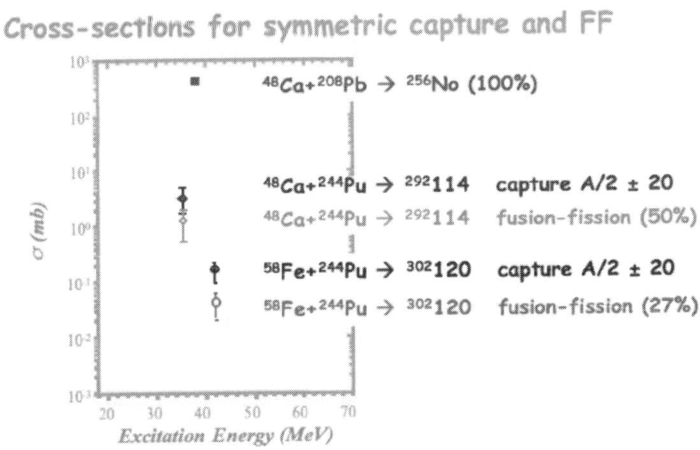

Figure 11. Cross sections for symmetric capture and fusion-fission as a function of excitation energy for ^{48}Ca +^{208}Pb, ^{244}Pu and ^{58}Fe + ^{244}Pu deduced with T$_H$OMATE from the experimental neutron observables.

2.4. Conclusions

Quasi-fission and fusion-fission can compete in the symmetric fragment mass region. The two processes can be discriminated by the neutron pre-scission distribution obtained thanks to the T$_H$OMATE analysis. Cross sections could be deduced. It appears that in the superheavy mass region, the quasi-fission dominates the fusion-fission. For the first time it was possible to show that in this mass region one-body dissipation reproduces well the data in the superheavy region and that there is no need for two-body dissipation.

3. Perspectives

3.1. Entrance channel effects

The different TKE-fragment mass distributions presented above show that entrance channel asymmetry and shell effects have a large influence on the competition between quasi-fission and fusion-fission. Figure 12 gives a compilation of systems induced by ^{48}Ca. It shows the evolution of the capture on quasi-fission ratio as a function of the mass of the composite system. This ratio first decreases with the mass of the composite system, presents a minimum for the ^{208}Pb target system and then increases drastically for targets heavier than Pb. A probable explanation for this non-trivial behaviour is the corresponding probability of formation of the different spherical shells during the decay of the composite system.

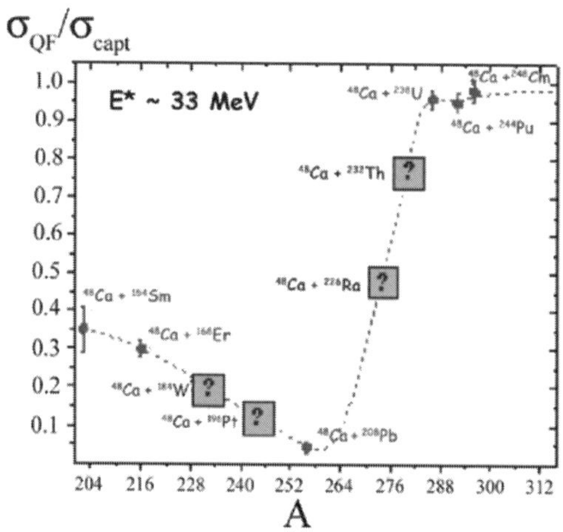

Figure 12. Ratio of quasi-fission to capture cross sections as a function of the composite nuclei masses for ^{48}Ca induced reactions at an excitation energy of 33 MeV.

One has to note that the curve in figure 12 has been obtained by measuring the scission fragments alone, considering thus as true fusion-fission all mass symmetric fragments whereas quasi-fission fragments correspond to asymmetric fragments. The neutron results presented above have shown that this is not true. This will not affect the superheavy region where the asymmetric quasi-fission dominates but it might have an influence on the lighter systems where the competition between quasi-fission and fusion-fission is unknown.

Obviously the behaviour of the competition is driven by shell effects as can be seen also on the mass distributions of figure 13 which shows TKE-fragment mass and mass distributions for different systems induced by ^{48}Ca. For systems with a target heavier than Pb (right), a large cross section corresponding to a quasi-fission process leading to Pb and the corresponding light fragment is observed. For targets lighter than Pb (left), the quasi-fission process is concentrated around nuclei with Z=28 and/or N=28,50 and their corresponding heavy fragment. The symmetric splitting region is characterized by the occurrence of a peak around mass A= 132.

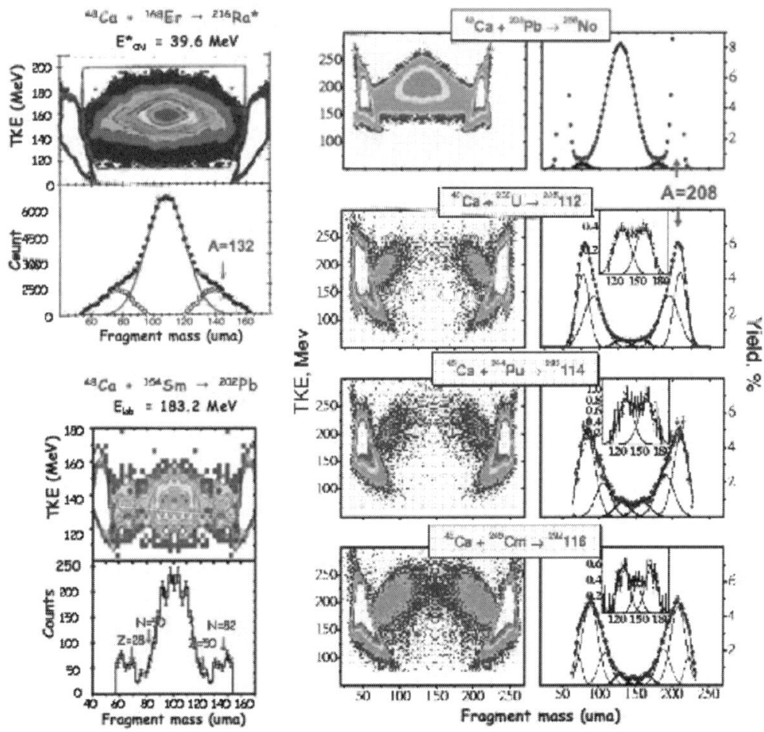

Figure 13. TKE-fragment mass and fragment mass distributions for different systems induced by a ^{48}Ca beam. The peaks in the mass distributions are centred on closed shells.

3.2. γ information

A huge amount of data concerning the scission fragments have been accumulated. The neutron information has shown to be a powerful tool to separate the quasi-fission from the fusion-fission. It is now important to investigate the capture dynamics with these means over a wide mass range in a more systematic way. In order to perform a more complete investigation the γ information will be included in future experiments.

The GDR may be a complementary way to separate quasi-fission from fusion-fission. The angular momentum may behave differently for quasi-fission and fusion-fission. In figure 14, calculations performed by Nasirov et al [13] in the frame of the DNS model for $^{48}Ca + ^{154}Sm$ at two excitation energies show two components for fusion-fission (dashed curve) and quasi-fission (dotted curve) which might be separated.

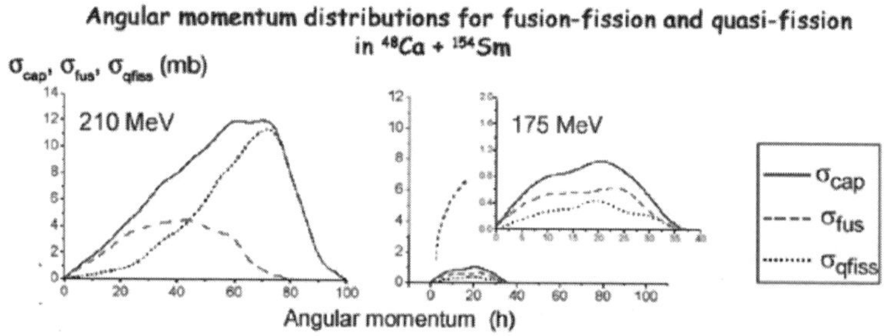

Figure 14. Angular momentum distributions calcuted by Nasirov et al in the frame of the DNS model for $^{48}Ca + ^{154}Sm$ at two excitation energies.

There exists little data on simultaneous neutron and γ measurements. The existing data measured separately often disagree as can be seen on figure 15 which gives lifetimes for ^{232}Th [14]. Thus it would be of great importance to measure the n- γ correlation.

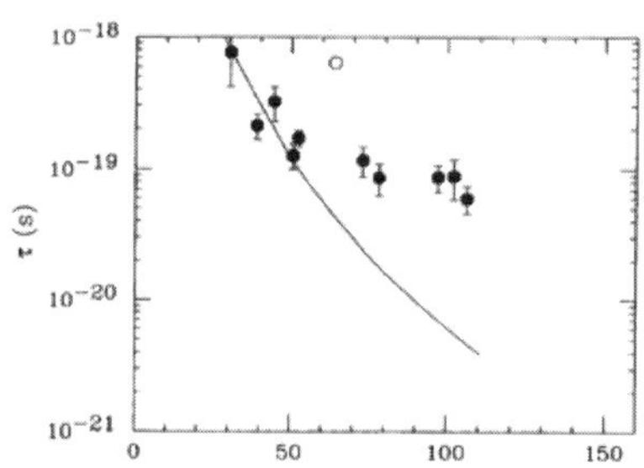

Figure 15. Fission lifetimes extracted from pre-scission neutron multiplicities for compound nuclei between Z=85 and Z=91 (full circles) and from γ-ray measurements of ^{232}Th (open circles) versus initial energy. From [14].

A systematic investigation of the systems presented in figure 12 associating the experimental setup CORSET for the scission fragments and DEMON for the neutrons with a set of BaF2 as HELENA from Milano (Italy) or the Château de Cristal of GANIL (France) will be performed.These measurements are planned at Legnaro in Italy for the lighter systems and at Dubna in Russia of the superheavy systems.

Acknowledgments

This work has been supported by INTAS 97-11929 and 00-655.

References

1. J. Wilczynski et al, *Phys. Rev.* **C54**, 325 (1996).
2. K.Siwek-Wilczynska et al, *Phys. Rev.* **C51**, 2054 (1995).
3. D. Hinde et al, *Phys. Rev.* **C45**, 1229 (1992).
4. M.G. Itkis et al, Proceedings the International Workshop of Fusion Dynamics at the Extremes, Dubna, Russia, may 25-27, 2000. Ed. By Yu.Ts. Oganessian and V.I Zagrebaev, World Scientific, p. 93.
5. Y. Aritomo et al, *Nucl. Phys.* **A738**, 221 (2004).
6. M. Morjean et al, *Nucl. Phys.* **A630**, 200c (1998).
7. P. Désesquelles et al, *Nucl. Phys.* **A604**, 183 (1996).
8. E. de Góes Brennand, PhD Thesis, Univ. of Brussels, Brussels, 2000.
9. K. Pomorski et al, *Nucl. Phys.* **A679**, 25 (2000).
10. L. Donadille et al, *Nucl. Phys.* **A656**, 259 (1999).
11. B. Benoit, PhD Thesis, Univ. of Brussels, Brussels, 2000.
12. T. Materna, PhD Thesis, Univ. of Brussels, Brussels, 2003 and *Nucl. Instr. Meth. in Phys. Res.* **A544**, 679 (2005).
13. N. Amar, PhD Thesis, Univ. of Caen, Caen, 2003.
14. P.Paul and M.Thoennessen,. *Annu. Rev. Part. Sci.* **44**, 65 (1994).

THE PROCESSES OF FUSION-FISSION AND QUASI-FISSION OF SUPERHEAVY NUCLEI

M. G. ITKIS, A. A. BOGACHEV, I. M. ITKIS, J. KLIMAN, G. N. KNYAZHEVA, N. A. KONDRATIEV, E. M. KOZULIN, L. KRUPA, Yu. Ts. OGANESSIAN, I. V. POKROVSKY, E. V. PROKHOROVA, A. Ya. RUSANOV, R. N. SAGAIDAK

Flerov Laboratory of Nuclear Reactions, JINR, 141980 Dubna, Russia

B. R. BEHERA, L.CORRADI, E. FIORETTO, A. GADEA, A. LATINA, A. M. STEFANINI, S. SZILNER

INFN, Laboratori Nazionali di Legnaro, Legnaro (Padova), Italy

S. BEGHINI, G. MONTAGNOLI, F. SCARLASSARA

INFN and Universita di Padova, Padova, Italy

M. TROTTA

Dipartimento di Fisica and INFN, Napoli, Italy

V. BOUCHAT, F. HANAPPE, T. MATERNA

Universite Libre de Bruxelles, Bruxelles, Belgium

O. DORVAUX, N. ROWLEY, C. SCHMITT, L. STUTTGE

Institut de Recherches Subatomiques, Strasbourg, France

Results of the experiments aimed at the study of fission and quasi-fission processes in the reactions ^{48}Ca+^{144,154}Sm, ^{168}Er, ^{208}Pb, ^{238}U, ^{244}Pu, ^{248}Cm; ^{50}Ti+^{208}Pb, ^{244}Pu;^{58}Fe+^{208}Pb, ^{244}Pu, ^{248}Cm, and ^{64}Ni+^{186}W, ^{242}Pu leading to the formation of heavy and super-heavy systems with Z=82-122 are presented. Cross sections, mass-energy and angular distributions for fission and quasi-fission fragments have been studied at energies close and below the Coulomb barrier. The influence of the reaction entrance channel properties such as mass asymmetry, deformations, neutron excess, shell effects in the interacting nuclei and producing compound nucleus, the mechanism of the fusion-fission and the competitive process of quasi-fission are discussed.

1. Introduction

One of the main problems of nuclear physics is the finding of extremal conditions for the existence of atomic nuclei. The synthesis of new superheavy elements in heavy ion induced reactions is an important part of these investigations. The survivability of superheavy elements (SHE) with respect to the fission process is defined by the fission barrier height, which completely depends on nuclear shells in the superheavy region. At approaching the shell closure the fission barrier grows and the survivability of superheavy nuclei increases. The recent success in the synthesis of new elements with Z=112-118, produced in the reactions with ^{48}Ca ions and actinide targets ^{238}U, ^{244}Pu, ^{248}Cm and ^{249}Cf [1,2] confirms the existence of "the island of stability" of SHE near the proton shell closure with Z=114, 120, 122 and neutron shell N=184.

However, the fusion process of massive nuclei with heavy ions (A≥27) differs greatly from the reactions with more light ions, in which the fusion always occurs after the capture stage. An increase in the Coulomb repulsion forces between interacting massive nuclei can lead to the decay of the composite nuclear system before it reaches a compact shape near the ground state. In this case the quasi-fission (QF) process [3,4] takes place. Quasi-fission is a fast process and its experimental features differ from those of the true fusion-fission (FF) process: large width of mass distributions, asymmetries in mass angle correlation of fragments and large angular anisotropies [5]. The neutron [6] and γ-multiplicities [7,8] are also different in both processes. It has been found recently [9] that the total kinetic energy (TKE) of fragments is higher in the QF process than that in the FF-process.

Different factors influence the FF/QF competition, such as the entrance channel mass asymmetry, N/Z ratio, excitation energy and angular momentum, the deformation and orientation of colliding nuclei, shell structure of the entrance channel and the formed compound nucleus (CN).

It is well known that in heavy ion induced reactions leading to the formation of SHE the deep-inelastic and quasi-fission processes are the dominant reaction channels, whereas the fusion probability is much lower. In the reactions with ^{48}Ca-ions and actinide targets the probability of fusion relative to QF is less than 10%, and the ratio decreases for more symmetrical target-projectile combinations used in cold fusion reactions. The last SHE which can be produced in the reaction with ^{48}Ca-ions is the element with Z=118 since the heaviest possible target is Cf .Thus to produce more heavy elements more heavy projectiles such as ^{50}Ti, ^{54}Cr and ^{58}Fe should be used. However, as mentioned above, a decrease in the mass asymmetry in the reaction entrance channel leads to an increase in the QF and a decrease in the FF contributions into the capture cross sections. That is why a more detailed study of both processes is needed. The distinctive features of both processes in MED of fragments and the relative contribution into the total reaction cross-section have been studied in order to find the optimal conditions for the compound nucleus formation.

Results of the experiments devoted to the study of fusion-fission and quasi-fission processes in the reactions leading to the formation of the superheavy nuclei with Z = 102-122 are presented. The heavy ions ^{48}Ca, ^{50}Ti, ^{58}Fe and ^{64}Ni were used as projectiles. The experiments were carried out at the U-400 accelerator of the Flerov Laboratory of Nuclear Reactions (JINR, Russia) using the time-of-flight spectrometer of fission fragments CORSET [10]. Mass-energy distributions (MED) of the fragments were obtained using a kinematic coincidence method. The capture (σ_{cap}) and fusion–fission (σ_{FF}) cross-sections were deduced.

The latest results of the investigation of the fusion-fission process of superheavy nuclei and prospects of the "hot" fusion reactions are considered.

2. The fusion-fission process of superheavy nuclei with Z=102-122

2.1. *Reactions with ^{48}Ca-projectiles*

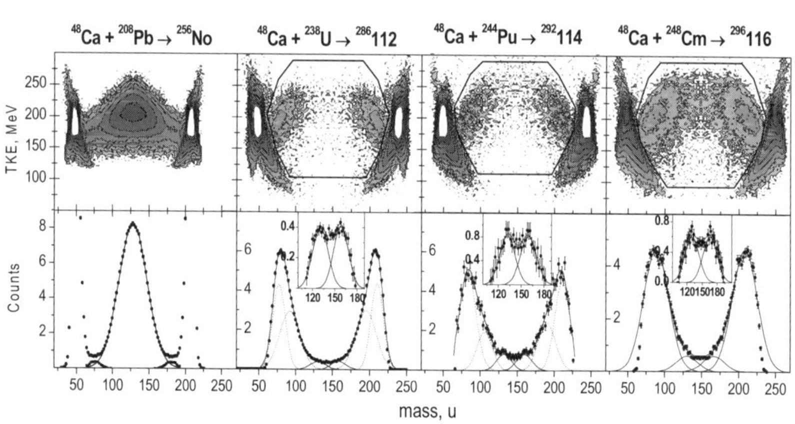

Figure 1. Mass-energy distributions of the fission fragments of ^{256}102-^{296}116 nuclei produced in "hot" fusion reactions with ^{48}Ca-projectiles.

The neutron rich isotope ^{48}Ca is used in experiments of SHE synthesis due to its undoubted advantages. Firstly, high neutron excess allows one to reach in the reaction with actinide targets N_{CN}≈170-180 in contrast to cold fusion reaction in which N_{CN}≈150-160. Secondly, the doubly magic structure of ^{48}Ca leads to the excitation energy E*≈ 30 MeV at the Coulomb barrier for elements Z= 112-116, that is by 10-15 MeV lower than in classical hot fusion reactions.

Figure 1 presents MED of fragments of the elements with Z = 102-116, produced in the ^{48}Ca induced reaction on the targets ^{208}Pb, ^{238}U, ^{244}Pu and ^{248}Cm at the excitation energy E*≈ 33 MeV. On the top of Fig.1 two-dimensional matrixes of (TKE/Mass) are presented, the mass yields of the reaction products are shown in the bottom. The main peculiarity of the data is the sharp change of the MED triangular shape in the reaction $^{48}Ca+^{208}Pb$, in which the fusion-fission process dominates, to the quasi-fission shape of MED in the $^{286}112$-$^{296}116$ nuclei. The distinctive feature of the quasi-fission process for these superheavy nuclei is the wide two-humped mass distribution with a high peak of heavy fragments near the doubly magic lead (M_H≈208). In spite of the dominating role of the quasi-fission process for these reactions we assume that in the symmetric region of the fragment masses (A/2±20u) the FF process coexists with QF. In a frame at the bottom of Fig. 1 the mass yield of the fusion-fission process is shown. It is obtained as a difference between experimental spectra and quasi-fission peak descriptions. One can see there that the mass distribution of the fusion-fission process is asymmetric in shape, with a nearly constant mass of the light fragment M_L≈132-134 u. In the case of superheavy elements the light spherical fragment with M_L≈132-134 u plays a stabilizing role, whereas the heavy fragment with M_H≈140 u is playing the same role for actinide nuclei.

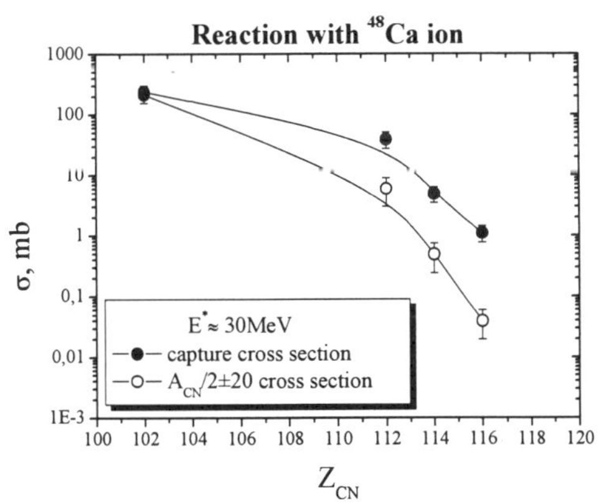

Figure 2. The capture (solid circles) and FF (open circles) cross-sections for the reactions with ^{48}Ca ions at excitation energy E*≈33 MeV.

The capture (solid circles) and FF (open circles) cross-sections for all these reactions are shown in Fig. 2 for the excitation energies E*≈16 and 30 MeV. We estimated the upper limit for the FF process as the mass region $A_{CN}/2 \pm 20$, in which the FF process, in our opinion, prevails.

2.2. *Reactions with ^{208}Pb target*

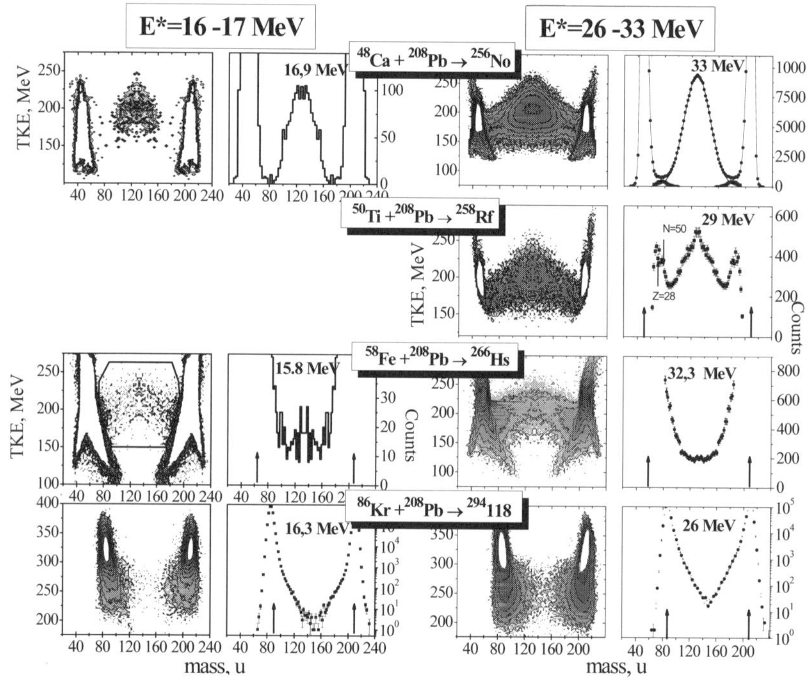

Figure 3. Two-dimensional matrixes (TKE/M) and mass yields for the reactions between ^{48}Ca, ^{50}Ti, ^{58}Fe and ^{86}Kr-projectiles and the ^{208}Pb-target. The left-hand side shows MED at the excitation energy E^*=16-17 MeV, the right-hand side – those at E^*=26-33 MeV.

One of possible ways to synthesize superheavy nuclei is "cold" fusion reactions in which one of the partners is the magic ^{208}Pb or ^{209}Bi nucleus and second one is a massive ion. Due to the magic structure of targets the excitation energy of compound nucleus at the Coulomb barrier reaches $E_x \approx$10-15 MeV, and the CN de-excitation occurs with the emission of only one neutron and γ-quanta. This strongly increases the survivability of CN at the de-excitation stage. However CN produced in these reactions have a small neutron excess and are shifted from the β-stability line by 10-15 u which results in decreasing their half-lives. Furthermore, the increase of projectile mass leads to a decrease in the entrance channel asymmetry and growth of Coulomb repulsion forces. Thus, QF contribution into the capture cross section increases. Mass-energy distributions of fission fragments for the reactions ^{48}Ca, ^{50}Ti, ^{58}Fe, ^{86}Kr+^{208}Pb at excitation energies E*≈ 16 and 30 MeV are presented in Fig. 3. It is clearly seen that at the transition from ^{48}Ca to ^{86}Kr the shape of fission fragment MED changes very strongly due to the QF process. In the case of ^{48}Ca the main process is FF of the compound nucleus ^{256}No; in the case of the the ^{50}Ti ion the contribution of the QF process into the capture cross-section is around 40%; whereas in the case of ^{86}Kr the QF is a dominant process.

At low excitation energies MEDs of the reactions with ^{48}Ca and ^{58}Fe-projectiles reveal the structure peculiarity caused by multimodal fission. The capture (solid circles) and FF (open circles) cross-sections for all these reactions are shown in Fig. 4 for the excitation energies E*≈16 and 30 MeV. We estimated the upper limit for the FF process as the mass region $A_{CN}/2 \pm 20$, in which the FF process, in our opinion, prevails. One can see that at both excitation energies the capture cross-section of the 118 element decreases by $\approx 5\times10^2$ times as compared with ^{256}No, while the FF cross-section drops by $\approx 10^4$ times. It is important to note that in the case of the reaction ^{86}Kr+^{208}Pb the ratio between the fragment yields in the region of asymmetric masses and that in the region of masses $A_{CN/2}$ exceeds by about 30 times a similar ratio for the reaction with ^{48}Ca and ^{58}Fe ions. It means that in the case of the ^{86}Kr+^{208}Pb →^{294}118 reaction in the region of symmetric masses the mechanism of QF prevails.

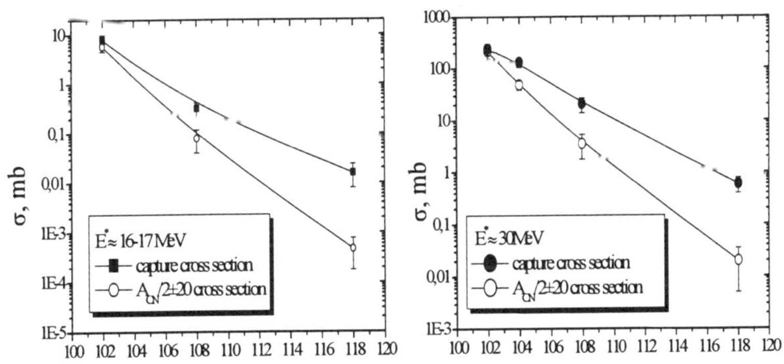

Figure 4. The capture (solid circles) and FF (open circles) cross-sections for the reactions leading to the production of elements with Z=102-118 on ^{208}Pb targets at excitation energies E*≈16 MeV and 33 MeV.

2.3. *Mass asymmetry in low energy fission of superheavy nuclei*

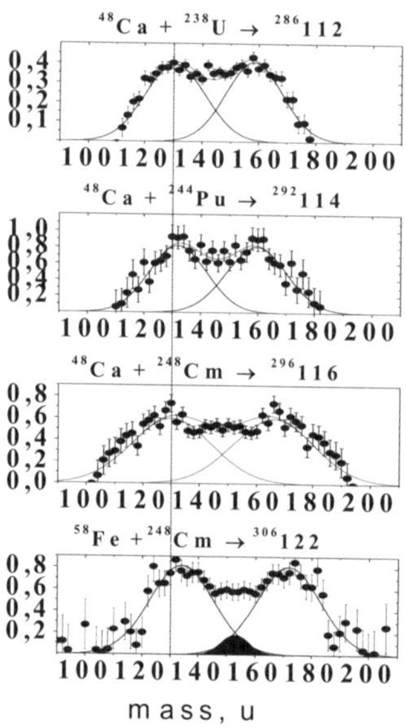

Figure 5. Mass distribution of fission-like fragments of $^{256}112$-$^{296}122$ nuclei for symmetric region.

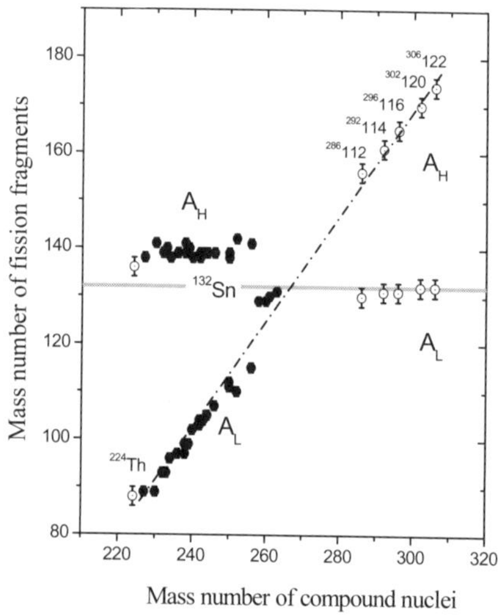

Figure 6. The dependence of the light and heavy fragment masses on the compound nucleus mass.

In Figure 5 mass distribution of fission-like fragments of $^{256}112$-$^{296}122$ nuclei for symmetric region is shown. Figure 6 show the dependence of the light and heavy fragment masses on the compound nucleus mass. It is very well seen that in the case of superheavy nuclei the light spherical fragment with mass 132-134 plays a stabilizing role, in contrast to the region of actinide nuclei.

2.4. *Shell effects manifestation*

Figure 7 shows the ratio of the QF to the capture cross-sections σ_{QF}/σ_{cap} as a function of the composite nucleus mass for the reactions with ^{48}Ca-projectiles at excitation energies E*=33-40 MeV. The solid circles represent the measured reactions; the question marks are the reactions to be investigated. Our prediction of this curve behavior is shown by the line. It is seen that the QF contribution increases for the targets lighter than ^{208}Pb and decreases at the ^{144}Sm target.

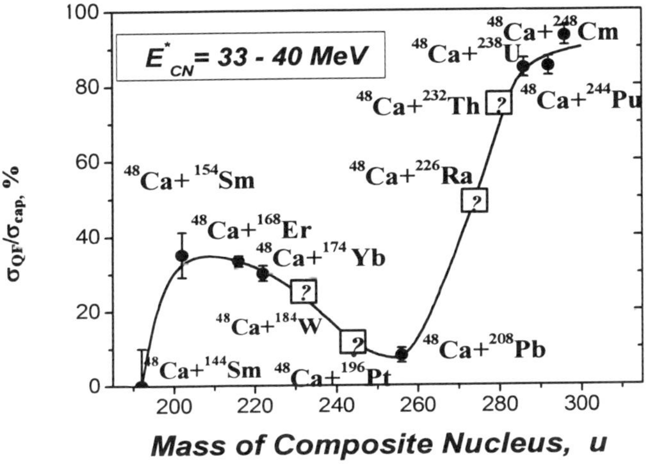

Figure 7. The ratio of σ_{QF}/σ_{cap} as function of the composite nucleus mass number for reactions with ^{48}Ca-ions and different targets.

The most probable explanation of such an unusual behavior of the ratio σ_{QF}/σ_{cap} is the influence of spherical shell closures in the entrance channel and in the nascent fragments. This means, that symmetric region of fission fragment masses $[(A_{CN}/2) \pm 20]$ seems to originate mainly from the regular fusion-fission process in the reaction $^{48}Ca+^{248}Cm \rightarrow ^{296}116$. However, as shown in [11], evolving from the initial configuration of two nuclei in contact into the state of spherical or near-spherical compound nucleus, the system goes through the same configurations through which a compound nucleus goes in regular fission, i.e., configurations close to the saddle point. In such configuration and in a state of complete thermodynamic equilibrium, the nuclear system is much likely to go into the fission channel without overcoming the saddle point, and producing a spherically symmetric compound nucleus.

The graphic example of the shell effect manifestation in the MED of the fission fragments is shown in Fig. 8. One can see (Fig. 7) that in the case of the heaviest targets ^{238}U, ^{244}Pu and ^{248}Cm the ratio σ_{QF}/σ_{cap} changes slightly, and the tendency was also observed in the σ_{ER} excitation functions for the superheavy elements [11].

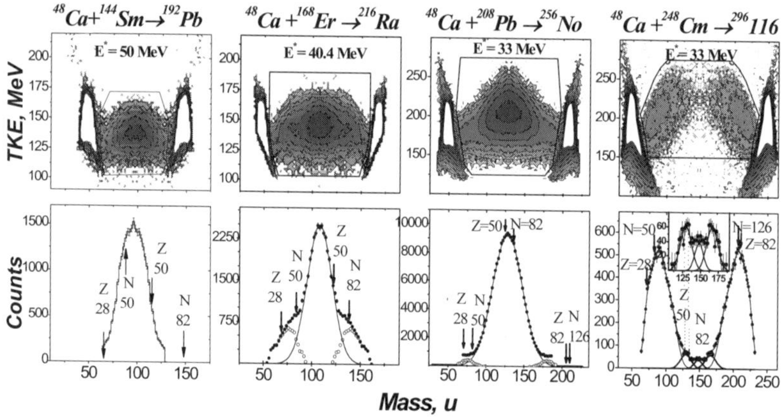

Figure 8. Two-dimensional matrices TKE-Mass (top panels) and mass yields (bottom panels) of fission fragments of ^{192}Pb, ^{216}Ra, ^{256}No, ^{286}112 nuclei produced in the reactions with ^{48}Ca.

2.5. *Transition from ^{48}Ca to ^{58}Fe-ions*

In order to investigate the dynamics of the FF/QF processes in the reactions with ^{48}Ca and ^{50}Ti-projectiles, the MED of fission fragments and excitation functions were measured in the reactions ^{48}Ca +^{246}Cm and ^{50}Ti + ^{244}Pu, leading to the formation of the ^{294}116 nucleus. The experiments with ^{48}Ca-ions showed that in the region of SHE with Z=112-116 the contribution of the QF component in the total reaction cross-sections σ_{QF}/σ_{cap} was approximately constant and higher than 90%. On the other hand, the heaviest element, which can be produced using ^{48}Ca-ions, is the nucleus with Z=118 (in the reaction with a Cf target). The production of more heavy elements demands the use of projectiles with a higher Z. Thus, the spherical neutron magic nucleus ^{50}Ti (N=28) seems to be a promising candidate for a projectile in the reactions of synthesis of superheavy nuclei.

In Figure 9 and 10 two-dimensional matrixes TKE-Mass (top panels) and mass yields of the fragments (bottom panels) for the reactions ^{48}Ca, ^{50}Ti, ^{58}Fe + ^{244}Pu and for the reactions ^{48}Ca, ^{50}Ti, ^{58}Fe + ^{208}Pb are shown.

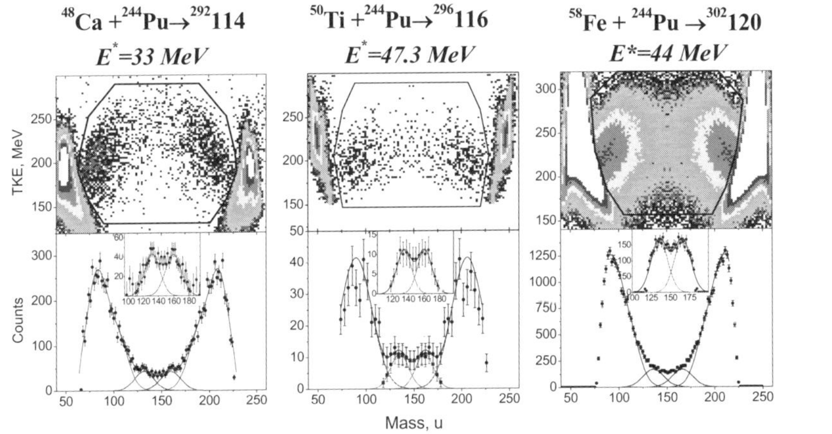

Figure 9. Two-dimensional matrixes TKE-Mass (top panels) and Mass yields of the fragments (bottom panels) for the reactions ^{48}Ca, ^{50}Ti, ^{58}Fe + ^{244}Pu.

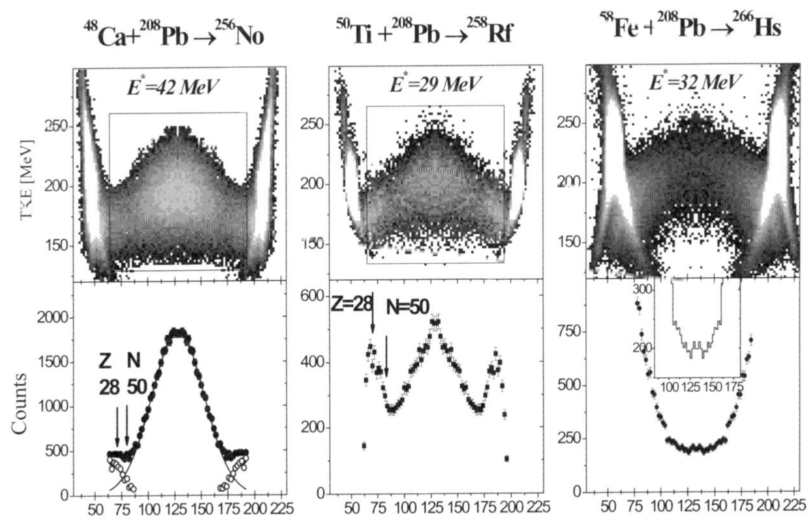

Figure 10. Two-dimensional matrixes TKE-Mass (top panels) and Mass yields of the fragments (bottom panels) for the reactions ^{48}Ca, ^{50}Ti, ^{58}Fe + ^{208}Pb.

In the formation of ^{294}116 in the reactions with ^{48}Ca and ^{50}Ti ions at energies close the Coulomb barrier it was observed that the shapes of the fragment mass distribution were alike, and the ratio σ_{FF}/σ_{cap} was approximately the same for both reactions. Figure 11 shows the capture and FF cross sections for these reactions. The experimental results show that in the transition from ^{48}Ca to ^{50}Ti ions the capture cross sections σ_{cap} and hence the fusion-fission cross sections σ_{FF} decrease by ≈ 3 times at E*=45-50 MeV. At the same time, the capture cross-sections σ_{cap} as a function of the excitation energy above the Coulomb barrier ($E^*-E^*_B$) are very similar in both reactions. That is why ^{50}Ti is a promising projectile for the SHE synthesis. However, for a more detailed study of the FF–QF competition in these reactions some new high statistics experiments are needed.

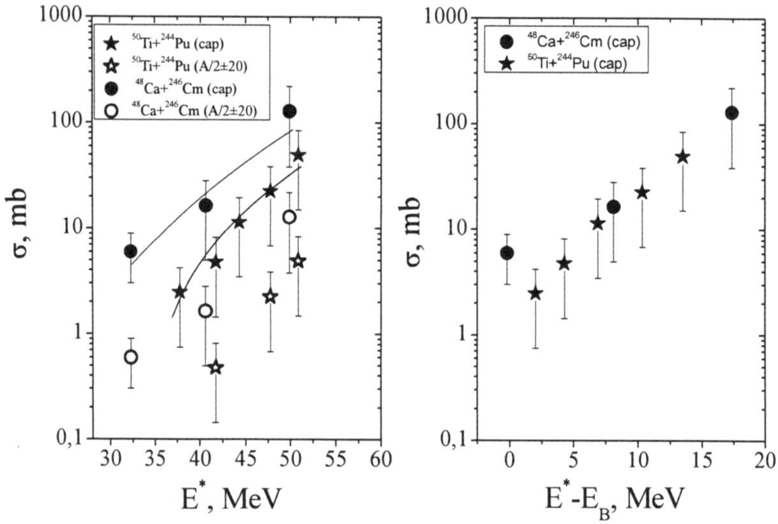

Figure 11. The capture (σ_{cap}) and fusion-fission ($\sigma_{A/2\pm20}$) cross sections as function of excitation energy E*(left-hand side) and excitation above the barrier $E^*-E^*_B$ (right-hand side).

2.6. *Reactions with ^{58}Fe and ^{64}Ni-Ions*

Figure 12 shows the data for the reactions between the ^{58}Fe and ^{64}Ni projectiles and ^{232}Th, $^{242,244}Pu$ and ^{248}Cm targets, leading to the formation of the nuclei from $^{290}116$ to $^{302}120$ and $^{306}122$ (where N = 182-184), i.e. to the formation of spherical CN, predicted by theory [12]. As seen from Fig. 12, we observe here an even stronger manifestation of asymmetric mass distributions for $^{306}122$ and $^{302}120$ fission fragments with the light fragment mass M_L≈132 u.

The corresponding structures are seen well in the ⟨TKE⟩(M) dependence. Only for the reaction $^{58}Fe + ^{232}Th \rightarrow ^{290}116$ (E*= 53 MeV) the valley in the region of M=A/2 disappears – this can be seen from the mass distribution as well as from the ⟨TKE⟩ (M) and σ^2_{TKE} (M) dependences. This fact is connected with a damping of the shell effects with an increase in the excitation energy. On the right-hand side of Fig. 12 the characteristics of $^{306}122$ fission fragments formed in the reactions $^{64}Ni+^{242}Pu$ and $^{58}Fe+^{248}Cm$ are demonstrated. Despite the fact that the compound nucleus $^{306}122$ undergoes fission at approximately the same excitation energy E*=31.5 MeV the form of the energy distributions changes to a more flat ⟨TKE⟩(M) dependence in the case of the ^{64}Ni-induced reaction.

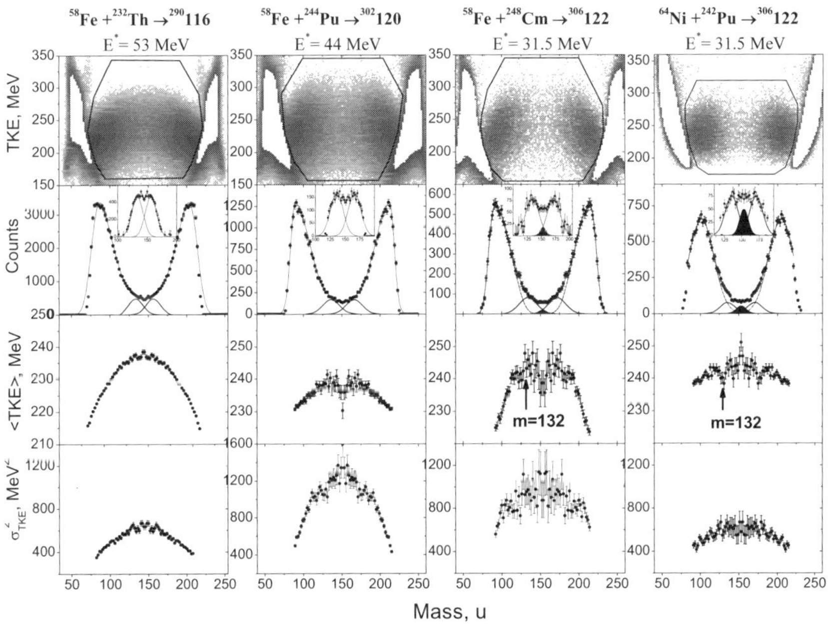

Figure 12. Two-dimensional TKE-Mass matrixes, the mass yields, ⟨TKE⟩(M) and σ_{TKE}(M) for $^{290}116$, $^{302}120$ and $^{306}122$ nuclei.

3. Capture and fusion-fission cross section

Figure 13 shows the results of measurements of the capture cross section σ_c and the fusion-fission cross section σ_{ff} for the studied reactions as a function of the initial excitation energy of the compound systems.

Comparing the data on the cross sections σ_{ff} at $E^* \approx$ 14-15 MeV (cold fusion) for the reactions ^{58}Fe + ^{208}Pb and ^{86}Kr + ^{208}Pb, one can obtain the following ratio: $\sigma_{ff}(108)/\ \sigma_{ff}\ (118) \geq 10^{\ 2}$. In the case of the reactions from ^{48}Ca + ^{238}U to ^{58}Fe + ^{248}Cm at $E^* \approx$ 33 MeV (warm fusion) the value of Z changes by the same 10 units as in the first case, and the ratio $\sigma_{ff}\ (112)/\ \sigma_{ff}\ (122)$ is $\approx$ 4-5 which makes the use of asymmetric reactions for the synthesis of spherical superheavy nuclei quite promising.

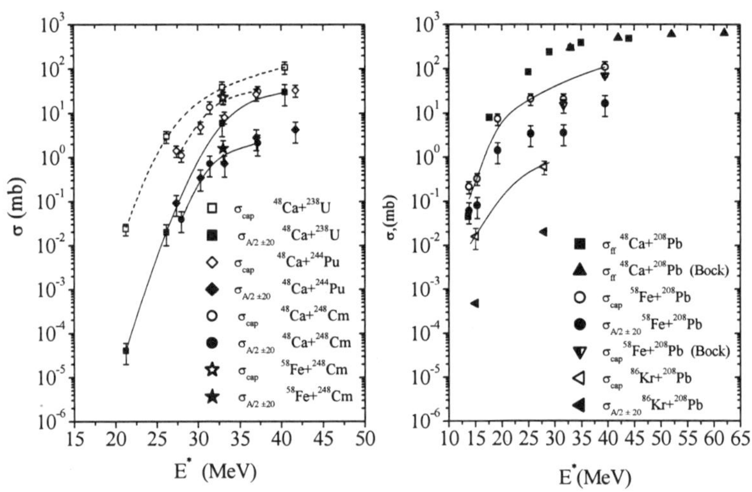

Figure 13. The capture cross section σ_{cap} and the fusion-fission cross section σ_{ff} for the production of elements 102-122 as a function of the excitation energy

Another interesting result is connected with the fact that the values of σ_{ff} for ^{256}102 and ^{266}108 at E^* = 14-15 MeV are quite close to each other, whereas the evaporation residue cross sections σ_{xn} [13] differ by almost three orders of magnitude ($\sigma_{ff}/\ \sigma_{xn}$) which is evidently caused by a change in the Γ_f/Γ_n value for the above mentioned nuclei. At the same time, for the ^{294}118 nucleus formed in the reaction ^{86}Kr+^{208}Pb, the compound nucleus formation cross section is decreasing at an excitation energy of 14 MeV by more than two orders of magnitude according to our estimations (σ_{ff} $\approx$ 500 nb is the upper limit) as compared with σ_{ff} for ^{256}102 and ^{268}108 produced in the reactions ^{48}Ca+ ^{208}Pb and ^{58}Fe +^{208}Pb at the same excitation energy. But when using the value of $\approx$ 2.2 pb for the cross section σ_{ev}(1n) from work [14], one obtains the ratio $\sigma_{xn}/\sigma_{ff} \approx 4{\cdot}10^{-6}$ for ^{293}118, whereas for ^{266}108 the ratio is $\sigma_{xn}/\sigma_{ff} \approx 10^{-6}$.

In one of recent works [15] it has been proposed that such unexpected increase in the survival probability for the $^{294}118$ nucleus is connected with the sinking of the Coulomb barrier below the level of the projectile's energy and, as a consequence, leads to an increase in the fusion cross section. However, our data do not confirm this assumption.

4. Multimodal fission

4.1. *Bimodal fission of ^{256}No and ^{250}No*

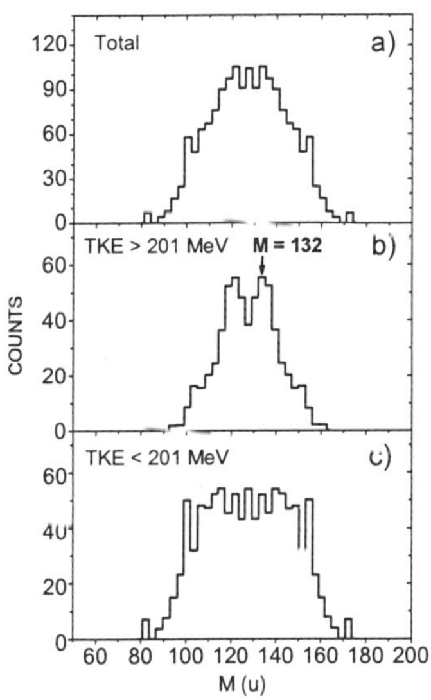

Figure 14. Mass yields of the fission fragments for the reaction ^{48}Ca+ ^{208}Pb→ ^{256}No at E*= 17.6 MeV. a) total mass yield; b) for TKE>201 MeV c) for TKE < 201 MeV.

The total and differential mass distributions for TKE > 201 MeV and TKE < 201 MeV are presented in Fig. 14 for the projectile energy E_{lab} = 211 MeV. In the case of TKE>201 (Fig 14b) a narrow two-humped structure is distinctly seen for the region of heavy masses $M_H \cong$ 131-135 u. The shape of this distribution is very similar to that in the MD of spontaneous fission of neighboring nuclei ^{259}Lr (Z = 103) [16]. Figure 14c shows a wide flat MD in the region of heavy fragment masses $M_H \cong$ 140-150 u for TKE < 201 MeV. Such a behavior of the low-energy component of MD is also typical of the superheavy nucleus fission, for example for the spontaneous fission of ^{259}Md (Z=101) [17] or ^{260}Md [18]. Thus, one can see that the ^{256}No mass distribution at E* = 17,6 MeV consists of at least two components.

Figure 15 presents energy distributions of ^{256}No for the energies of ^{48}Ca ions E_{lab}=211-242 MeV. The TKE distributions for all mass range are shown in the left-hand side of the Figure; TKE distributions for symmetric mass range M = 124-132 u (i.e. for the masses whose MD is two-humped, Fig. 14) are shown in the right-hand side. The TKE distributions of the selected symmetric masses practically do not differ from integral distributions at energies E_{lab} = 220-242

MeV. However, at the lowest projectile energies E_{lab}=217 and 211 MeV two components of the TKE appear in the region of symmetric masses, i.e., a low-energy one with $\langle TKE \rangle_{low} \sim 200$ MeV, and a high energy one with $\langle TKE \rangle_{high} \sim$ 233 MeV. These TKE values as well as mass yields are typical of the standard and SS modes in the spontaneous fission of superheavy nuclei [16, 17]. Thus, we observed here both in the mass and energy distributions the bimodality (first –LDM mode and SS-mode)of the ^{256}No induced fission, although on the whole the contribution of the SS-mode is quite small and equals ≈ 2.5 % for symmetric masses M=124-132 u at E_{lab}=211 MeV.

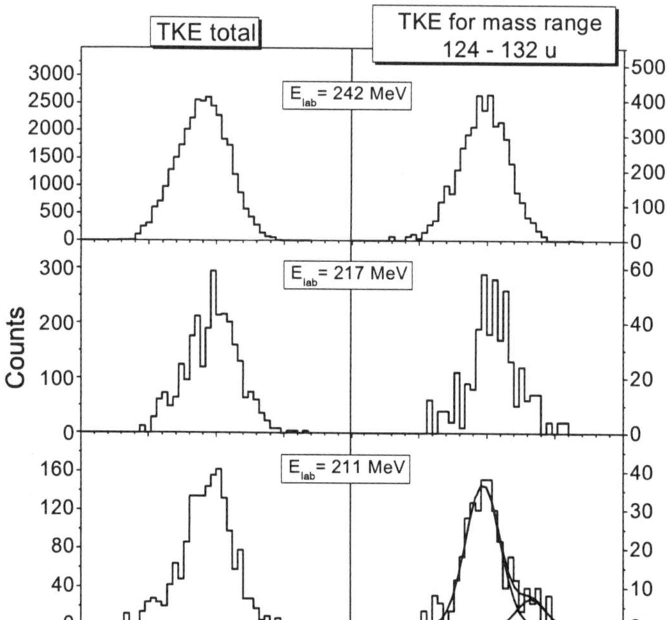

Figure 15. TKE distributions for the energies of ^{48}Ca ions E_{lab}=211-242 MeV. The integral distributions for all masses are shown in the left-hand side, for masses M = 124-132 u – in right hand side. For the lowest energies E_{lab}= 211 and 217 MeV the decompositions of the TKE into 2 modes are presented for the symmetric mass region.

Figure 16 (a) and (b) shows the mass distributions, (c) – the ⟨TKE⟩ distributions as a function of the fragment mass for ^{44}Ca + ^{206}Pb and ^{64}Ni + ^{186}W at an energy close to the Bass barrier (the compound nucleus excitation energy is about 30 MeV). One can see that mass-energy distributions for these systems are very different. In the case of ^{44}Ca, the mass distribution has a complicated structure: i) the asymmetric fission connected with the formation of the deformed shell near the heavy fission fragment mass 140; ii) the symmetric fission component determined by the effect of the Z = 50 proton shell; and iii) the quasifission component, visible around Z = 20, 28 and N = 28, 50. In contrast to this reaction, the contribution of the quasifission component into the total mass distribution in the case of ^{64}Ni + ^{186}W increases greatly. This observation is confirmed by the different behavior of the ⟨TKE⟩ distributions for fission fragments in the systems (see Fig. 16c). In the mass region $A_{CN}/2\pm20$ the ⟨TKE⟩ distributions are similar in both reactions, while in the asymmetric mass region ⟨TKE⟩ for ^{64}Ni + ^{186}W is higher than that in ^{44}Ca + ^{206}Pb. Our analysis shows that only a small part (~25%) of the fission cross section can be associated with complete fusion for the ^{64}Ni + ^{186}W system, the remainder should be attributed to quasi-fission. In the case of ^{44}Ca + ^{206}Pb, the contribution of CN-fission component into the total mass distribution is ~70%.

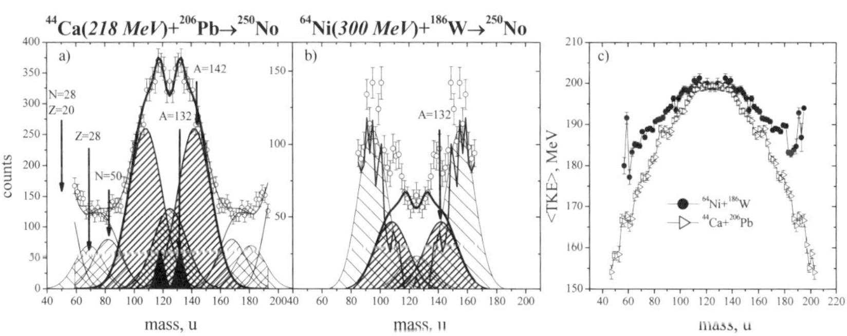

Figure 16. Mass distributions for the reactions $^{44}Ca+^{206}Pb$ (a) and $^{64}Ni+^{186}W$ (b) and average kinetic energies <TKE> as a function of fragment mass (c) for these reactions at an excitation energy of about 30 MeV.

5. Conclusions

Mass and energy distributions of fragments, fission and quasifission cross sections, have been studied for a wide range of nuclei with Z=102-122 produced in reactions with ^{48}Ca, ^{50}Ti, ^{58}Fe and ^{64}Ni ions at energies close and below the Coulomb barrier.

In the case of the fission process as well as in the case of quasifission, the observed peculiarities of mass and energy distrubutions of the fragments, the ratio between the fission and quasifission cross sections, in dependence of the nucleon composition and other factors, are determined by the shell structure of the formed fragments.

Entrance channel effect plays important role in the fusion fission dynamics and competition between Fusion-Fission and Quasi-Fission processes. The target deformation has a dominant role on the evolution of the comopsite system, whereas shell effects in exit channel determine the main characteristics of reaction fragments just as in the case of superheavy systems.

The dependence of the capture (σ_c) and fusion-fission (σ_{ff}) cross sections for nuclei ^{256}No, ^{286}112, ^{292}114, ^{296}116, ^{294}118 and ^{306}122 on the excitation energy in the range 15–60 MeV has been studied. It should be emphasized that the fusion-fission cross section for the compound nuclei produced in reaction with ^{48}Ca and ^{58}Fe ions at excitation energy of ≈30 MeV depends only slightly on reaction partners, that is, as one goes from ^{286}112 to ^{306}122, the σ_{ff} changes no more than by the factor 4–5. This property seems to be of considerable importance in planning and carrying out experiments on the synthesis of superheavy nuclei with Z>114 in reaction with ^{48}Ca and ^{58}Fe ions. In the case of the reaction ^{86}Kr+^{208}Pb, leading to the production of the composite system ^{294}118, contrary to reactions with ^{48}Ca and ^{58}Fe, the contribution of quasi-fission is dominant in the region of the fragment masses close to $A_{CN}/2$.

A further progress in the field of synthesis of superheavy nuclei can be achieved using hot fusion reactions between actinide nuclei and ^{48}Ca ions as well as actinide nuclei and ^{50}Ti, ^{54}Cr, ^{58}Fe ions. Of course, for planning the experiments on the synthesis of superheavy nuclei of up to Z=122, new research and more precise quantitative data obtained in the processes of fusion-fission and quasifission of these nuclei are required.

Acknowledgments

This work was supported by the Russian Foundation for Basic Research under Grant 03-02-16779 and INTAS grant 03-51 6417.

References

1. Yu. Ts. Oganessian, et al. Eur.Phys.J, A5(1999) 63 ; Nature 400(1999) 242
2. Yu. Ts. Oganessian, et al.; Phys. Rev. C 70, 064609 (2004) ; Phys. Scr. 125 (2006) 57
3. R. Bock, et al., Nucl. Phys. A 388 (1982) 334.
4. J.Töke et al., Nucl.Phys. A 440, 327 (1985)
5. W. Q. Shen et al., Phys. Rev. C 36, 115 (1987).
6. D.J.Hinde et al., Phys.Rev.C 45 (1992) 1229

7. B.B.Back et al., Phys.Rev.C 41 (1990) 1495
8. M.G.Itkis et al, Nucl.Phys. A 734 (2004) 136
9. A. Yu. Chizhov, et al., Phys. Rev. C 67 (2003) 011603(R).
10. E. M. Kozulin, et al., Instrum. and Exp. Techniques Vol.51 (2008) p44.
11. V.I. Zagrebaev, *Phys. Rev.* **C64**, 034606 (2001) ; *J. Nucl. Radiochem. Sci.*, **3,** No 1, 13 (2001).
12. Z. Patyk, A. Sobiszevski, Nucl. Phys. A 533, 132 (1991).
13. Hofmann S., Münzenberg G., *Reviews of Modern Physics*, 72 (2000) №3.
14. Ninov V. *et al, Phys. Rev. Lett.* 83 (1999) 1104.
15. Myers W. D. and Swiatecki W.J., *Phys. Rev. C*, **62** (2000) 044610.
16. T. M. Hamilton, et al., Phys. Rev. C 46 (1992) 1873.
17. E. K. Hulet, et al., Phys. Rev. Lett, 56 (1986) 313; Phys. Rev. C 40 (1989) 770; Phys. At. Nucl. 57 (1994) 1099.
18. J. F. Wild, et al., Phys. Rev. C 41 (1990) 640.

FISSION AND QUASIFISSION IN THE REACTIONS ^{44}CA+^{206}PB AND ^{64}NI+^{186}W*

G.N. KNYAZHEVA, A. YU. CHIZHOV, M.G. ITKIS, E.M. KOZULIN
Flerov Laboratory of Nuclear Reaction, JINR, 141980, Dubna, Russia

V.G. LYAPIN, V.A. RUBCHENYA, W.H. TRZASKA
Department of Physics, University of Jyvaskyla, FIN-40014 Jyvaskyla, P.O. Box 35, Finland

S.V. KHLEBNIKOV
V.G. Khlopin Radium Institute, 194021, St. Petersburg, Russia

The mass-energy and angular distributions of binary fission-like fragments produced in the reactions ^{44}Ca+^{206}Pb and ^{64}Ni+^{186}W, leading to the same compound nucleus ^{250}No have been measured at two CN excitation energies 30 and 40 MeV. The presence of quasifission component was observed for the both systems. But in the case of ^{64}Ni-ion the quasifission process dominates, while in the case of ^{44}Ca-ion the main process is fission of the compound nucleus ^{250}No. From measured angular distributions the reaction times for quasifission and fission were found for both reactions.

1. Introduction

The study of nuclear reactions with heavy ions is of great interest for understanding of nuclear interactions. The collision of two heavy nuclei can lead to different reaction channels such as elastic, quasielastic, deep-inelastic, fast-fission, quasifission (QF), compound nucleus (CN) -fission, formation of evaporation residue.

For the CN-fission process, the projectile is completely absorbed by the target, and the resulting compound nucleus reaches its equilibrium (near spherical) deformation before fission. For this to occur, it is necessary that the system has fission barrier. If the angular momentum is very high, the fission barrier is reduced to zero. Such fission-without-barrier is generally called fast-

* This work is supported by the Russian Foundation for Basic Research (Grant Number 03-02-16779).

fission, and this process should be faster that CN-fission. One of possible processes in heavy-ion induced reactions is QF. It has been observed in reactions between nuclei with larger Coulomb energy ($Z_1Z_2 \geq 1600$). Although, such systems have the fission barriers, they also show evidence for fission occurring in fast time scale. It has been suggested that due to the high Coulomb repulsion the fission trajectory does not pass inside the true (unconditional) fission barrier. In other words, true fission does not occur.

In resent years a big progress in synthesis of new syperheavy nuclei was made [1, 2]. All these elements were formed in the reactions with ^{48}Ca-ion. It is known that deep-inelastic and QF processes are dominating channels in this type of reaction, whereas fusion probability is small fraction of the capture cross section [3]. The competition between the formation of CN and QF is, probably, determined by the properties of di-nuclear configuration at contact point, where entrance-channel effects are expected to play the major role in the reaction dynamics [4]. The relative orientation of the symmetry axis of the deformed nuclei changes the Coulomb barrier and the distance between the centers of colliding nuclei. Decreasing the entrance-channel mass-asymmetry $\eta=(M_1-M_2)/(M_1+M_2)$ with increasing compound nucleus fissility are responsible for the appearance of the QF effect manifested in the suppression of the fusion cross section for combinations leading to strongly fissile compound nucleus [4, 5, 6, 7].

This paper presents the investigation of the role of entrance-channel mass-asymmetry on the fusion probability in the reactions ^{44}Ca + ^{206}Pb and ^{64}Ni + ^{186}W leading to the same ^{250}No*- CN. The investigation of the CN-fission of nuclei with Z>100 obtained in the reactions with Ca, Ti, Fe, Ni ions is very important for further planning of new superheavy nuclei synthesis, since these nuclei belong to the class of transfermium elements, the stability of which is mainly determined by the shell effects as it is in the case of superheavy elements.

2. Experiment

Experiments were carried out at the K-130 accelerator of the University of Jyväskylä. Beam intensity on the target was ≈2-5pnA, depending on the experimental conditions.

The targets were placed in the center of a ∅ = 150 cm scattering chamber. They were produced by metal evaporation of ^{206}Pb (150 μg/cm^2) and of $^{186}WO_3$ (150 μg/cm^2) on carbon backing (40 μg/cm^2). In experiment the backings faced the beam.

Four silicon detectors monitored continuously the beam intensity and position. They detected Rutherford yields from the target and were placed above and below, and to the left and right of the beam line at the same scattering angle Θ_{lab}=16°. Small corrections to measured cross sections were made according to observed variations of the relative yields in the monitors, due to possible changes of beam focusing and its position during the various experimental runs.

Precise mass-energy distributions of binary reaction events were measured using the ToF-ToF spectrometer CORSET [8] consisted of compact start detectors and position-sensitive stop detectors. The arms of the spectrometer were installed at angles 60°-60 with respect to the beam axis that corresponds to 180° in the center of mass system for fission fragments. The distance between start and stop detectors is 15 cm. Start detectors were placed at the distance of 5 cm from the target. The angular acceptance for both arms was 25^0 in-plane and $\pm 10^0$ out-of-plane, the mass resolution was about 2–3 amu. The efficiency of registration of each arm was determined with α -source and it was ≈86%. It is mainly depends on the transparency of electrostatic mirror of start detector.

To measure mass-angular distributions of fission fragments we also installed ToF-E telescopes at the angles of 5^0, 10^0, 20^0, 30^0 and 60^0 to the beam line. The distance between start and stop detectors for these arms is 18 cm. Starts detectors were placed at the distance of ~30 cm from the target. The angular acceptance of each ToF-E telescope was $\pm 1^0$ and mass resolution corresponded to 3amu. The registration efficiency of each arm also was obtained with α -source and it was ≈75%.

3. Results and analysis

3.1. *Mass-energy distributions of the binary reaction products for the $^{44}Ca+^{206}Pb$ and $^{64}Ni+^{186}W$*

Mass-energy distributions of fission fragments have been measured in the ^{44}Ca+^{206}Pb $\rightarrow ^{250}$No*, ^{64}Ni+^{186}W $\rightarrow ^{250}$No* at the excitation energies of the compound nucleus 30 and 40 MeV. Figure 1 displays the main characteristics of fission fragment mass-energy distributions for all these reactions (from top to bottom: two-dimensional matrix of counts as a function of mass and total kinetic energy; mass distribution for fission events involved into the contour line; average total kinetic energy of fission fragments involved into the contour line as a function of mass). Table 1 contains the information about some entrance channel characteristics for these reactions.

In Fig. 1 (upper panels), the reaction products with masses close to those of the projectile and the target are identified as elastic, quasielastic and deep-inelastic events in the two-dimensional TKE-mass matrix, and it will not be considered in this paper. The reaction products in the mass range A=60÷180 a.m.u. can be identified as totally relaxed events, i.e. as fission-like events. Mass distributions for fission-like fragments have the complicated structure: the symmetric component is typical for the fission of excited CN; the asymmetric fission is connected with the formation of the deformed shell near the heavy fission fragment mass 140 and the asymmetric "shoulders", visible around Z = 28 and N = 50, 88. In the study of the spontaneous fission properties of heavy actinide nuclei (Z > 98) it was found that the transition from asymmetric to symmetric fission in the No isotopes takes place somewhere at N = 154 [9], mass distribution of No-isotopes which have less neutrons than 154 is asymmetric and its properties mainly determined by the heavy fragment, peaked around A=140.

Table 1. The main characteristics of studied reactions.

Reaction	Z_1Z_2	η	Elab, MeV	E_{CN}^*, MeV	<L>,
$^{44}Ca+^{206}Pb$	1640	0.648	217	30	15
			227	40	28
$^{64}Ni+^{186}W$	2072	0.488	300	30	12
			311	40	30

For the composite systems which are similar to $^{44}Ca+^{206}Pb$ it was shown [5] that the main process in these reactions is CN-fission. To extract the CN-fission from all fission-like products for this reaction we made the decomposition of observed mass distribution on the symmetric and the asymmetric (with the mass of the heavy fragment A_H=140) components. This decomposition is given in Fig. 1 by solid lines. It is clearly seen that the symmetric component increases with increasing of the CN-excitation energy that should be observed for the fission of excited CN. The shaded area is the difference between experimental mass distribution and our selection of CN-fission events.

The theoretical calculation for heavy and superheavy region of nuclei [10] predicts the value for the height of fission barrier for ^{250}No around 4 - 5MeV. This fission barrier doesn't disappear for all angular momentum brought into the composite systems. It means that this asymmetric "shoulder" may be explained in the term of QF and we may exclude the fast-fission process from our consideration of possible reaction channels in studied reactions.

In contrast to the reaction with ^{44}Ca, the contribution of the asymmetric "shoulders" into the total mass distribution in the case of ^{64}Ni + ^{186}W greatly increases, the QF is dominating process. The angular momentum for the ^{44}Ca+^{206}Pb and ^{64}Ni+^{186}W systems are similar, so, they should not reveal the significant difference between the mass distributions of CN-fission for both systems. We suggest that the main process for symmetric mass split of the ^{64}Ni+^{186}W system is CN-fission. In order to estimate the upper limit of CN-fission for this reaction, we inscribe the mass distribution for the CN-fission extracted from the ^{44}Ca+^{206}Pb reaction at the same excitation energies in the experimental mass distribution of the ^{64}Ni+^{186}W reaction.

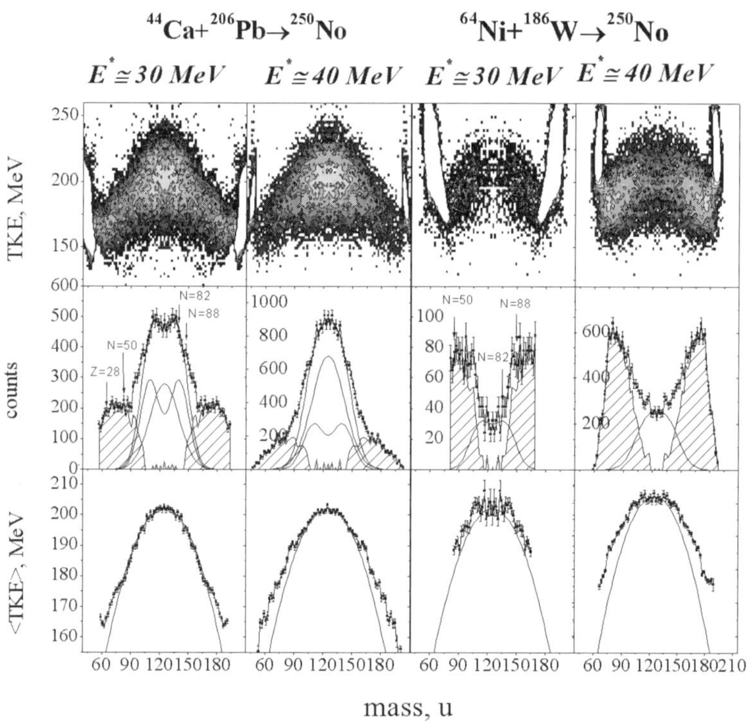

Figure 1. Two-dimensional TKE-mass matrixes (upper panels), yields of fragments and their <TKE> as a function of the fragment mass (middle and bottom panels, respectively) in the ^{44}Ca+^{206}Pb (coulombs 1 and 2) and ^{64}Ni+^{186}W (coulombs 3, 4) at CN excitation energies 30 and 40 MeV.

This observation is confirmed by the different behavior of the <TKE> distributions for the fission fragments in the systems. In the mass region

$A_{CN/2}\pm 20$ the <TKE> distributions are similar for both reactions, while for the asymmetric mass region <TKE> for $^{64}Ni + ^{186}W$ is higher than that for the $^{44}Ca + ^{206}Pb$ reaction.

The arrows in Figure 1 show the positions of the spherical closed shells with Z=28 and N=50, 82 and deformed neutron shell N=88 [11], derived from the simple assumption on the N/Z equilibration. In the case of the $^{44}Ca+^{206}Pb$ the major part of the QF component fits into the region of these shells, and its maximal yield is a "compromise" between Z=28 and N=50. In the case of the $^{64}Ni+^{186}W$ the closed shell N=50 and deformed shell with N=88 play important role in the formation of the QF asymmetric component and the drift of mass to the symmetry is more pronounced.

3.2. *Mass-angular distributions*

The analysis of the mass-angular distributions of the fission fragments allow one to derive the QF and FF components from all fission-like products detected in the experiment. According to standard formalism [12], the angular distribution of the fission fragments for the CN-fission in the centre-of-mass system is given by the expression

$$W(\theta)=\sum_{I=0}^{\infty}(2I+1)T_I\frac{\sum\limits_{K=-I}^{K=I}\frac{1}{2}(2I+1)\left|d_{0K}^{I}(\theta)\right|^2\exp\left\{-\frac{K^2}{2K_0^2(I)}\right\}}{\sum\limits_{K=-I}^{I}\exp\left\{-\frac{K^2}{2K_0^2(I)}\right\}} \qquad (1)$$

where I is the spin of the CN, d_{0K}^{I} is the symmetric top wave function, K is the projection of the spin I on the axis of symmetry, K_0 is the variance of the K distribution and T_I is the transmission coefficient for the I-th partial wave. Within the framework of this model the fragment angular distribution depends only on the spin of the CN via the transmission coefficient T_I and the parameter K_0, where T is the temperature, J_{eff} is the effective moment of inertia. From average γ-ray multiplicity for the system $^{48}Ca+^{208}Pb$ the following relation was obtained $J_0/J_{eff}=0.79$ [6] where J_0 is the moment of inertia of sphere of the same mass. In our estimation we take the same value for the effective moment of inertia.

The angular distribution for the asymmetric (where we expected the QF process) and symmetric (where we assume the domination of CN-fission) mass split for both reactions was extracted. In Figure 2 the angular distributions for the selected mass bins of fission-like fragments are shown. The solid curves are fits to the experimental data which are given by

$$d\sigma / d\theta = 2\pi \sin\theta \cdot (a + b e^{\beta(\theta-\pi/2)}) \cdot W(\theta), \tag{2}$$

where β is a slope parameter in the exponential decay function reproducing the evident forward - backward asymmetry, and a, b are normalization parameters corresponding to the symmetrical and asymmetrical parts of angular distributions. The value of slope parameter β was fixed on -0.02 for all mass bins. Approximately the same value for this slope parameter was found in [13]. One can see, that for both reactions angular distributions are symmetrical for all symmetrical masses and could be described very well with eq. (7) with b=0, while for the asymmetric mass region the significant forward-backward asymmetry in angular distribution is observed and fitted well by Eq. (2) with parameter a=0. The parameter of K_0 obtained from this fitting is listed in Table 2 for all cases.

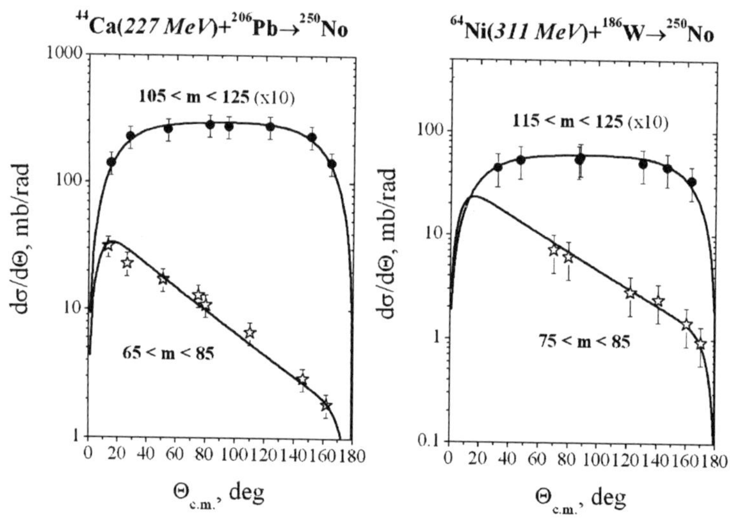

Figure 2. Differential cross section for fission-like fragments for the reactions $^{44}Ca+^{206}Pb$ and $^{64}Ni+^{186}W$ for different mass bins. The solid lines are the best fits to the data using Eq. (2).

3.3. *Time scale for CN-fission and QF*

From the mass-angle correlation it is evident that the composite system, which suffers the asymmetric mass split (A $\approx$ 80 a.m.u) for both reactions, rotates less than one turn. Consequently, it is possible to estimate the reaction time using these angular distributions in terms of the angle of rotation $\Delta\theta$ of the composite system during the reaction, provided that the relevant angular momentum and moments of inertia are known [7].

For the symmetric mass split the angular distribution is isotropic for the reactions with ^{44}Ca and ^{64}Ni and the value for the value of K_0 agrees well with expected for the CN-fission. It means that typically the nucleus probably rotates several times before scission and the main process, leading to the symmetric mass split, is CN-fission for both reactions. Experimentally, evidence for long CN-fission time scales comes from fission angular distributions, pre-scission neutron measurements also indicate long fission time scale of several 10^{-20} s for CN-fission process [14, 15]. In the paper [7] it was found that the characteristic relaxation time for the mass-asymmetry degree of freedom is $(5.2\pm0.5)\times10^{-21}$s. Besides, it appears that there is time lag of about 2×10^{-21}s before the mass drift starts.

For the asymmetric masses where the QF process dominates, the reaction time is shorter in order of magnitude that for CN-fission and the value of K_0 is less than it should be for the CN-fission. The angles of rotations and reaction times for CN-fission and QF are given in Table 2.

Table 2. Angles of rotation and reaction time for CN-fission and QF processes in the reactions ^{44}Ca+^{206}Pb and ^{64}Ni+^{186}W.

Reaction	<l>,	<m>, a. u.	TKE, MeV	K_0,	$\Delta\tau$, deg.	$\Delta\tau$, 10^{-21}s
^{44}Ca+^{206}Pb	28	75	177.5	6.0	231	3.6
		116	200	16.7	>270	>32.4
^{64}Ni+^{186}W	30	80	187	7.0	226	4.4
		120	200	16.7	>270	>30.2

3.4. *Fusion probability for the ^{44}Ca+^{206}Pb and ^{64}Ni+^{186}W*

From the mass-energy and angular distributions it follows that the main process for the symmetric mass split is CN-fission of ^{250}No for both studied reactions and our selection of the fission events from TKE-mass matrixes is correct. In that way we can estimate the fusion probability for the reactions $^{44}Ca+^{206}Pb$ and $^{64}Ni+^{186}W$, so as the cross section of the evaporation residues formation is around 1nb, and fission is the main decay channel for this compound nucleus. The Table 3 represents the capture and CN-fission cross sections for studied reactions. Only a small part around 30% of the fission cross section can be associated with CN-fission for the $^{64}Ni + ^{186}W$ system, the remaining part should be attributed to QF. In the case of $^{44}Ca + ^{206}Pb$, the contribution of CN-fission component into the total mass distribution is ~70%.

Table 3. The capture and CN-fission cross sections for $^{44}Ca+^{206}Pb$ and $^{64}Ni+^{186}W$ reactions.

Reaction	E_{lab}, MeV	σ_{cap}, mb	σ_{CN}/σ_{cap},%	σ_{CN}, mb
$^{44}Ca+^{206}Pb$	217	-	60±5	-
	227	135±20	80±5	108±22
$^{64}Ni+^{186}W$	300	-	<30	-
	311	86±13	25±5	22±6

4. Summary

The mass-energy and angular distributions of binary fission-like fragments produced in the reactions $^{44}Ca+^{206}Pb$ and $^{64}Ni+^{186}W$, leading to the same CN – ^{250}No, have been measured at two CN excitation energies of 30 and 40 MeV.

Pronounced QF component was observed for the more symmetric combination $^{64}Ni + ^{186}W$, when compared with the $^{44}Ca + ^{206}Pb$ system. Our analysis shows that only a small part (~30%) of the fission cross section can be associated with CN-fission for the $^{64}Ni + ^{186}W$ system, the remaining part should be attributed to QF. In the case of $^{44}Ca + ^{206}Pb$, the contribution of CN-fission component into the total mass distribution is ~70%.

Preferential forward-peaking of light-mass fission fragment (A≈80a.m.u) demonstrated that QF bypass the CN stage and occurs in the scales shorter than rotation period for the ^{250}No. From the mass-angle correlation it was found that the reaction time for QF component is about $(3\text{-}5)\times10^{-21}$s, while for the CN-fission the reaction time is greater than 30×10^{-21}s.

Acknowledgments

These experiments were performed by the FLNR and the Academy of Finland under Physics Program at JYFL. The authors are grateful to the staff of the JYFL cyclotron for their careful work.

References

1. Yu Ts. Oganessian et al., *Phys. Rev.* **C70,** 064609 (2004).
2. Yu. Ts. Oganessian et al., *Phys. Rev.* **C74,** 044602 (2006).
3. M.G. Itkis et al., *Nucl. Phys.* **A734**, 136 (2004).
4. W.Q. Shen et al., *Phys. Rev.* **C36**, 115 (1987).
5. B.B. Back, *Phys. Rev.* **C31**, 2104 (1985).
6. R. Bock et al., *Nucl. Phys.* **A388**, 334 (1982).
7. J. Toke et al., *Nucl. Phys.* **A440**, 327 (1985).
8. N.A. Kondratiev et al., in *Dynamical Aspects of Nuclear Fission, Proc. of 4th Intern. Conf.*, Časta-Papiernička 1998, p.431 (WS, Singapore, 1999).
9. E.K. Hulet, *Yad. Fiz.* **57**, 1165 (1994).
10. Z. Patyk, R. Smolanczuk and A. Sobiczewski, *Nucl. Phys.* **A626** , 337 (1997).
11. B.D. Wilkins, E.P. Steiberg and R.R. Chasman, *Phys. Rev.* **C14**, 1832 (1976).
12. I. Halpern and V.M. Strutinski, in *Proceedings of the Second UN International Conference on the Peaceful Uses of Atomic Energy*, Geneva, Switzerland, 1957, (United Nations, Geneva, Switzerland, 1958), p. 408.
13. B. B. Back et al., *Phys. Rev.* **C53** (1996) 1734.
14. W.P. Zank et al., *Phys. Rev.* **C33**, 519 (1986).
15. D.J. Hinde et al., *Nucl. Phys.* **A452**, 550 (1986).

MASS-ENERGY CHARACTERISTICS OF THE REACTIONS $^{58}Fe+^{208}Pb \to ^{266}Hs$ AND $^{26}Mg+^{248}Cm \to ^{274}Hs$ AT COULOMB BARRIER

E.V. PROKHOROVA, L. KRUPA, A.A. BOGACHEV, I.M. ITKIS, M.G. ITKIS, M. JANDEL, J. KLIMAN, G.N. KNYAZHEVA, N.A. KONDRATIEV, E.M. KOZULIN, YU.TS. OGANESSIAN, I.V. POKROVSKY, A.YA. RUSANOV, V.M. VOSKRESENSKI

Flerov Laboratory of Nuclear Reaction, JINR, 141980, Dubna, Russia

V. BOUCHAT, F. HANAPPE, T. MATERNA

Universite Libre de Bruxelles, 1050 Bruxelles, Belgium

O. DORVAUX, N. ROWLEY, C. SCHMITT, L. STUTTGE

Institut de Recherches Subatomiques, F-67037 Strasbourg Cedex, France

The mass-energy distributions of the fission fragments, fusion-fission (σ_{FF}) and capture (σ_{cap}) cross-sections in the reactions $^{58}Pb+^{208}Pb$ and $^{26}Mg+^{248}Cm$ were measured at the U400 accelerator of Flerov Laboratory of Nuclear Reactions (JINR, Russia) with use of double-arm time-of-flight spectrometer CORSET. The influence of reaction entrance channel on the competition between fusion-fission and quasi-fission processes was studied. Strong manifestation of the shell effects in mass distributions of fusion-fission and quasifission fragments was observed. The bimodal fission phenomena were found for both ^{266}Hs and ^{274}Hs nuclei at low excitation energies. For several beam energies of ^{26}Mg and ^{58}Fe studied in this work, the average neutron multiplicities $\langle M_n \rangle$ and γ–ray multiplicities $\langle M_\gamma \rangle$ in coincidence with fission fragments were also measured. Local minima are observed in $\langle M_\gamma \rangle$ as a function of mass suggesting the great influence of nuclear structure of fission fragments on $\langle M_\gamma \rangle$. The different dependence of $\langle M_n \rangle$ and $\langle M_\gamma \rangle$ as a function of fission fragment mass, total kinetic energy and excitation energy for fusion-fission and quasifission processes were observed.

1. Introduction

One of the main motivations in great interest of studying the entrance-channel dependence of the fusion-fission reactions between heavy ions from experimental [1,2,3,4] and theoretical [5] point of view, during the last years, is connected with successful Dubna experiments on the synthesis of 113-118 element isotopes [6]. These experiments on competition between the fusion-fission and quasi-fission extended the number of reactions investigated previously [7,8,9]. The main difficulty in their experimental study stems from the fact that the two-body exit channels result predominantly from nonequilibrium fission-like processes, namely, deep inelastic collision (DIC) and quasi-fission (QF). The compound nucleus (CN) fission in more symmetric heavy systems is the minor reaction mechanism. The quasi-fission conceptually bridges the gap between DIC, where the reactions partners get into close contact to exchange many particles without altering their average mass and charge and the complete fusion process where the reactions partners lose their identity after forming the compound nucleus. The mass distributions of the quasi-fission processes can extend toward mass symmetry, approaching $A_{CN}/2$, where $A_{CN}=(A_p+A_T)$ is the mass of the composite system and A_p and A_T are the projectile and target masses, respectively. The mass-symmetric region may therefore be populated by both the CN-fission process and the QF process. It thus becomes often difficult to separate the contributions of CN fission from the QF process on the basis of the fragment-mass distributions alone.

A number of systematic studies done so far have identified the following experimental signatures of quasifission: (1) fragment mass distributions is wider than the mass distributions resulting from fusion-fission reactions, (2) asymmetries in the mass-angle correlations increasing with the target mass, and (3) angular anisotropies significantly larger than those in fusion-fission reactions [9]. Later more differences between FF and QF were observed: higher TKE and lower neutron and γ–ray multiplicities for QF compared to CN fission [2,3].

Competition of CN fission and QF and its manifestation in mass-energy characteristics of reactions $^{58}Fe+^{208}Pb\rightarrow^{266}Hs$ and $^{26}Mg+^{248}Cm\rightarrow^{274}Hs$ close to and below the Coulomb is one of two main subjects studied in the present work. Additional information on this competition is obtained from study on neutron and γ–ray emission measured in coincidence with fission fragments.

A second factor influencing the study of reactions $^{58}Fe+^{208}Pb\rightarrow^{266}Hs$ and $^{26}Mg+^{248}Cm\rightarrow^{274}Hs$ at low excitation energies in the present work is related to manifestation of bimodal fission in this nuclei region (Super-Short mode). If to consider the properties of true fission process of superheavy elements, today it is well known that MEDs of fission fragments have bimodal nature in the spontaneous fission [10, 11, 12] in Fm-Rf region of nuclei chart. Bimodality means the co-existence of two different types (modes) of fission with special mass (symmetric or asymmetric shapes) and TKE characteristics in the same nucleus. It deals with valley structure of Potential Energy Landscape due to shell effects. The contribution of various modes can be very different in neighboring nuclei and isotopes [12, 13]. At moderate excitation energies (E^* = 10-20 MeV) shell effects begin to fade away, however the modal structure is still observed. Recently the bimodality was observed for ^{256}No at excitation energies E^* = 17.6 - 40.0 MeV [14] and also for ^{270}Sg (Z_{CN} = 106), at E^* = 28 MeV [15].

Theoretical calculations made for the superheavy nuclei show that at the potential energy surface in the multidimensional deformation space of fissioning nuclei there exist at least two fission valleys. One of the valleys, named by U.Brosa *et al.* [16] the Super-Short (SS) valley is connected with a possibility for these nuclei to have in both fragments the number of neutrons and protons close to magic numbers N = 82 and Z = 50. This possibility will fast disappear at moving from Fm to more heavy nuclei. However the calculations show that the SS-valley should exist with the superheavy nuclei up to ^{270,272}Hs (Z = 108) [17]. Since $A_{CN}/2$=133 is close to double magic nucleus ^{132}Sn one could expect manifestation of SS mode at low excitation energies also in our experiment.

2. Experiment

The experiment was carried out on the U-400 accelerator of the Flerov Laboratory of Nuclear Reactions (Dubna). The beam currents of ^{26}Mg ^{58}Fe ions were 10-20 nA and the targets were 180-220 μg/cm^2 layer of ^{208}Pb evaporated onto 50 μg/cm^2 carbon backing and 125 μg/cm^2 layer of ^{248}Cm (125 мкг/см^2) evaporated onto 20 μg/cm^2 ^{27}Al backing and covered by carbon layers (40 μg/cm^2) from both sides. The measurements were made in the energy range E_{lab}=282-324 MeV (excitation energy E^*=14-46.8 MeV) for ^{58}Fe and E_{lab}=125-160 MeV (excitation energy E^*=31.7–63.4 MeV) for ^{26}Mg. A kinematic coincidence method [18, 19, 7, 9] was used for the registration of reaction products using a double-arm time-of-flight spectrometer CORSET [20]. The arms of the CORSET spectrometer were set symmetrically to the beam axis at the angle $\Theta_{lab1}=\Theta_{lab2}=55°$. A schematic drawing of the experimental arrangement is shown in Fig.1.

Each arm of the CORSET spectrometer consists of a Start micro-channel plate (MCP) detector with an electrostatic mirror and an assembly of position-sensitive Stop MCP-detectors. The coordinate system of Stop MCP-detectors was constructed with the delay-line readout technique and allowed one to measure coordinates in (X) and perpendicular (Y) to the horizontal plane. The size of each detector in the assembly is 6x4 cm^2. Two vertical and two horizontal copper strip masks with thickness 1 mm were mounted on the surface of each Stop detector as X and Y coordinate references. Start detectors were installed at a distance of 4 cm away from the target. The converter foils of electrostatic mirrors of Start detectors were made from mylar (170 μg/cm^2) with evaporated gold (30 μg/cm^2). Each Stop detector array included 4 microchannel plate detectors. The distance between the target and Stop detectors was 18.0 cm. The solid angle of the spectrometer was 560 msr.

The detected reaction products were recorded into the files on the computer hard disk event by event. Each event consists of two coincident fragments and includes the following parameters: time interval between Start1 and Stop1 detectors (time-of-flight Tof1), between Start2 and Stop2 (Tof2), between Start1 and the Stop2 detector (long time-of-flight, LTof2), and between Start2 and the Stop1 (LTof1), the coordinates of particles in Stop detectors X1, Y1, X2, Y2, and amplitudes of signals from each MCP detector. The event was written if the signals from both Stop detectors and any Start detector were registered. The advantage of using of two Start detectors is the possibility to measure of normal TOF and long Tof's and hence to determine the efficiency of each of Start detector.

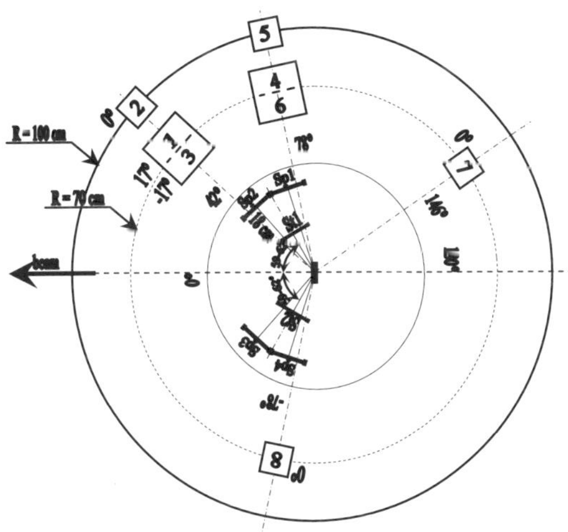

Figure 1. The scheme of experimental setup.

The data processing was performed in a usual procedure [9] assuming two-body kinematics. Primary masses, velocities, energies and angles in the center-of-mass (CM) system of reaction products were calculated from measured velocities and angles in the lab-system using the mass and momentum conservation laws. For the calculation of primary masses and energies the iteration cycle was used which takes into account a change in the velocities due to ionization losses [21] in the target, backing and converter foils of electrostatic mirrors. The time-of-flight calibration was performed by the energies of registered elastically scattered ^{58}Fe (^{26}Mg) and ^{208}Pb (^{248}Cm) ions. The spectrometer mass resolution (FWHM) was estimated by the peaks of elastic scattering events and was about 2–3 u. The selection of events from random coincidences and incomplete linear momentum transfer was performed event by event from the analysis of folding angle correlations in the center-of-mass system in- and out-of the reaction plane. Only events corresponding to the full linear momentum transfer (FLMT) were selected for a subsequent analysis. A circular contour "FoldXY" which is the circle with the radius $R = 5^\circ$ and center in $\Theta_{CM} = 180^\circ$ and $\Psi_{CM} = 180^\circ$ was used for the selection of FLMT events. $\Theta_{CM} = \theta_{1CM} + \theta_{2CM}$ is the sum of angles in-plane, while $\Psi_{CM} = \psi_{1CM} + \psi_{2CM}$ is the sum of angles out-of-plane of both reaction products in the CM system.

For detection of neutrons and prompt γ-rays a time-of-flight neutron spectrometer DEMON [22], consisted of 8 detectors, and four 7.62×7.62 cm NaI(Tl) detectors were used in the experiment (see Fig. 1). Neutron detectors are indicated in Fig. 1 by N1 to N8. They were placed both in plane and out of plane at distances ranging from 70 to 100 cm. Neutrons were separated from γ-rays by pulse shape discrimination. Their energy is determined by time-of-light measurement over a flight path. The time resolution of DEMON is about 1.5 ns. The discrimination thresholds were set using ^{22}Na, ^{137}Cs, ^{60}Co and ^{241}Am. The efficiencies of the neutron detectors were determined using a ^{252}Cf source mounted at the target position, supplemented for high neutron energies by calculated values from a Monte Carlo code.

The decomposition of the total neutron multiplicities into its pre- and post-scission components was obtained by means of a three moving source fit. Neutrons were supposed to be emitted isotropically in corresponding rest frames with Maxwellian energy distributions from the compound nucleus and two fully accelerated fission fragments. The whole set of DEMON detectors was taken into account in the χ^2 minimization using the MINUIT code [23]. The detailed description of neutron data processing can be found for example in [24].

The NaI(Tl) detectors were placed in the lower hemisphere at 35 cm distance from the target. The threshold for γ-ray registration was set to 100 keV and upper range to 5 MeV. The energy calibration was carried out by using discrete γ-transitions from standard calibrated sources such as ^{22}Na, ^{60}Co and ^{152}Eu, before, in the middle, and after experiment. No significant shifts were observed in the amplitude and threshold positions. To reduce the accidental coincidences and contribution from neutrons the time between pulses from fission fragments and γ-rays were measured. The time resolution of the γ ray detectors was 10 ns. A contribution from neutrons to γ-ray multiplicities was estimated to be about 5 percent and the accidental coincidences were less than 1 percent.

The response matrix techniques [25] was used to carry out the data processing of γ-rays. The data from all detectors have been summed in order to obtain good statistics. We analyzed the γ-rays as functions of both fragments. Thus, we do not distinguish between the gammas from light and heavy fragments and we do not into account the angular distribution of γ-rays with respect to the fission axis. All results are then presented per fission event. The total efficiency as a function of energy and response matrix of the NaI(Tl) detectors was taken from experimental calibration (by ^{60}Co γ–ray source placed at the target position) and simulation using EGSnrc code [26].

3. Experimental results and analysis

3.1. *Mass-energy distributions of the binary reaction products for the $^{58}Fe+^{208}Pb$ and $^{26}Mg+^{248}Cm$*

The main experimental results of mass-TKE distributions of fission like fragments for both studied reactions are presented in Fig. 2 and 3. The figures show the formation of the isotopes of ^{266,274}Hs in the "cold" fusion (^{58}Fe+^{208}Pb → ^{266}Hs) and "hot" fusion (^{26}Mg+^{248}Cm→^{274}Hs) reactions. Although in the former reaction ^{266}Hs is produced at lower excitation energies which should increase the CN survivability in the de-excitation process, the quasi-fission process is nevertheless the main reaction mechanism at the asymmetry of the entrance channel η=0.56. In the ^{26}Mg-induced reaction a more n-rich isotope ^{274}Hs is formed with the asymmetry coefficient η=0.81.

The two-dimensional TKE/Mass matrixes (left-hand side panels) and mass yields (right-hand side panels) for the reaction ^{58}Fe+^{208}Pb→ ^{266}Hs at the excitation energies of the compound nucleus E*= 14-39.5 MeV are shown in Figure 2. It is clearly seen from Figure 2 that the fragment distribution is two-humped, similar to that observed earlier in the reactions ^{48}Ca+^{238}U, ^{244}Pu, ^{248}Cm [1], but with the mass of the heavy fragment m_H≈200 u. It is difficult to separate in this case the quasi-fission events from those of elastic scattering and deep inelastic transfer. It is seen nevertheless that the quasi-fission peak of the heavy fragment shifts from m_H≈208 u to m_H≈198-200 u in contrast to the reactions studied before [1]. If one uses the assumption on the mass-charge independence, the spherical shell with Z_L=28 manifests itself in the light fragment. Both the lead shell in the heavy fragment and the nickel shell in the light fragment obviously influence the shape of the mass distribution of quasi-fission peaks.

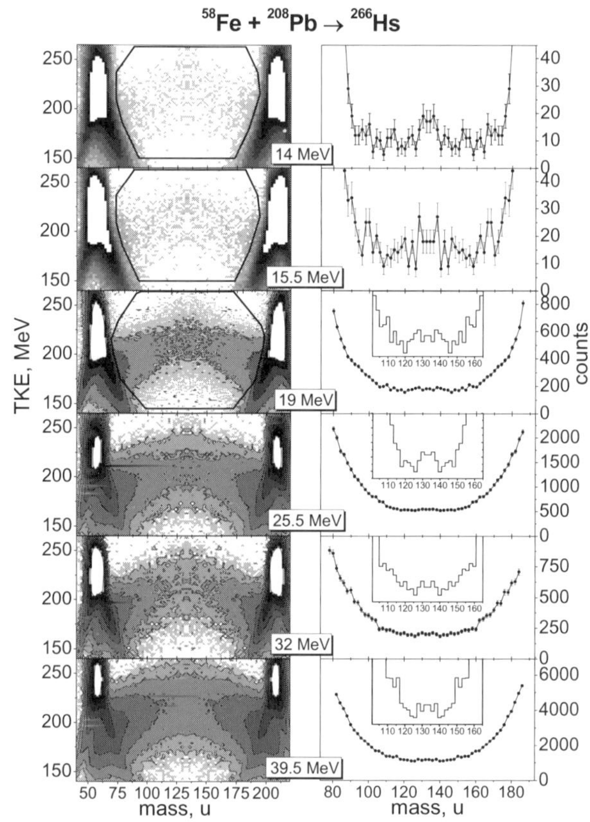

Figure 2. Two-dimensional distributions of the reaction products (TKE, Mass) and mass yields, for reaction $^{58}Fe+^{208}Pb \rightarrow ^{266}Hs$ at excitation energies E*= 14-39.5 MeV. In the right panels the mass yields of symmetric mass division in enlarged scale (insertions, $M=A_{CN}/2\pm35$) are also presented. The black solid lines separate the fission-like events (see text).

As one can see from the central part of Fig. 3 mass-TKE matrixes in the case of reaction $^{26}Mg+^{248}Cm \rightarrow ^{274}Hs$ have triangular shapes typical of the fission of heated nuclei, described by the Liquid Drop Model (LDM) [27]. Only at lower excitation energies E* = 31.7 - 35.3 MeV there appears some difference in the matrix shape from the LDM predictions.

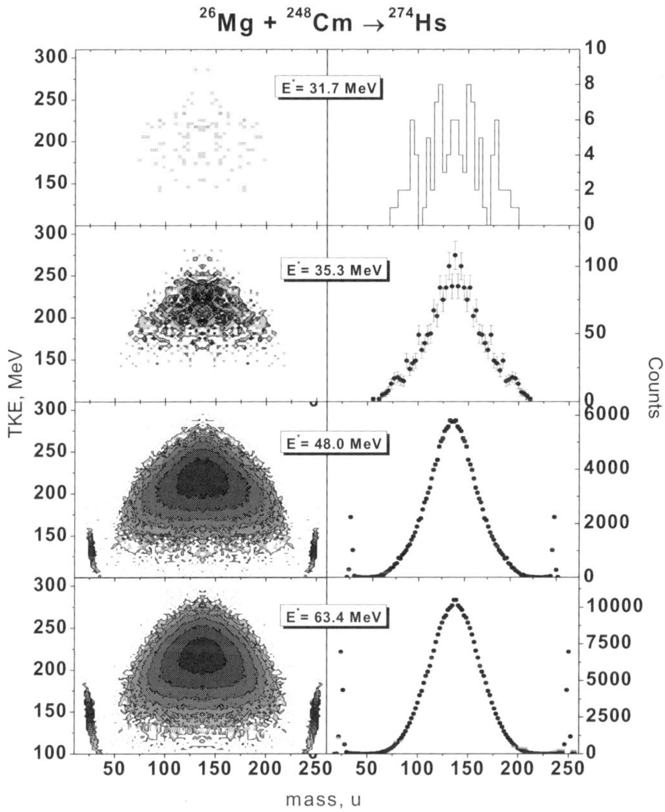

Figure 3. Two-dimensional distributions of the reaction products (TKE, Mass) and mass yields (right panels), for reaction $^{26}Mg+^{248}Cm \rightarrow ^{274}Hs$ at excitation energies E*= 31.7-63.4 MeV.

The arrows in Figure 5 show the positions of the spherical closed shells with Z=28 and N=50, 82 [28], derived from the simple assumption on the N/Z equilibration. In the case of the $^{28}Mg+^{248}Cm$ the major part of the QF component fits into the region of these shells, and its maximal yield is a "compromise" between Z=28 and N=50. In the case of the $^{58}Fe+^{208}Pb$ the closed shell N=50 and deformed shell with N=88 play important role in the formation of the QF asymmetric component and the drift of mass to the symmetry is more pronounced (see Fig. 2).

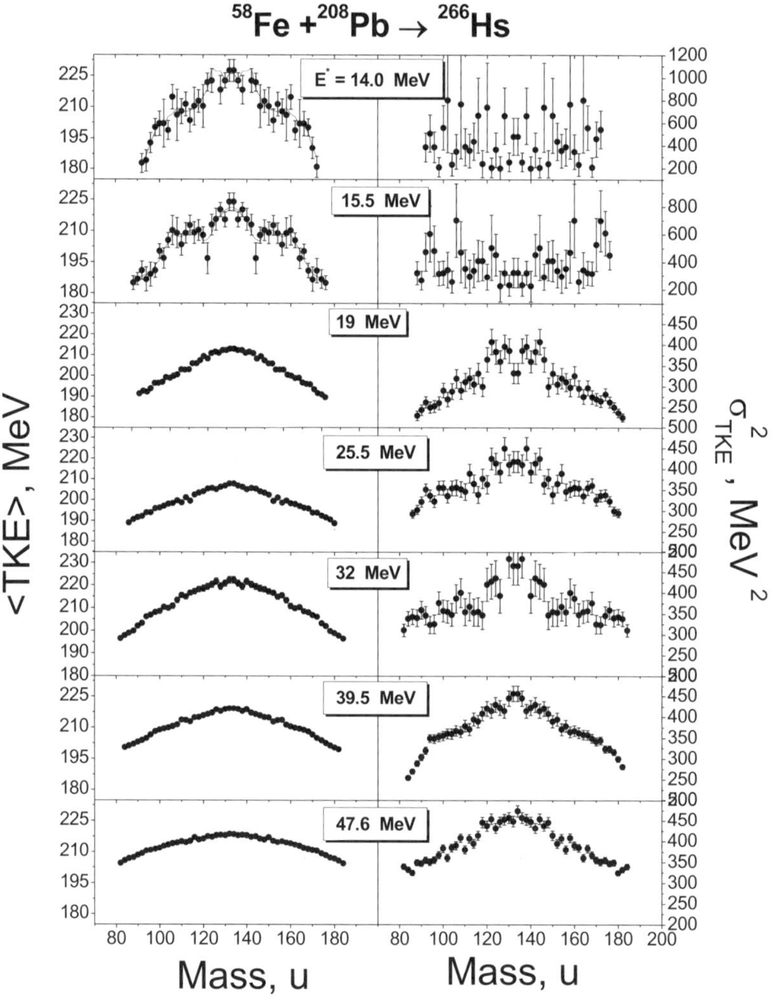

Figure 4. Average total kinetic energy <TKE> and its dispersion σ^2_{TKE} as a function of mass for reaction ^{58}Fe+^{208}Pb→ ^{266}Hs at excitation energies E*= 14-47.6 MeV.

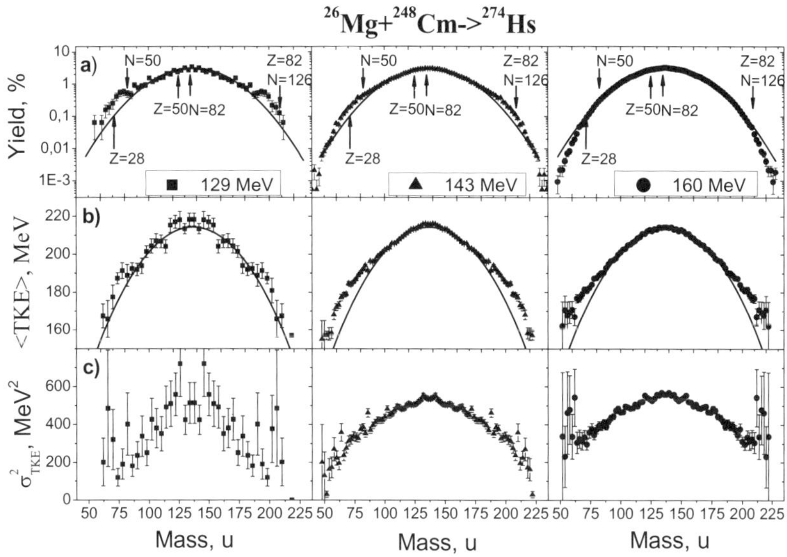

Figure 5. Mass yield, average total kinetic energy ⟨TKE⟩ and TKE dispersion $\sigma_{TKE}(M)$ for the reaction ^{26}Mg+^{248}Cm at energies E_{lab}=129, 143 and 160 MeV.

3.2. *Bimodal fission of ^{266}Hs and ^{274}Hs*

Mass yields for the reaction ^{58}Fe+^{208}Pb are presented in Fig. 2 on the right-hand side (mass regions m=A/2±45 are enlarged and framed). The increase in the mass yields caused by shell effects is observed in the region of the symmetrical fission (m=126-140) at a low excitation energy ($E^* \leq$ 19 MeV). In this case the spherical proton shell ($Z_L \approx 50$) manifests itself in the light fragment with $m_L \approx 126$, whereas the spherical neutron shell ($N_H \approx 82$) is manifested in the heavy fragment $m_H \approx 140$.

The dependences of the average TKE and variances σ^2_{TKE} on the fragment mass for all excitation energies of the compound nucleus are shown in Figure 4. At low excitation energies (E*≤ 19 MeV) the structures are observed on the <TKE>/Mass curve, and they get smoother with an increase in the excitation energy. Some structure was also found in the dependence of the variance on the mass at low excitation energies of up to 32 MeV in the mass region m ≈ 126-140 u. The parabolic dependece of ⟨TKE⟩(M) at E_{lab}=282 MeV becomes very narrow. Such a behavour of mass-energy distribution of fragments presented in

Fig. 2 and 4, in our opinion, is due to shell effects manifestation in the low energy fission of ^{266}Hs.

Let us look now at the energy distribution of fission-like fragments. The TKE distributions of ^{266}Hs at excitation energies E*=14, 15.5 и 19 MeV for symmetric mass range M= $A_{CN}/2\pm20$ u (i.e. for the masses whose MD is two-humped, see 2) are shown in Fig. 6. One can see that two components of the TKE appear in the region of symmetric masses, i.e., a low-energy one with $<TKE>_{low}$ ~ 208 MeV, and a high energy one with $<TKE>_{high}$ ~ 233 MeV. These TKE values as well as mass yields are typical of the standard and SS modes in the spontaneous fission of superheavy nuclei.

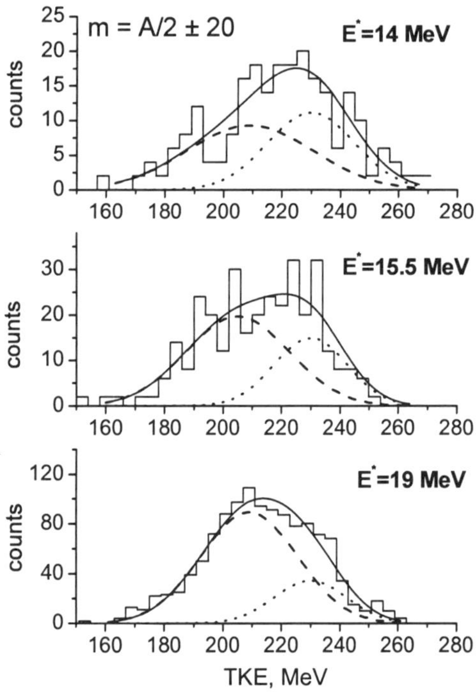

Figure 6. Total kinetic energy of fission fragments for reaction $^{58}Fe+^{208}Pb\rightarrow^{266}Hs$ in symmetric mass division (M= $A_{CN}/2\pm20$) at excitation energies E* = 14, 15.5 и 19 MeV.

Thus, it has been experimentally shown that at medium excitation energies the MED of ^{266}Hs (Z=108) has multimodal character. In the total distribution of the fission fragment masses at E^*<19 MeV a narrow symmetric mode (Super Short) has been found that is connected with the manifestation of the magic neutron number N=82. The yield of the SS mode in the central region (M=

$A_{CN}/2\pm20$) ranges from 45% at E_{lab}=282 MeV to 22% at 289 MeV. We can state that the investigation of fission of superheavy nuclei, deep under the Coulomb barrier at very low excitation energies, provides abundant information on the properties of the fission process.

Figure 5 present the mass yields, ⟨TKE⟩(M) and variances $\sigma^2_{TKE}(M)$ and $\sigma^2_M(TKE)$ for the reaction ^{26}Mg+^{248}Cm at energies E_{lab}=129, 143 and 160 MeV (E^*=35.3, 48 and 64.3 MeV). Fig 5a shows the shoulders in mass yield for $M_H\approx$193-220 u and additional light fragment masses for the E_{lab}=129 and 143 MeV. The fit of the mass yield by the Gaussian according to the LDM model is shown by solid line. The experimental dependence ⟨TKE⟩(M) was fitted by parabola (thin line in Fig. 5b).

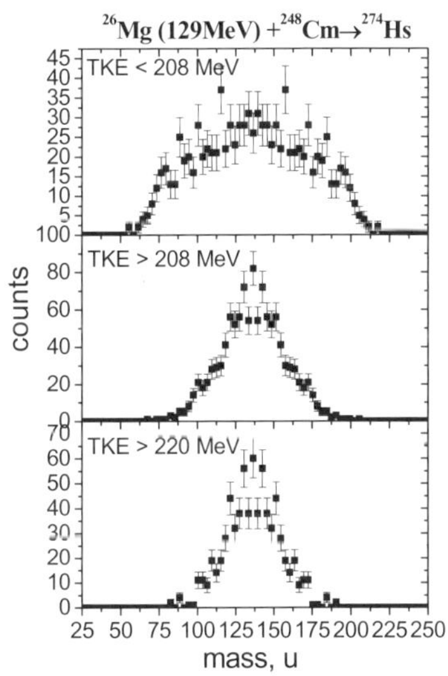

Figure 7. Mass yields of the fission fragments of ^{274}Hs (E^*=35.3MeV) for different TKE ranges.

One can see from Fig.5a for all three excitation energies ⟨TKE⟩(M) is higher for the masses $M_H\approx$193-220 u than the LDM parabolic dependence. The similar increase of ⟨TKE⟩(M) was already observed for QF component of ^{256}N [14]. The variance $\sigma^2_{TKE}(M)$ also increases for the mass region $M_H\approx$200-220 u revealing indirectly the presence of QF and FF processes. QF component on the edges of the mass distribution is caused by the closed shells with Z=28 in light fragment and Z=82, N=126 in the heavy fragment (Fig. 5a). Figure 7 shows the mass yields of the fragments at E^*=35.3 MeV for different TKE ranges. For TKE>220 and TKE>208 MeV one can observe that mass yield consists of two fission modes – symmetric narrow and wider one, which manifests itself as a pediment in mass yield. In case of symmetric fission of ^{274}Hs (N_{CN}=166) both fragments are close to the magic shell N=82, that results to narrow mass distribution. The bottom panels of Fig. 7 exhibits the wide mass distribution for TKE<208 MeV. The TKE distributions for all masses and for the symmetric mass region (A/2±20) are shown in Fig. 8. It consists from two components with ⟨TKE⟩ ≈ 198 MeV and 227 MeV. The strong increasing of the yield of high-energetic symmetric fission mode when neutron number in both fragments is close to the shell N=82 is the general tendency for the superheavy nuclei [29]. The effect of bimodal fission of ^{274}Hs (N=166) is appeared even more obviously than for ^{270}Sg (N=164) [30], probably due to pre-fission neutrons.

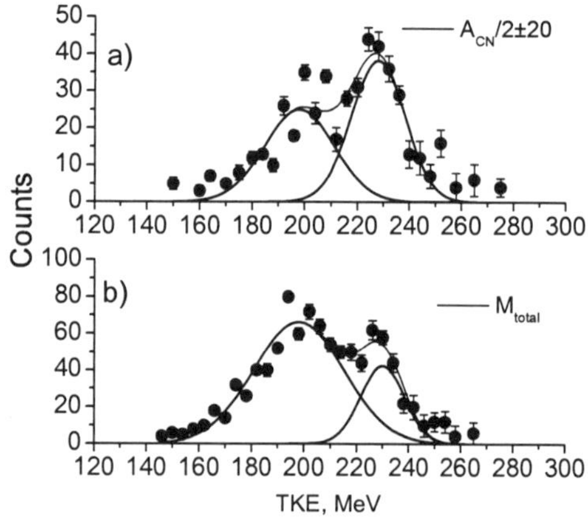

Figure 8. TKE distributions of the ^{274}Hs fission fragments for a) symmetric fragmentation M=A/2±20 u; b) for all masses.

3.3. *Neutron and gamma-ray emission*

In Fig. 9 for reaction $^{58}Fe+^{208}Pb \rightarrow ^{266}Hs$ at two excitation energies E^*= 32 and 47.6 MeV the average total $<M_n^{tot}>$ (middle panels) and for E^*= 46.8 MeV (due to much higher statistics) also pre-scission $<M_n^{pre}>$ and post-scission $<M_n^{post}>$ neutron multiplicities (lower panel) as a function of fragment mass are shown. The neutron multiplicities were obtained for (Mass, TKE) regions indicated in Fig.9 (upper panels) by solid polylines. In fig 10 for $^{26}Mg+^{248}Cm \rightarrow ^{274}Hs$ the average total neutron multiplicity $<M_n^{tot}>$ at excitation energy E^*= 45 MeV is shown. A higher average total neutron multiplicity $<M_n^{tot}>$ is seen for FF region (M= A_{CN}/2±20) compared to QF one in both reactions. The same is valid for $<M_n^{pre}>$ and $<M_n^{post}>$.

The average γ-ray multiplicities $<M_\gamma>$ obtained from the present experiment are shown as square points as a function of fragment mass and TKE for different excitation energies in Fig. 11. Some trends in the experimental data are immediately obvious and worth noticing. We observe that the γ-ray multiplicity curves as a function of mass is *n* shaped, the highest γ-ray multiplicities being found for mass symmetric mass divisions. This behavior is similar to that of average total neutron multiplicity $<M_n^{tot}>$. In addition we see the expected trend that the γ-ray multiplicities for fusion-fission region $<M_\gamma^{FF}>$ (M = $A_{CN}/2\pm20$) increase with bombarding energy (Fig. 12). Both the total angular momentum and the excitation energy of the system increase with the bombarding energy. The increasing angular momentum implies an increase in the amount of aligned spin transferred to the nuclei due to tangential friction. At the same time the increasing excitation energy implies an increase in the spin fluctuation generated by diffusion [31]. On the other hand γ-ray multiplicities for quasi-fission region $<M_\gamma^{QF}>$ increase more slowly with excitation energy (see Fig. 12).

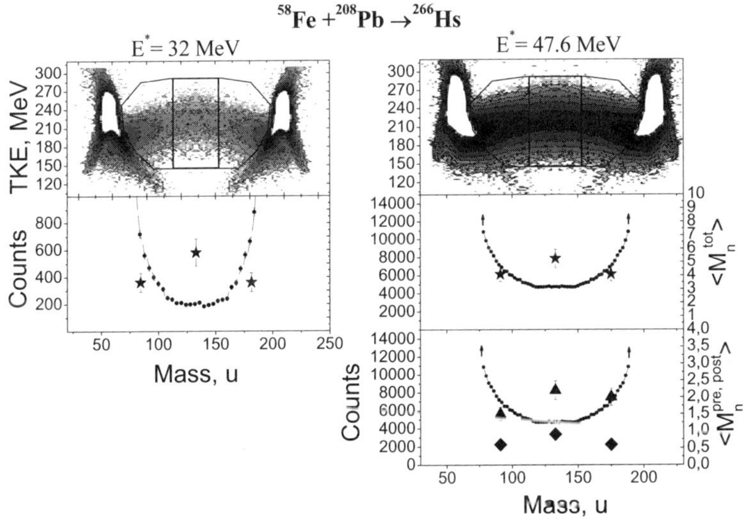

Figure 9. Two-dimensional matrices (Mass, TKE) (upper panels), average total $<M_n^{tot}>$ (middle panels), pre-scission $<M_n^{pre}>$ and post-scission $<M_n^{post}>$ neutron multiplicities (lower panel) as a function of fragment mass obtained for two excitation energies for reaction $^{58}Fe+^{208}Pb \rightarrow ^{266}Hs$.

For all excitation energies some local minima are observed in $<M_\gamma>$ as a function of mass, suggesting the influence of nuclear structure of fission fragments. Evidently the minima occur for these cases when both fragments are near a closed shell, namely for A=132. The similar structures were observed in previous experiments [2,7]. With increasing excitation energy E^* these minima are washed out. Since the Super Shot mode, discussed in section 3.4, is manifestation of double closed shell ^{132}Sn, the behavior of $<M_\gamma>(M)$ is another suggestion that for lower excitation energies the SS mode should manifest itself in mass-energy distribution of fission fragments what was observed in this paper.

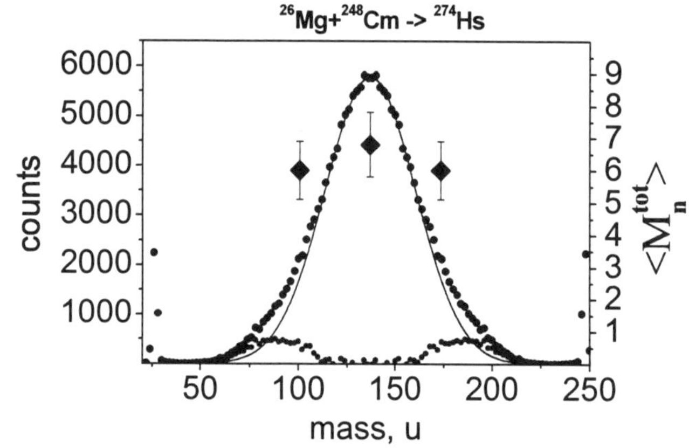

Figure 10. Average total $< M_n^{tot} >$ as a function of fragment mass obtained for excitation energy E^*=45 MeV for reaction $^{26}Mg+^{248}Cm$.

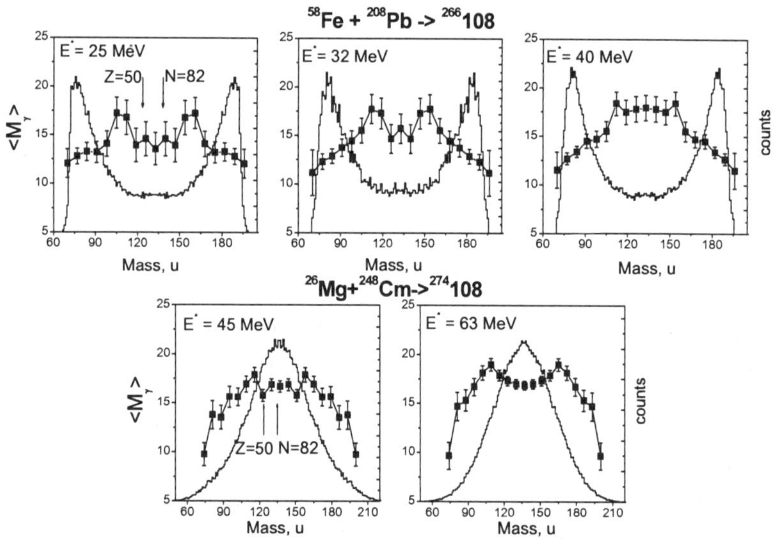

Figure 11. Average γ–ray multiplicities $<M_\gamma>$ as a function of fragment mass for ^{58}Fe (upper panels) and ^{26}Mg (lower panels) induced reactions with indicated excitation energies.

Average γ-ray multiplicities as a function of TKE are shown in Fig. 13. One sees that $<M_\gamma>$ decreases with increasing TKE for symmetric mass splits ($M=A_{CN}/2\pm20$), where the fusion-fission process dominates. On the other hand in the case of asymmetric mass distribution, where quasifission dominates $<M_\gamma>$ is almost constant as a function of TKE. This trend is apparent for all excitation energies (Fig. 13).

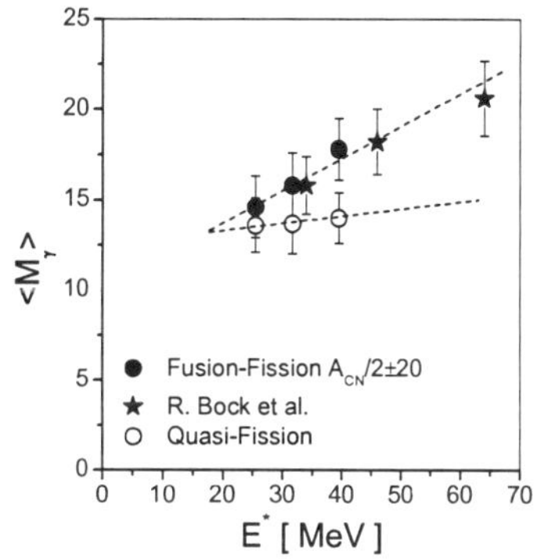

Figure 12 Average γ–ray multiplicities for fusion-fission ($M=A_{CN}/2\pm20$, solid circles) and quasi-fission regions (open circles) as a function of the excitation energy in the reaction $^{58}Fe+^{208}Pb\to{}^{266}108$. The data from the work by Bock *et al* [7] are shown as stars.

The analysis of neutron and γ-ray emission of fission fragments has shown that the total neutron and γ-ray multiplicities in the symmetric mass division, where the compound nuclei are formed, are considerably higher than in the asymmetric one, where the quasifission is the dominant reaction mechanism. Along with the higher TKE for QF in comparing with that expected for FF process this behavior is the suggestion that the QF is probably much colder process than classical fusion-fission.

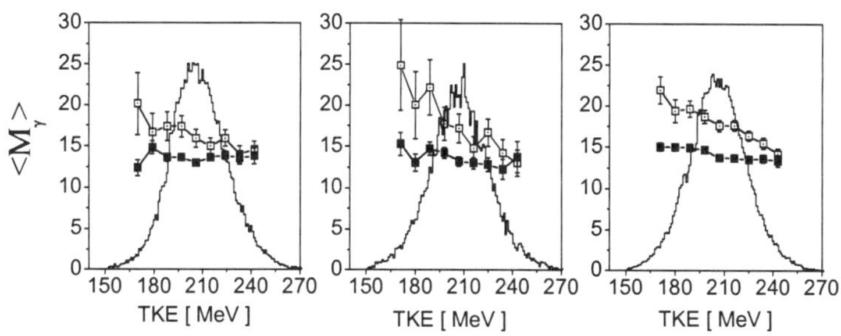

Figure 13. Average γ–ray multiplicities as a function of TKE for ^{58}Fe-induced reactions with indicated excitation energies.

4. Summary

Mass and energy distributions of fragments have been studied in ^{26}Mg and ^{58}Fe ion induced reactions at energies close and below the Coulomb barrier. It has been observed that MED of the fragments at energies near the Coulomb barrier consists of two parts, namely, the classical fusion-fission process of compound nucleus ^{266}Hs and the quasi-fission corresponding to the light fragment masses ~50-80 u and their complimentary heavy fragment masses 186-216 u. From MED of fragments we concluded that spherical shells Z = 82 and N = 126 play significant role in QF. In addition, it has been found that the quasifission has a higher total kinetic energy as compared with that expected for the classical fusion-fission.

For the first time the phenomenon of multimodal fission was observed and studied for superheavy element ^{266}Hs and ^{274}Hs. A high-energy Super-Short mode has been discovered in the region of heavy fragment masses M = 130-135 and TKE ≈ 233 MeV. This nucleus is the one with the highest charge Z=108 where SS mode was revealed so far.

Local minima are observed in $\langle M_\gamma \rangle$ as a function of mass suggesting the great influence of nuclear structure of fission fragments on $\langle M_\gamma \rangle$. The analysis of neutron and γ-ray emission of fission-like fragments has shown that the total neutron and γ-ray multiplicities in the symmetric mass division, where the compound nuclei are formed, are considerably higher than in the asymmetric one, where the quasifission is the dominant reaction mechanism. That means the QF is much colder process than classical fusion-fission and this is probably the one of main reasons why the influence of the shell effects on the observed characteristics of QF process is much stronger than in the case of classical fission of CN.

References

1 M.G. Itkis et al., Nucl.Phys.A 734 (2004) 136,
2 B.B. Back et al., Phys. Rev. C 41 (1990) 1495;
3 A.Yu. Chizhov, et al., Phys. Rev. C 67 (2003) 011603(R).
4 P.K. Sahu, et al., Phys. Rev. C 72 (2005) 034604.
5 G.G. Adamian, N.V. Antonenko and W. Scheid, Phys. Rev. C 68 (2003) 034601, and references in it.
6 Yu.Ts. Oganessian, et al., Nature, 400 (1999) 242; Phys. Rev. Lett. 83 (1999) 3154; Phys. Rev. C 62 (2000) 041604(R); C 63 (2001) 011301(R); C 69 (2004) 054607.

7 R. Bock, et al., Nucl. Phys. A 388 (1982) 334.
8 G. Guarino et al., Nucl. Phys. A 424 (1984) 157.
9 J. Töke, et al., Nucl. Phys. A 440 (1985) 327 ; W.Q. Shen, et al., Phys. Rev. C 36 (1987) 115.
10 E. K. Hulet, et al., Phys. Rev. Lett, 56 (1986) 313; Phys. Rev. C 40 (1989) 770; Phys. At. Nucl. 57 (1994) 1099.
11 M.R. Lane Phys. Rev. C 53 (1996) 2893.
12 D. C. Hoffman, and M. R. Lane, Radiochim. Acta 70/71 (1995) 135; D. C. Hoffman, T. M. Hamilton, and M. R. Lane, Nuclear Decay Modes, edited by D. N.Poenaru (Institute of Physics Publishing, Bristol, 1996) p. 393.
13 D. C. Hoffman, et al., Phys. Rev. C 41 (1990) 631.
14 E.V. Prokhorova et all., Nucl. Phys. A 802 (2008) 45.
15 M. G. Itkis et al., Phys. Rev. C 59 (1999) 3172.
16 U. Brosa et al., Phys. Reports 197 (1990) 167; P. Möller, et al., Nucl. Phys. A 492 (1989) 349.
17 P. Möller et al., Nucl. Phys. A 469 (1987) 1; S. Čwiok et al., Phys. Part. Nucl. 25 (1994) 119.
18 T. Sikkeland, E.L . Haines and V.E. Viola, Phys . Rev. 125 (1962) 1350.
19 G. G. Chubaryan, M. G. Itkis, S. M. Lukyanov, V. N. Okolovich, Yu. E. Penionzhkevich, V. S. Salamatin, A. Ya. Rusanov, and G. N. Smirenkin, Phys. At. Nucl. 56 (1993) 286
20 E. M. Kozulin, et al., Instrum. and Exp. Techniques Vol.51 (2008) p44.
21 E. V. Benton, and R. P. Henke, Nuclear Instruments and Methods 67 (1969) 87; G.N.Knyazheva, S.V.Khlebnikov, E.M. Kozulin, T.E.Kuzmina, V.G.Lyapin, M.Muterrer, J.Perkowski, W.H.Trzaska, NIM B248 (2006) 7.
22 S. Mouatassim et al, *Nucl. Instr. and Meth.* **A359** (1995) 330.
23 http://seal.web.cern.ch/seal/snapshot/work-packages/mathlibs/minuit/
24 Hinde et al. Nucl. Phys. A452 (1986) 550.
25 M. Guttormsen et al, *Nucl. Instr. and Meth.* **A374** (1996) 371.
26 http://www.irs.inms.nrc.ca/EGSnrc/EGSnrc.html.
27.J. R. Nix and W. J. Swiatecki, Nucl. Phys. 71, 1 (1965)
28.B.D. Wilkins, E.P. Steiberg and R.R. Chasman, *Phys. Rev.* **C14**, 1832 (1976).
29 D.C. Hoffman et al., *Radiochim.Acta* **70/71**, 135 (1995)
30 M.G.Itkis et al., *Phys.Rev.* **C59**, 3172 (1999)
31 L. Moreto and R.P. Schmitt, Phys. Rev. C 21 (1980) 204; R.P. Schmitt and A. J. Pacheco, Nucl. Phys. A379 (1982) 313.

FUSION OF HEAVY IONS AT EXTREME SUB-BARRIER ENERGIES

Ş. MIŞICU

National Institute for Nuclear Physics-HH,
Bucharest-Magurele, P.O.Box MG6, Romania
** E-mail:misicu@theor1.theory.nipe.ro*
http://theor1.theory.nipne.ro/ misicu/

H. ESBENSEN

Physics Division, Argonne National Laboratory,
Argonne, Illinois 60439, USA
E-mail: esbensen@phy.anl.gov

After shortly reviewing some essential facts related to the sub-barrier fusion like the problem of the inner part of the Coulomb barrier, enhancement of fusion cross sections due to coupling to excited channels, the far-bellow the barrier data relevant for nuclear reactions in stars, we present calculations performed for the cases ^{58}Ni+^{58}Ni, ^{64}Ni+^{64}Ni, ^{64}Ni+^{74}Ge and ^{64}Ni+^{100}Mo where we were able to confirm the steep falloff of the cross sections. Along with the cross sections we present a diagnosis of the deep sub-barrier fusion using specific tools such as the S-factor and the logarithmic derivative L.

Keywords: heavy-ion fusion; coupled channels; astrophysical factor; nuclear equation of state.

1. Introduction

The topic of sub-barrier fusion presents a particular interest in heavy-ion physics at low energy due to several reasons among which we would like to quote the following four main reasons:

1) It represents a tool to test the heavy-ion potential on the inner flank of the Coulomb barrier. The outer shape of the potential and the positions of the fusion barriers are known from a number of experiments like the elastic scattering, fusion near the barrier, etc., when the ions are at most barely touching their tails. However little is known about the evolution of the interaction when the projectile and the target are overlapping more and more.

2) It is a fact, established long time ago, that the sub-barrier fusion is enhanced when one takes into account the coupling to the vibrational or rotational channels in the target or the projectile, or to the neutron transfer channels. Thus, it is also a tool to confirm the nuclear single-particle or collective structure. When one talks about the vibrational nuclei, inclusion of two-phonon or three-phonon couplings turns out to be crucial in explaining the enhancement of the cross sections (see[1] and references therein). The inclusion of quadrupole or higher order deformations is on the other hand necessary in explaining the enhancement of cross sections when the target is a rotational nucleus.

3) In last decades it was possible to synthesize heavy and super-heavy elements using bombarding energies also below the Coulomb barrier. Thus, it is also a gate to the archipelago of unknown nuclei.

4) Recalling the still open problem of extrapolating the near-barrier data to lower energies for the reaction cross sections of light nuclei like $^{12}C+^{12}C$, $^{12}C+ ^{16}O$, $^{12}C+ ^{13}C$, $^{16}O+ ^{16}O$, $^{16}O+ ^{24}Mg$, it is then easy to realize the relevance of the far below the barrier fusion problem for astrophysical applications.

5) An unexpected trend for the excitation function to decrease steeply was very recently disclosed by C. L. Jiang et al.[2] Among the most conspicuous cases reported in the past are $^{58}Ni+^{58}Ni$[3] , where the departure from the expected behavior takes places already at cross sections ~ 0.1 mb, whereas the new fusion data reported by Jiang et al. are even more spectacular because the reported cross sections are measured down to 10 nb : $^{60}Ni+ ^{89}Y$[2] ($\sigma_f \geq 100$ nb), $^{64}Ni+ ^{64}Ni$[4] ($\sigma_f \geq 10$ nb), $^{64}Ni+^{100}Mo$[5] ($\sigma_f > 10$ nb). The hindrance of fusion was first reported as a suppression of the measured low-energy fusion cross sections with respect to model calculations.[2] This newly discovered phenomenon could imply that the synthesis of heavy elements is hindered below a certain energy threshold.

Very recently we proposed a mechanism that could explain this new phenomenon in sub-barrier fusion[6,7] . Essential in getting a good description of the data was to take into account the saturation of nuclear matter and to use realistic neutron and proton distributions of the reacting nuclei. These two ingredients are naturally incorporated in a potential calculated via the double-folding method with tested effective nucleon-nucleon forces and with realistic charge and nuclear densities, a fact which is often overlooked or only indirectly included in the Woods-Saxon parametrization. In subsequent publications we confirmed this scenario for other combinations: $^{58}Ni+^{58}Ni$, $^{64}Ni+^{100}Mo$[7] and $^{28}Si+^{64}Ni$[8] .

2. Coupled-Channels Approach

We use the same approach as in previous publications (see[9] and references therein), i.e. coupled-channels calculations performed in the so-called iso-centrifugal or rotating-frame approximation, where it is assumed that the orbital angular momentum L for the relative motion of the dinuclear system is conserved. The rotating frame approximation (RFA) allows a drastic reduction of the number of channels used in the calculations. If, for example, we consider the phonon structure for quadrupole excitations in one of the participating nuclei with account of up to N=3 phonons, then we are facing 33 channels whereas after applying RFA we end-up with only 10 channels.

The set of coupled channels reads:

$$\left(\frac{\hbar^2}{2M_0}\left[-\frac{d^2}{dr^2}+\frac{L(L+1)}{r^2}\right]+\frac{Z_1Z_2e^2}{r}+V(r)+\sum_{n_1,n_2}\varepsilon_{n_1,n_2}-E\right)u_{n_1n_2}(r)$$

$$=-\sum_{m_1m_2}\langle n_1n_2\mid\delta V_C+\delta V_N\mid m_1m_2\rangle u_{m_1m_2}(r), \tag{1}$$

where E is the relative energy in the center of mass frame, L is the conserved orbital angular momentum, and M_0 is the reduced mass of the dinuclear system. The C. C. equations (1) are written for two coupled vibrators of eigenenergy ε_{n_1,n_2} and consequently the radial wave function $u(r)$ is labeled by the quantum numbers n_1 and n_2.

As for the spherical part of the potential, $V(r)$, the "proximity" approximation allows us to express it as a function of the shortest distance between the nuclear surfaces of the reacting nuclei:

$$s=r-R_1-R_2-\delta R\ , \tag{2}$$

where

$$\delta R=R_1\sum_{\lambda\mu}\alpha^{(1)}_{\lambda\mu}Y^*_{\lambda\mu}(\hat{r})+R_2\sum_{\lambda\mu}\alpha^{(2)}_{\lambda\mu}Y^*_{\lambda\mu}(-\hat{r}), \tag{3}$$

and $\hat{r}$ specifies the spatial orientation of the projectile-target system in the laboratory frame and $\alpha^{(i)}_{\lambda\mu}$ are the deformation parameters. In the RFA the direction of $\boldsymbol{r}$ defines the z-axis. The only vibrational excitations that can take place are therefore the $\mu=0$ components, since $Y_{\lambda\mu}(\hat{z})\propto\delta_{\mu,0}$.

The explicite form of the nuclear potential is given by double folding integral of two nuclei with one-body deformed densities ρ_1 and ρ_2, subjected to vibrational fluctuations, and center of masses separated by distance $\boldsymbol{r}$,

$$V_N(\boldsymbol{r})=\int d\boldsymbol{r}_1\int d\boldsymbol{r}_2\ \rho_1(\boldsymbol{r}_1)\ \rho_2(\boldsymbol{r}_2)\ v(\boldsymbol{r}_{12}), \tag{4}$$

where $\boldsymbol{r}_{12} = \boldsymbol{r} + \boldsymbol{r_2} - \boldsymbol{r}_1$. The density of the ion i is parametrized by a two parameter Fermi function

$$\rho_i(\boldsymbol{r}) = \frac{\rho_{0i}}{1 + e^{\frac{r - R_i}{a_i}}} \tag{5}$$

For the effective realistic nucleon-nucleon interaction we take the M3Y in the Reid parametrization. This interaction is independent of the density of nuclear matter in which the two nucleons are embedded[10] . Such a potential is providing a good starting point to evaluate fusion barriers but its interior part is unphysically deep. Because of the lack of the density dependence in the $N - N$ force the nuclear matter is collapsing instead of saturating! In order to account for the effect of the density of the surrounding medium we add to the above double folding integral a second one which has the role to simulate the weakening of the nuclear attraction as the two density start to overlap. For this reason we refer to the used potential as the M3Y+repulsion.

When deriving the properties of the short-range repulsive core, we must keep in mind that an overlapping region with doubled nucleon density is formed once the distance R between the nuclei becomes less than $R_p + R_t$, where R_p and R_t are the nuclear radii along the collision axis. The doubling of the density increases the energy of the nucleons in the overlapping region. In the case of the complete overlap (for $R = 0$) the increase of the interaction energy per nucleon is, up to quadratic terms in the normal density,

$$\frac{\Delta V}{2A_p} = B(2\rho) - B(\rho) \approx \frac{1}{2}\rho_0^2 \left(\frac{\partial^2 B(\rho)}{\partial \rho^2} \right) \Bigg|_{\rho=\rho_0} = \frac{1}{18} K \tag{6}$$

where in the last equality we use the proportionality of the incompressibility K of normal nuclear matter to the curvature of the energy per nucleon, $B(\rho)$,[11] . Since K is usually not measured directly but deduced from isoscalar giant monopole or dipole resonances,[12] and since there are conflicting results coming from the cross-sections of the corresponding experimental data we use instead of a universal value the predictions of the Thomas-Fermi model[13] as a function of the relative neutron excess $\delta = (\rho_n - \rho_p)/\rho$ of the compound (fused) nucleus. Eventually the strength of the repulsive core V_{cor} is determined by assuming that ΔV must be identified with the value of the heavy-ion interaction potential at the coordinate origin $R = 0$[14] .

To illustrate the shape of this potential for a case of interest we compare in Fig. 1 the spherical heavy-ion potential for the symmetric projectile-target combination ^{58}Ni+^{58}Ni calculated with the M3Y+repulsion and with Akyüz-Winther (AW).[15]

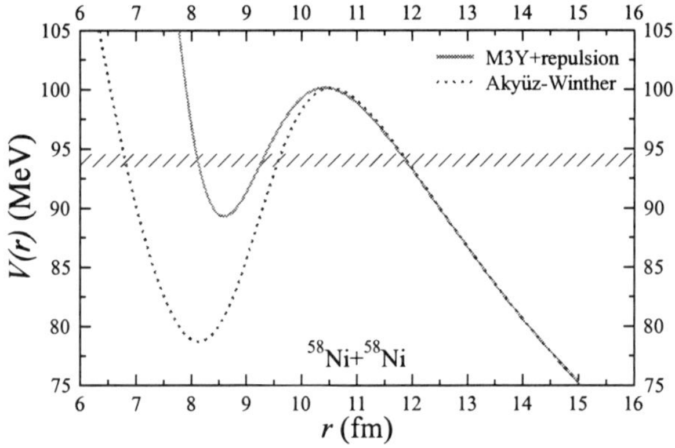

Fig. 1. Various spherical ion-ion potentials for ^{58}Ni+^{58}Ni. The solid curve is the potential employed in the present work. The curve with small dashes is the Akyüz-Winther potential used in[2]-.[5] The dashed strip corresponds to experimental boundaries of the threshold energy E_s.

The non-spherical part of the nuclear potential results from the difference between the total interaction and the potential in the elastic channel. Since linear and quadratic interactions are necessary and often sufficient to fit the data at least in the intermediate energy region (see[9] and references therein) $\delta V(r)$ is expanded up to second-order in the surface distortion (3),

$$\delta V_N(r) = -\frac{\partial V}{\partial r}\delta R + \frac{1}{2}\frac{\partial^2 V}{\partial r^2}\,[(\delta R)^2 - \langle 0|(\delta R)^2|0\rangle]. \qquad (7)$$

It is seen that the ground state expectation of this interaction, $\langle 0|\delta V_N|0\rangle$, is zero, but the second order term will give a non-zero contribution to the diagonal matrix element in an excited state, this prescription being exact for a harmonic oscillator up to second order in the deformation amplitudes. We include a similar expansion of the Coulomb field, δV_C, but only to first order in the deformation amplitudes[9] . Expressions for the matrix elements of δV in the double-oscillator basis are given in[16] . These expressions are inserted into the C. C. formalism in the rotating frame approximation which singles out only axially symmetric distortions ($\alpha_{\lambda\mu=0}$).

The C. C. equations, as written down in eq.(1), are solved with the usual boundary conditions at large distances whereas on the left side of the barrier the so-called in-going-wave boundary condition (IWBC) is imposed, more precisely, at the radial separation where the potential pocket attains its minimum. The IWBC ensures that there will be no reflected wave inside the barrier. In the same time on the outer flank of the barrier the quantum flux has ingoing as well as reflected components.

In the asymptotic region usual scattering boundary conditions are set[16]

$$u_n^{LM}(r) \longrightarrow \delta_{no} F_L(k_n r) + \mathcal{T}_n{}^L H_L^{(+)}(k_n r) \tag{8}$$

where the reaction matrix can be related to the S-matrix through the relation

$$\mathcal{T}_n{}^L = \frac{1}{2\pi}(\delta_{n0} - S_{n0}{}^L) \tag{9}$$

Eventually the transmission probabilities are obtained through the formula

$$T_L = 1 - \sum_n | S_{n0}{}^L |^2 \tag{10}$$

Then fusion cross section is given by the total in-going flux

$$\sigma_F = \frac{\pi}{k_0^2} \sum_L (2L+1) T_L \tag{11}$$

The calculations include one-phonon excitations of the lowest 2^+ and 3^- states in target and projectile, and all two-phonon and mutual excitations of these states up to a 7.2 MeV excitation energy. This energy cutoff was chosen so that all of the two-phonon states were included in the calculations for ^{64}Ni+^{64}Ni, whereas the two-phonon octupole states were excluded in the calculations for ^{58}Ni+^{58}Ni. The necessary structure input for ^{64}Ni, ^{74}Ge and ^{100}Mo is given in Ref.[1] The input for ^{58}Ni is from Ref.[16]

In Fig.2 the calculated as well the experimental fusion excitation functions are displayed for the systems ^{58}Ni+^{58}Ni (experimental data from[3]), ^{64}Ni+^{64}Ni (experimental data from[4]), ^{64}Ni+^{74}Ge (experimental data from[17]) and ^{64}Ni+^{100}Mo (experimental data from[5]). For the sake of comparison with previous C. C. calculations we present also the case when the AW potential is used and the no-coupling case (NOC). We see that when we use the potential with shallower depth the low-energy points are well and even very well reproduced.

The agreement with data, when using the M3Y+repulsion potential, is sensitively better than the one provided by the AW one starting at 90 MeV, for ^{64}Ni+^{64}Ni, not only for the 4 lowest experimental data points but also at higher energies. The excitation function obtained with the M3Y+repulsion potential has the right slope, not only because the potential attains a higher-lying pocket but also because the curvature of the barrier is different, with a thicker barrier in the overlapping region. For the reaction ^{64}Ni+^{64}Ni, for example, the best χ^2 per point is only $\chi^2/N = 0.86$. This value is obtained

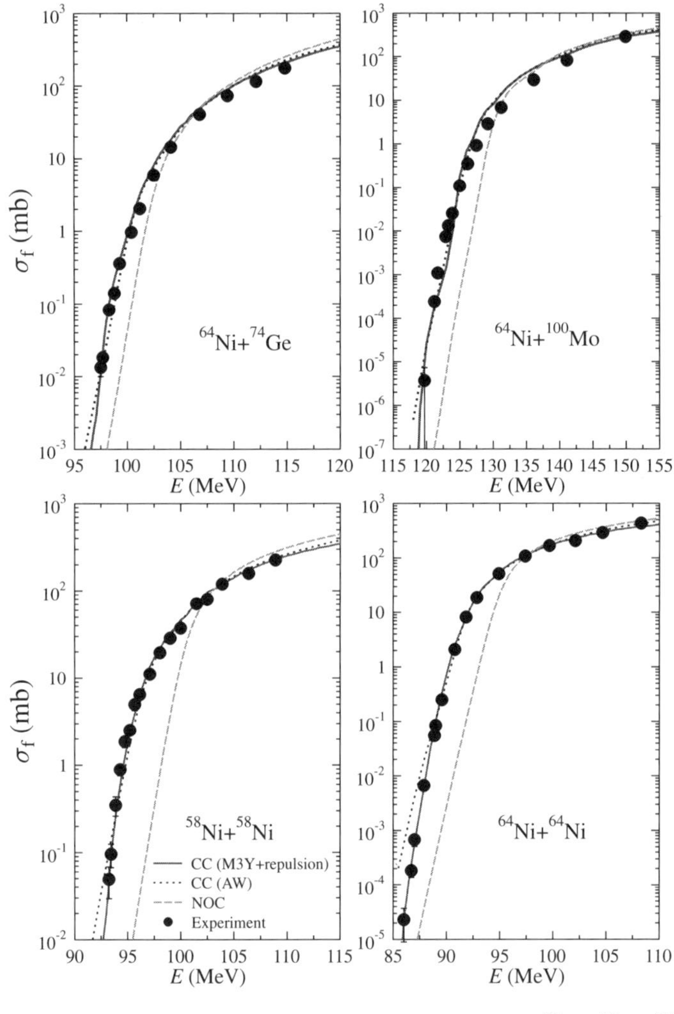

Fig. 2. Experimental fusion excitation functions for the systems ^{58}Ni+^{58}Ni, ^{64}Ni+^{64}Ni, ^{64}Ni+^{64}Ge and ^{64}Ni+^{100}Mo are compared to various C. C. calculations described in the text, and to the no-coupling (NOC) limit for the AW potential.

by applying the energy shift $\Delta E = 0.16$ MeV to the calculated excitation function. The best fit obtained with AW potential, on the other hand, gives a $\chi^2/N = 10$ and requires an energy shift of $\Delta E = 0.9$ MeV.

However, for the case ^{64}Ni+^{100}Mo the agreement with the data seen in the upper right panel of Fig. 2 is clearly not as good as in the other three cases shown in the same figure, although in order to improve the fit to the data we included up to three phonon excitations of the quadrupole mode in ^{100}Mo using the structure parameters given in Ref.[1] . The reason is that the C. C. effects are very strong for this heavy-ion system and the calculations have not fully converged with respect to multiphonon excitations, as discussed in Ref.[1] . Another problem is that the nuclear structure properties of multiphonon states are often poorly known, so we did not try to improve the fit to the data here.

In a first attempt to get a diagnosis for the various cases where the hindrance in sub-barrier fusion occurs, Ref.[18] proposed the use of two representations. The first one is the astrophysical S-factor,

$$S = E\sigma(E)\ \exp(2\pi\eta), \tag{12}$$

where E is the center-of-mass energy and η the Sommerfeld parameter. The experimental value of S increases with decreasing bombarding energy and has the tendency to develop a maximum for the systems of interest. The necessity to resort to this quantity comes from the fact that the reaction cross section varies by many orders of magnitude below the Coulomb barrier (7 orders of magnitude for ^{64}Ni+ ^{64}Ni). Using the S-factor representation we are able to magnify structures in the excitation function at energies below the barrier. From Fig.3 we can easily conclude which of the four cases under study is better reproduced by our calculations.

Using the M3Y+repulsion potential, for the case ^{64}Ni+ ^{64}Ni we get a very good description of the lowest experimental points. In the case of ^{58}Ni+ ^{58}Ni although we do not fit some of the data points we get the right location and curvature of the S-factor maximum. Also in the case of ^{64}Ni+ ^{74}Ge we get a better description than in the case of the AW potential, which is not able in any of the four cases to predict the bending of the S-factor curve. For the case ^{64}Ni+ ^{100}Mo we reproduce roughly the trend of the S-factor to develope a maximum, when using the potential with repulsive core.

Another diagnostic tool proposed in Ref.[18] is the logarithmic derivative,

$$L(E) = \frac{d[\ln(E\sigma)]}{dE} = \frac{1}{E\sigma}\frac{d}{d\,E}(E\sigma). \tag{13}$$

The fact that the early attempts to reproduce the low-energy data points of the measured fusion cross sections using the AW potential were unsuccesfull is most clearly seen from the inspection of the logarithmic derivative $L(E)$ (see Fig.4) . For energies below a certain threshold the experimental values of $L(E)$ increase steeply with decreasing energy, whereas the the-

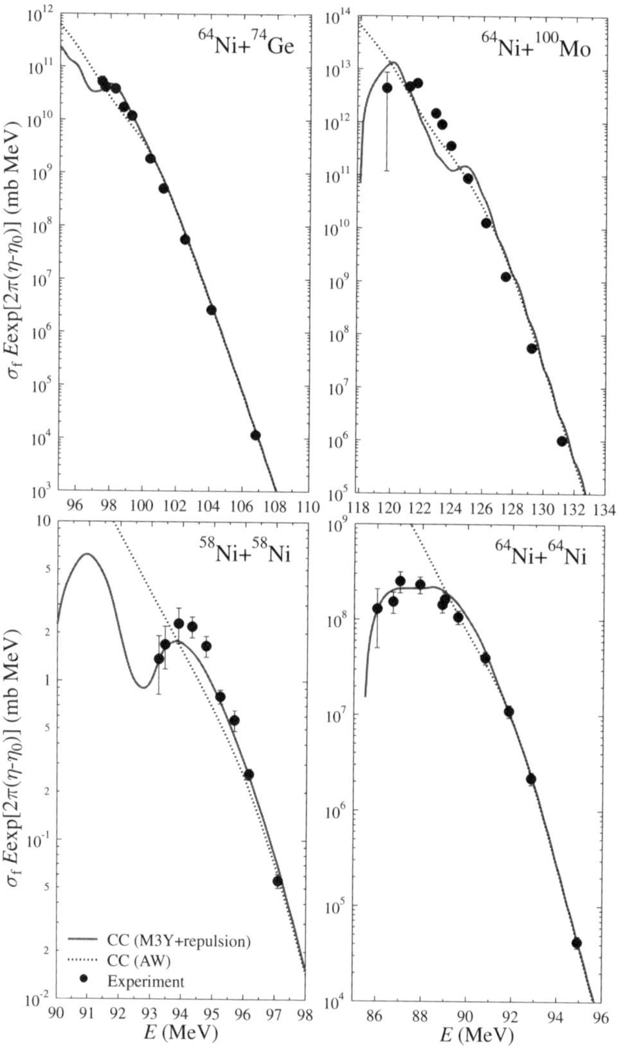

Fig. 3. Experimental S-factors for the systems under study are indicated by solid circles. They are compared to the coupled-channels calculations performed with the M3Y+repulsion (solid curve) and Akyüz-Winther (dashed curve) potentials.

oretical curve, corresponding to the AW potential, increases with a much smaller slope. Once we include the repulsive core in the dynamic model, the divergency in $L(E)$, as indicated by the experimental data, is explained for all four fusion reactions investigated by us.

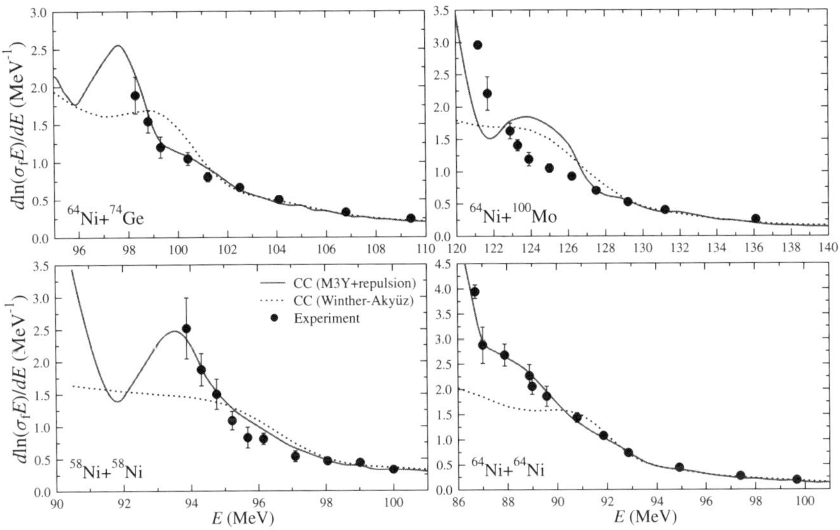

Fig. 4. Logarithmic derivatives of the energy-weighted cross sections for the systems under study.

For the case ^{64}Ni+ ^{100}Mo a local maximum at 124 MeV, predicted by our calculations, is most likely caused by a poor convergence with respect to multiphonon excitations.

Another tool to investigate the sub-barrier fusion is the spin distribution whose first moment (average angular momentum) is given by

$$\langle L \rangle = \frac{1}{\sigma_f} \sum_L L\sigma_f(L) \tag{14}$$

In the past it has always been believed that $\langle L \rangle$ for fusion would approach a constant at low energy. However if we use the potential with shallower depth we obtain a narrowing of the spin distribution for fusion as the center-of-mass energy decreases and approaches the minimum pocket energy. An example is shown in Fig. 5, where the measurements of the γ-ray multiplicity from the compound nucleus formed in the fusion of ^{64}Ni+^{64}Ni have been converted into an average angular momentum for fusion[19] . The thin dashed curve shows the prediction based on the AW potential in the no-coupling limit (NOC). It approaches a constant value at low energy but the data are always above that limit. The solid curve in Fig. 5 shows the results we obtain in the C. C. calculations we discussed earlier, which were based on the M3Y+repulsion potential.

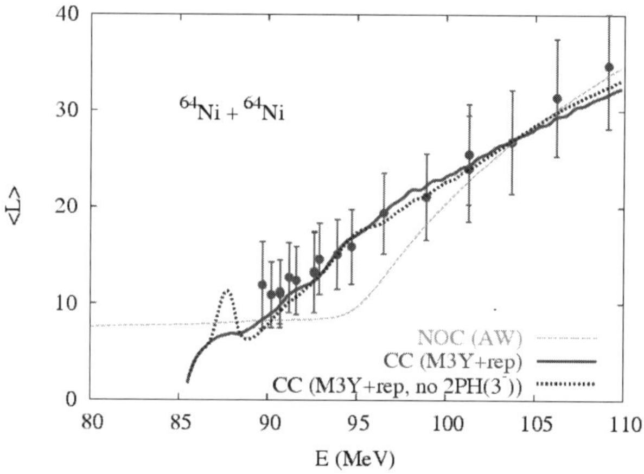

Fig. 5. (Color online). Average angular momentum for the fusion of $^{64}Ni+^{64}Ni$ obtained in C. C. calculations based on the M3Y+repulsion potential with (solid curve) and without (thick dotted curve) the effect of couplings to the two-phonon octupole states. The thin dashed curve was obtained in the no-coupling limit (NOC) using the AW potential. The data are from[19] .

It is also important to mention that the low-energy behavior of several observables can have a strong sensitivity to the couplings to multi-phonon states. Although the fusion only occurs in the elastic channel at energies close to the minimum of the pocket, the polarization of inelastic channels can still have a large effect. We found, in particular, that couplings to the two-phonon octupole states are very important. This is illustrated in Fig. 5 by the dotted curve which was obtained without any couplings to the two-phonon octupole states. It is seen that the calculation in this case develops a rather sharp peak at 87.7 MeV. The peak disappears when the coupling to the two-phonon octupole states is included, as illustrated by the solid curve. Unfortunately, the data cannot tell us which of these two calculations is the most realistic.

Before ending it is worthwile to mention that the repulsive core mechanism which seems to explain the hindrance in sub-barrier fusion was also confirmed for the asymmetric combination $^{64}Ni+^{28}Si$[8] .

Acknowledgements

One of the authors (Ş.M.) is grateful to the Fulbright Commission for financial support and for the hospitality of the Physics Division at Argonne National Laboratory. H.E. acknowledge the support of the U.S. Department of Energy, Office of Nuclear Physics, under Contract No. W-31-109-ENG-38. We are also grateful to C. L. Jiang, B. B. Back, R. V. F. Janssens and K. E. Rehm for usefull discussions.

References

1. H. Esbensen, Phys. Rev. C **72**, 054607 (2005).
2. C. L. Jiang *et al.*, Phys. Rev. Lett. **89**, 052701 (2002).
3. M. Beckerman, J. Ball, H. Enge, M. Salomaa, A. Sperduto, S. Gazes, A. Di Rienzo and J. D. Molitoris, Phys. Rev. C **23**, 1581 (1982).
4. C. L. Jiang *et al.*, Phys. Rev. Lett. **93**, 012701 (2004) .
5. C. L. Jiang *et al.*, Phys. Rev. C **71**, 044613 (2005).
6. Ş. Mişicu and H. Esbensen, Phys. Rev. Lett. **96**, 112701 (2006).
7. Ş. Mişicu and H. Esbensen, Preprint ANL, PHY-11523-TH-2006.
8. C. L. Jiang *et al.*,Phys. Lett. **B640**, 18 (2006).
9. H. Esbensen, Prog. Theor. Phys. (Kyoto), Suppl. **154**, 11 (2004).
10. M. E. Brandan and G. R. Satchler, Phys. Rep. **285** (1997) 143.
11. J. Eisenberg and W. Greiner, *Nuclear Theory, vol.I, Phenomenological Models*, (North-Holland, Amsterdam 1988).
12. M. N. Harakch and A. van der Woude, *Giant Resonances*, (Clarendon Pressm, Oxford, 2001)
13. W. D. Myers and W. J. Swiatecki, Phys. Rev. C **57**, 3020 (1998).
14. A. Sandulescu, Ş. Mişicu, F. Carstoiu and W. Greiner, Phys. Part. Nucl. **30**, 386 (1999).
15. R. A. Broglia and A. Winther, *Heavy Ion Reactions, Lecture Notes, Volume I: The Elementary Processes*, (Addison-Wesley Pub. Co., CA, 1991), p.114
16. H. Esbensen and S. Landowne, Phys. Rev. C **35**, 2090 (1987).
17. M. Beckerman, M. Salomaa, A. Sperduto, J. D. Molitoris and A. Di Rienzo, Phys. Rev. C **25**, 837 (1982).
18. C. L. Jiang, H. Esbensen, B. B. Back, R. V. F. Janssens, and K. E. Rehm, Phys. Rev. C **69**, 014604 (2004).
19. D. Ackerman *et al.*, Nucl. Phys. **A609**, 91 (1996).

FUSION AND FISSION DYNAMICS OF HEAVY NUCLEAR SYSTEM

Valery ZAGREBAEV[1,*] and Walter GREINER[2]

[1] *Flerov Laboratory of Nuclear Reaction, JINR, Dubna, Moscow Region, Russia,*
[2] *FIAS, J.W. Goethe-Universität, Frankfurt, Germany*
**E-mail: zagreb@jinr.ru*

The paper is focused on reaction dynamics of superheavy nucleus formation and decay at beam energies near the Coulomb barrier. The aim is to review the things we have learned from recent experiments on fusion-fission reactions leading to the formation of compound nuclei with $Z \geq 102$ and from their extensive theoretical analysis. The choice of collective degrees of freedom playing a principal role and finding the adiabatic multi-dimensional driving potential regulating the whole process are discussed. Dynamics of heavy-ion low energy collisions is studied within the realistic model based on multi-dimensional Langevin equations. Theoretical predictions are made for synthesis of SH nuclei in symmetric and asymmetric fusion reactions as well as in damped collisions of transactinides.

Keywords: damped collisions; fusion; fission; superheavy elements.

1. Introduction

To describe properly and simultaneously the strongly coupled deep inelastic (DI), quasi-fission (QF) and fusion-fission processes of low-energy heavy-ion collisions we have to choose, first, the unified set of degrees of freedom playing the principal role both at approaching stage and at the stage of separation of reaction fragments. The number of the degrees of freedom should not be too large so that one is able to solve numerically the corresponding set of dynamic equations. On the other hand, however, with a restricted number of collective variables it is difficult to describe simultaneously DI collision of two separated nuclei and QF of the highly deformed mono-nucleus. Second, we have to determine the unified potential energy surface (depending on all the degrees of freedom) which regulates in general all the processes. Finally, the corresponding equations of motion should be formulated to perform numerical analysis of the studied reactions.

The distance between the nuclear centers R (corresponding to the elongation of a mono-nucleus), dynamic spheroidal-type surface deformations δ_1 and δ_2, mutual in-plane orientations of deformed nuclei φ_1 and φ_2, and mass asymmetry $\eta = \frac{A_1-A_2}{A_1+A_2}$ are probably the relevant degrees of freedom in fusion-fission dynamics. Note that we take into consideration all the degrees of freedom needed for description of all the reaction stages. Thus, in contrast with other models, we need not to split artificially the whole reaction into several stages when we consider strongly coupled DI, QF and CN formation processes. Unambiguously defined initial conditions are easily formulated at large distance, where only the Coulomb interaction and zero-vibrations of the nuclei in their ground states determine the motion.

2. Adiabatic potential energy

The interaction potential of separated nuclei is calculated rather easily within the folding procedure with effective nucleon-nucleon interaction or parameterized, e.g., by the proximity potential [1]. Of course, some uncertainty remains here, but the height of the Coulomb barrier obtained in these models coincides with the empirical Bass parametrization [2] within 1 or 2 MeV. Dynamic deformations of colliding spherical nuclei and mutual orientation of statically deformed nuclei significantly affect their interaction changing the height of the Coulomb barrier for more than 10 MeV. It is caused mainly by a strong dependence of the distance between nuclear surfaces on the deformations and orientations of nuclei [3].

After contact the mechanism of interaction of two colliding nuclei becomes more complicated. For fast collisions ($E/A \sim \varepsilon_{\text{Fermi}}$ or higher) the nucleus-nucleus potential, V_{diab}, should reveal a strong repulsion at short distances protecting the "frozen" nuclei to penetrate each other and form a nuclear matter with double density (diabatic conditions, sudden potential). For slow collisions (near-barrier energies), when nucleons have enough time to reach equilibrium distribution (adiabatic conditions), the nucleus-nucleus potential energy, V_{adiab}, is quite different (Fig. 1). Thus, for the nucleus-nucleus collisions at energies well above the Coulomb barrier we need to use a time-dependent potential energy, which after contact gradually transforms from a diabatic potential energy into an adiabatic one: $V = V_{\text{diab}}[1 - f(t)] + V_{\text{adiab}} f(t)$. Here t is the time of interaction and $f(t)$ is a smoothing function with parameter $\tau_{\text{relax}} \sim 10^{-21}$ s, $f(t=0) = 0$, $f(t >> \tau_{\text{relax}}) = 1$. The calculation of the multidimensional adiabatic potential energy surface for heavy nuclear system remains a very complicated physical problem, which is not yet solved in full.

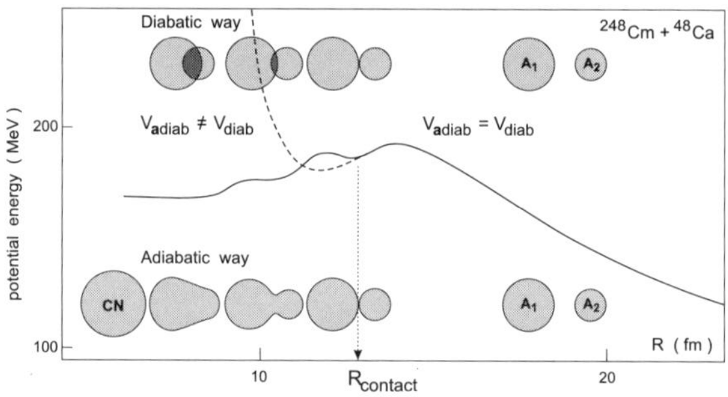

Fig. 1. Potential energy for ^{48}Ca+^{248}Cm for diabatic (dashed curve) and adiabatic (solid curve) conditions (zero deformations of the fragments).

In this connection the two-center shell model [4] seems to be most appropriate for calculation of the adiabatic potential energy surface. However, the simplest version of this model with restricted number of collective coordinates, using standard parametrization of the macroscopic (liquid drop) part of the total energy [5, 6] and overlapping oscillator potentials for calculation of the single particle states and resulting shell correction, does not reproduce correctly values of the nucleus-nucleus interaction potential for well separated nuclei and at contact point (depending on mass asymmetry). The same holds for the value of the Coulomb barrier and the depth of potential pocket at contact. No doubt, within an extended version of this model all these shortcomings may be overcome (see the talk of A. Karpov).

2.1. *Two-core model*

In Refs. [7, 8, 9] for a calculation of the adiabatic potential energy surface the "two-core approximation" was proposed based on the two-center shell model idea and on a process of step-by-step nucleon collectivization. It is assumed that on a path from the initial configuration of two touching nuclei to the compound mono-nucleus configuration and on a reverse path to the fission channels the nuclear system consists of two cores (z_1, n_1) and (z_2, n_2) surrounded with a certain number of common (shared) nucleons $\Delta A = A_{CN} - a_1 - a_2$ moving in the whole volume occupied by the two cores, see Fig. 2. Denote by ΔA_{CN} the number of collectivized nucleons at which the two cores a_1 and a_2 fit into the volume of CN ("dissolve" inside it and lose completely their individuality), i.e., $R(a_1, \delta_1) + R(a_2, \delta_2) = R(A_{CN}, \delta_{CN}^{g.s.})$, where δ_1 and δ_2 are the dynamic deformations of the cores. It is clear that $\Delta A_{CN} < A_{CN}$ and the compound nucleus is finally formed when the elongation of the system becomes shorter than a saddle point elongation of CN.

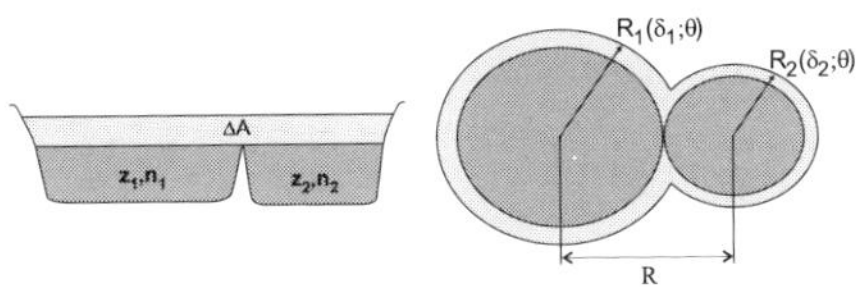

Fig. 2. Schematic view of nuclear system in the two-core approximation

Adiabatic driving potential can be defined in the following way

$$V_{\rm adiab}(R; z_1, n_1, \delta_1; z_2, n_2, \delta_2) = \tilde{V}_{12}(r, z_1, n_1, \delta_1; z_2, n_2, \delta_2)$$
$$-[\tilde{B}(a_1) + \tilde{B}(a_2) + \tilde{B}(\Delta A)] + B(A_P) + B(A_T). \qquad (1)$$

Here $\tilde{B}(a_1) = \tilde{\beta}_1 a_1$, $\tilde{B}(a_2) = \tilde{\beta}_2 a_2$ and $\tilde{B}(\Delta A) = 0.5(\tilde{\beta}_1 + \tilde{\beta}_1)\Delta A$ are the binding energies of the cores and of common nucleons. These quantities depend on the number of shared nucleons. Define the range of collectivization as $x = \Delta A/\Delta A_{CN}$, then $\tilde{\beta}_{1,2}$ can be roughly approximated as $\tilde{\beta}_{1,2} = \beta_{1,2}^{exp}\varphi(x) + \beta_{CN}^{exp}(1 - \varphi(x))$, where $\beta_{1,2}^{exp}$ and β_{CN}^{exp} are specific binding energies of the isolated (free) fragments, which can be derived from the experimental nuclear masses or calculated rather accurately within the macroscopic-microscopic model [6]. $\varphi(x)$ is an appropriate monotonous function satisfying the conditions $\varphi(x = 0) = 1$, $\varphi(x = 1) = 0$. Thus the specific binding energies of the cores $\tilde{\beta}_{1,2}$ approach the specific binding energy of CN with increasing ΔA. All the shell effects enter the total energy (1) by means of $\beta_{1,2}^{exp}$ and β_{CN}^{exp}. Interaction of the fragments $\tilde{V}_{12}$ is defined in usual way at $R \geq R_{\rm cont}$ as a sum of the Coulomb and nuclear potentials. This interaction weakens gradually with increasing the number of shared nucleons ΔA at $R_{CN} < R < R_{\rm cont}$, i.e. with gradual dissolving the two cores inside the compound mono-nucleus. The interaction energy just transforms to the binding energy of CN. Thus, once the compound nucleus has been formed ($\Delta A = \Delta A_{CN}$), the total energy of the system $V_{\rm adiab} = Q_{gg}^{\rm fus} = B(A_1^0) + B(A_2^0) - B(A_{CN})$.

Calculations of adiabatic potential energy within the two-core approximation and within the two-center shell model give rather similar results at $R < R_{\rm cont}$ [9, 10]. At the same time, there are several advantages of the two-core model. To get a reasonable value for the fission barrier we used the shell corrections at zero and ground state deformations (calculated according to Ref. [6]) and the parametrization of the liquid drop energy proposed in Ref. [11]. Based on these values the adiabatic potential was calculated for small deformations. Then it was joined together with the potential of two touching nuclei as it was proposed in Ref. [7, 9]. Experimental binding energies of two cores were used, thus giving us the "true" values of the shell corrections. As a result, the two-core model gives automatically an explicit (experimental) value of the nucleus-nucleus interaction energy in the asymptotic region for well separated nuclei where it is known (the Coulomb interaction plus nuclear masses). It gives also quite realistic heights of the Coulomb barriers, which is very important for description of near-barrier heavy ion reactions. Note that the proposed driving potential is defined in the whole region $R_{CN} < R < \infty$, it is a continuous function at $R = R_{\rm cont}$ and, thus, may be used for simultaneous description of the whole fusion-fission dynamics of heavy nuclear systems.

2.2. *Clusterization and shape-isomeric states*

Within the two-core model the processes of compound nucleus formation, fission and quasi-fission may be described both in the space of $(R, \eta, \delta_1, \delta_2)$ and in the space $(a_1, \delta_1; a_2, \delta_2)$, because for a given nuclear configuration $(R, \eta, \delta_1, \delta_2)$ we may unambiguously determine the two cores a_1 and a_2. It is extremely important for interpretation of physical meaning of some deep minima on the potential energy surface.

Adiabatic driving potential (1) is shown in Fig. 3 as a function of z_1 and z_2 at $R \leq R_{\rm cont}$ (minimized over n_1 and n_2) and also as a function of elongation and mass asymmetry at fixed deformations of both fragments. It is easily to see that the shell structure, clearly revealing itself in the contact of two nuclei is also retained at $R < R_{\rm cont}$ (see the deep minima in the regions of $z_{1,2} \sim 50$ and $z_{1,2} \sim 82$ in Fig. 3b). Following the fission path (dotted curves in Fig. 3a,b) the nuclear system goes through the optimal configurations (with minimal potential energy) and overcomes the multi-humped fission barrier (Fig. 3c). These intermediate minima correspond to the shape isomer states. Now, from analysis of the driving potential (see Fig. 3a,b), we may definitely conclude that these isomeric states are nothing else but the two-cluster configurations with magic or semi-magic cores surrounded with a certain number of shared nucleons.

It would be interesting to estimate the adiabatic potential energy also for the three-center configuration. Such clusterization may play a role in vicinity of scission point, where the shared nucleons ΔA may form a third cluster located between the two heavy cores a_1 and a_2. Such calculation need at least two more degrees of freedom and is difficult to be performed.

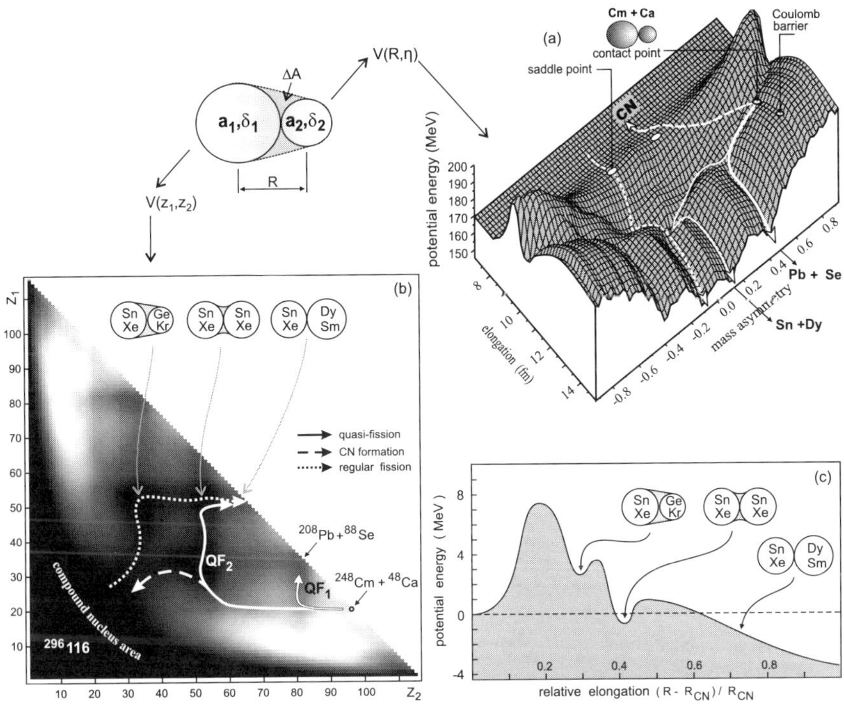

Fig. 3. Driving potential of nuclear system $^{296}116 \leftrightarrow {}^{48}\mathrm{Ca} + {}^{248}\mathrm{Cm}$. (a) Potential energy in the "elongation – mass asymmetry" space. (b) Topographical landscape of the driving potential on (z_1, z_2) plane. Dashed, solid and dotted curves show most probable trajectories of fusion, quasi-fission and regular fission, respectively. Diagonal corresponds to the contact configurations ($\Delta A = 0$). (c) Three-humped barrier calculated along the fission path (dotted curve).

2.3. *Orientations effects*

It is well known that the orientation effects play very important role in sub-barrier fusion of deformed nuclei by significantly increasing the capture probability due to decreasing the Coulomb barrier for nose-to-nose collisions. There is some evidence that the orientation effects could be very important also at the stage of CN formation [12] (especially in the synthesis of SH nuclei [13]) by significantly decreasing the fusion probability for the nose-to-nose configurations which lead the nuclear system preferably into the QF channels. Up to now this effect was not taken into account explicitly in theoretical models, only the empirical parametrization of it had been used [9]. The main difficulty here is a calculation of the adiabatic potential energy for (in principle, unknown) subsequent shapes of the nuclear system starting from the configuration of arbitrary oriented two touching deformed nuclei and up to more or less spherical CN. Moreover, additional degrees of freedom are definitely needed to describe these complicated shapes. The standard two-center shell model, as well as other macro-microscopic models, deal only with axially symmetric shapes.

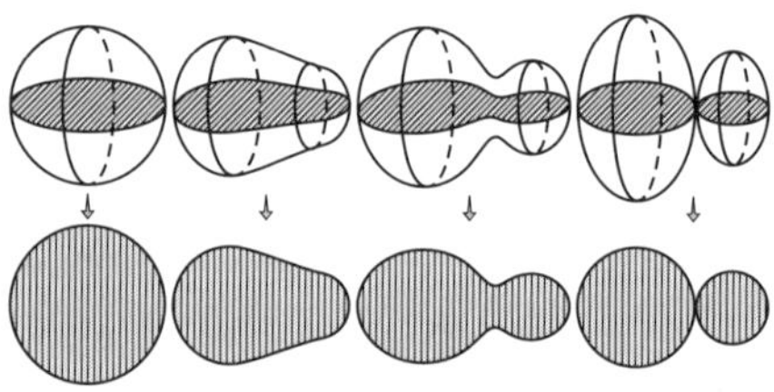

Fig. 4. Back and top views (upper and bottom figures) of subsequent shapes of the nuclear system evolving from the configuration of two equatorially touching statically deformed nuclei to the configuration of spherical CN.

Within the two-core model we may calculate the adiabatic potential energy not only for axially symmetric shapes but also for the side-by-side initial orientation assuming that on the way to CN only the equatorial dynamic deformations of both fragments may change, whereas the static deformations of the cores (along axes perpendicular to the line connecting two centers) gradually relax to zero values with increasing equatorial deformation and mass transfer, see schematic Fig. 4. This assumption seems quite reasonable because there are no forces which may change the "perpendicular" deformations of the fragments. In that case we need no additional degrees of freedom. The same variables δ_1 and δ_2 may be used for dynamic deformations along the axis between nuclear centers ($\delta_{1,2} = 0$ at contact). We assume that the static deformations of the nuclei just gradually disappear with increasing mass transfer and dynamic equatorial deformations: $\delta^{\perp}_{1,2} = \delta^{\perp}_{1,2}(0) \cdot exp[-(\frac{\eta-\eta_0}{\Delta_\eta})^2] \cdot exp[-(\frac{\delta_{1,2}}{\Delta_\delta})^2]$. Thus, they are not independent variables. Here η_0 is the initial mass asymmetry, $\delta^{\perp}_{1,2}(0)$ are the static deformations of the projectile and target, and $\Delta_\eta \sim \frac{10}{A_{CN}}$ and $\Delta_\delta \sim 0.2$ are the adjusted parameters which do not much influence the whole dynamics.

The calculated driving potentials for the two fixed orientations of statically deformed ^{248}Cm nucleus fusing with ^{48}Ca are shown in Fig. 5 in the space of mass asymmetry and elongation. For the side collision the Coulomb barrier in the entrance channel is significantly higher. However, at the contact point this configuration is much more compact and the path to formation of CN is much shorter comparing with the tip (nose-to-nose) collisions. So we may expect higher fusion probability in this case, and our calculations (see below) confirm that. It is rather difficult (if possible, at all) to derive adiabatic potential energy of the nuclear system evolving from the configuration of arbitrary oriented touching deformed nuclei. In contrast, the diabatic potential energy is calculated easily in that case by using the double folding procedure, for example. To take somehow into account the orientation effect in the cross section of CN formation we may simply average the results obtained for the two limiting orientations.

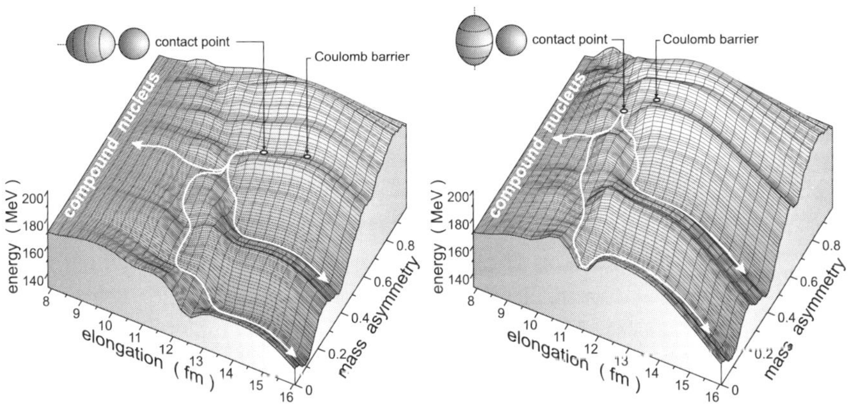

Fig. 5. Driving potentials for the nuclear system formed in $^{48}Ca+^{248}Cm$ collision at tip (top) and side (bottom) orientations of statically deformed ^{248}Cm. The solid lines with arrows show schematically (without fluctuations) the projections of the QF trajectories (going to lead and tin valleys) and the path leading to formation of CN..

3. Collision dynamics

3.1. *Equations of motion*

A choice of dynamic equations for the considered degrees of freedom is also not so evident. The main problem here is a proper description of nucleon transfer and change of the mass asymmetry which is a discrete variable by its nature. Moreover, the corresponding inertia parameter μ_η, being calculated within the Werner-Wheeler approach, becomes infinite at contact (scission) point and for separated nuclei. In Ref. [10] the inertialess Langevin type equation for the mass asymmetry

$$\frac{d\eta}{dt} = \frac{2}{A_{CN}} D_A^{(1)}(\eta) + \frac{2}{A_{CN}} \sqrt{D_A^{(2)}(\eta)} \Gamma(t), \tag{2}$$

has been derived from the corresponding master equation for the distribution function $\varphi(A,t)$ [14, 15]. Here $\Gamma(t)$ is the normalized random variable with Gaussian distribution, $\langle \Gamma(t) \rangle = 0$, $\langle \Gamma(t)\Gamma(t') \rangle = 2\delta(t-t')$, and $D_A^{(1)}$, $D_A^{(2)}$ are the transport coefficients. A is the number of nucleons in one of the fragments and $\eta = (2A - A_{CN})/A_{CN}$.

Assuming that sequential nucleon transfers play a main role in mass rearrangement, i.e. $A' = A \pm 1$, we have

$$\begin{aligned} D_A^{(1)} &= \lambda(A \to A+1) - \lambda(A \to A-1), \\ D_A^{(2)} &= \tfrac{1}{2}[\lambda(A \to A+1) + \lambda(A \to A-1)]. \end{aligned} \tag{3}$$

For nuclei in contact the macroscopic transition probability $\lambda(A \to A' = A \pm 1)$ is defined by nuclear level density [14, 15] $\lambda^{(\pm)} = \lambda_0 \sqrt{\rho(A \pm 1)/\rho(A)} \approx \lambda_0 \exp\left(\frac{V(R,\delta,A\pm1) - V(R,\delta,A)}{2T}\right)$. Here $T = \sqrt{E^*/a}$ is the local nuclear temperature, $E^*(R,\delta,\eta)$ is the excitation energy, a is the level density parameter, and λ_0 is the nucleon transfer rate ($\sim 10^{22}$ s^{-1} [14, 15]), which may, in principle, depend on excitation energy (the same holds for the diffuseness coefficient $D_A^{(2)}$). This feature, however, is not completely clear. Here we treat the nucleon transfer rate λ_0 as a parameter of the model. Later we hope to derive the temperature dependence of this parameter from a systematic analysis of available experimental data.

Nucleon transfer for slightly separated nuclei is also rather probable. This intermediate nucleon exchange plays an important role in sub-barrier fusion processes [16] and has to be taken into account in Eq. (2). It can be done by using the following final expression for the transition probability

$$\lambda^{(\pm)} = \lambda_0 \sqrt{\frac{\rho(A \pm 1)}{\rho(A)}} P_{tr}(R,\delta,A \to A \pm 1). \tag{4}$$

Here $P_{tr}(R,\delta,A \to A \pm 1)$ is the probability of one nucleon transfer depending on the distance between the nuclear surfaces. This probability goes exponentially to zero at $R \to \infty$ and it is equal to unity for overlapping nuclei. In our calculations we used the semiclassical approximation for P_{tr} proposed in Ref. [16]. Eq. (2) along with (4) defines a continuous change of mass asymmetry in the whole space (obviously, $\frac{d\eta}{dt} \to 0$ for far separated nuclei). Finally there are 13 coupled Langevin type equations for 7 degrees of freedom $\{R, \vartheta, \delta_1, \delta_2, \varphi_1, \varphi_2, \eta\}$ [10].

3.2. *Friction forces and nuclear viscosity*

A number of different mechanisms have been suggested in the literature for being responsible for the energy loss in DI collisions. A discussion of the subject and appropriate references can be found, e.g., in [2, 17]. The uncertainty in the strength of nuclear friction and in its form-factor is still very large. Because of that and for the sake of simplicity we use here for separated nuclei the phenomenological nuclear friction forces with the Woods-Saxon radial form-factor $F(\zeta) = (1 + e^{\zeta})^{-1}$, $\zeta = (\xi - \rho_F)/a_F$ and ξ is the distance between nuclear surfaces. The shift $\rho_F \sim 2$ fm serves to approach the position of the friction shape function to the strong absorption distance which is normally larger than the contact distance $R_{\rm cont}$. Thus $\gamma_R^F = \gamma_R^0 F(\xi - \rho_F)$, $\gamma_{\rm tang}^F = \gamma_t^0 F(\xi - \rho_F)$ and γ_R^0, γ_t^0, ρ_F and $a_F \sim 0.6$ fm are the model parameters.

For overlapping nuclei (mono-nucleus configuration) the two-body nuclear friction can be calculated within the Werner-Wheeler approach [18]. The corresponding viscosity coefficient μ_0 is however rather uncertain. From the analysis of fission-fragment kinetic energies it has been estimated to be of the order of several units of 10^{-23} Mev s fm^{-3} [18]. The one-body dissipation mechanism leads in general to stronger nuclear friction and some reduction coefficient for it is often used in specific calculations. Taking into account this uncertainty we use here the Werner-Wheeler approach [18] for calculating the form-factors of nuclear friction $\gamma_R^{WW}(R, \delta_1, \delta_2, \eta)$ and $\gamma_{\delta_1,\delta_2}^{WW}(R, \delta_1, \delta_2, \eta)$ with the viscosity coefficient μ_0 which is treated as a model parameter. To keep continuity of kinetic energy dissipation at contact point, where two colliding nuclei form a mono-nucleus, we switched the phenomenological friction γ_R^F to γ_R^{WW} by the "smoothed" (over 0.6 fm) step function $\theta_s(\xi) = (1 - e^{-\xi/0.3})^{-1}$. There is no problem at contact point for the nuclear friction $\gamma_{\delta_1,\delta_2}$ associated with the surface deformations.

The two strength parameters of nuclear friction, γ_R^0 for well separated nuclei and μ_0 for nuclear viscosity of the deformable mono-nucleus, reflect, from the one side, a possible difference in the mechanisms of dissipation of relative motion kinetic energy in DI collisions of two separated nuclei and nuclear viscosity of a mono-nucleus due to coupling of collective motion (shape parameters) with the particle-hole excitations. On the other side, these friction strength parameters are of the same order of magnitude. Using $\mu_0 = 0.2 \cdot 10^{-22}$ MeV s fm^{-3} proposed in [18] we get the nuclear friction coefficient $\gamma_R(\delta = 0) = 4\pi R_0 \mu_0 \approx 15$ MeV s fm^{-2} for a change in elongation of a spherical nucleus with radius $R_0 = 6$ fm. This value can be compared with the value of nuclear friction of two nuclei in contact $\gamma_R(\xi = 0) = 13$ MeV s fm^{-2} estimated from the "proximity theorem" (Ref. [2], p. 269). Nevertheless, as mentioned above, the uncertainty in the values of both parameters is very large. Moreover, microscopic analysis shows that nuclear viscosity may also depend strongly on nuclear temperature [19]. Analyzing experimental data on DI scattering of heavy ions we prefer to treat nuclear friction on a phenomenological base using appropriate strength parameters γ_R^0 and μ_0, which later could be compared with those calculated microscopically [19].

3.3. *Decay of primary fragments and cross sections*

The cross sections for all the processes can be calculated now in a simple and natural way. A large number of events (trajectories) are tested for a given impact parameter. Those events, in which the nuclear system overcame the fission barrier from the outside and entered the region of small deformations and elongations, are treated as fusion (CN formation). The other events correspond to quasi-elastic, DI and QF processes. Subsequent decay of the excited CN ($C \to B + xn + N\gamma$) is described then within the statistical model. The double differential cross-sections of all the processes are calculated as follows

$$\frac{d^2\sigma_\eta}{d\Omega dE}(E,\theta) = \int_0^\infty bdb \frac{\Delta N_\eta(b,E,\theta)}{N_{\text{tot}}(b)} \frac{1}{\sin(\theta)\Delta\theta\Delta E}. \tag{5}$$

Here $\Delta N_\eta(b,E,\theta)$ is the number of events at a given impact parameter b in which the system enters into the channel η (definite mass asymmetry value) with kinetic energy in the region $(E, E+\Delta E)$ and center-of-mass outgoing angle in the region $(\theta, \theta+\Delta\theta)$, $N_{\text{tot}}(b)$ is the total number of simulated events for a given value of impact parameter. In collisions of deformed nuclei averaging over initial orientations should be performed. It is made quite simply for DI and capture cross sections because the diabatic potential energy surface is easily calculated for any orientation of deformed nuclei. Probability of CN formation is determined mainly by the adiabatic potential, which, as mentioned above, can be calculated for the moment only for the two limiting orientations of touching deformed nuclei. Thus, the cross section of CN formation was calculated by averaging the results obtained for these two limiting orientations (nose-to-nose and side-by-side).

Expression (5) describes the mass, energy and angular distributions of the *primary* fragments formed in the binary reaction (both in DI and in QF processes). Subsequent de-excitation cascades of these fragments via fission and emission of light particles and gamma-rays were taken into account explicitly for each event within the statistical model leading to the *final* mass and energy distributions of the reaction fragments. The sharing of the excitation energy between the primary fragments was assumed to be proportional to their masses. For each excited fragment the multi-step decay cascade was analyzed taking into account a competition between evaporation of neutrons and/or protons and fission. At the final stage of the evaporation cascade ($E^* < E_n^{sep}$) a competition between γ-emission and fission was taken into account in the same way as for survival of CN. Due to rather high excitations of the fragments the analysis of this evaporation cascade needs the longest computation time. Mass, energy and angular distributions of the fission fragments (regular fission of CN) are also estimated within the statistical model using the dependent on mass asymmetry adiabatic potential energy surface at the scission point.

The used model allows us to perform also a time analysis of the studied reactions. Each tested event is characterized by the reaction time τ_{int}, which is calculated as a difference between re-separation (scission) and contact times. Those events, in which nuclei do not come in contact (e.g., for large impact parameters), are excluded from the analysis. In such a way, for all the channels we may calculate and analyze reaction-time distributions, which is very important for formation of giant quasi-atoms in collisions of heavy transactinide nuclei [20].

4. Deep inelastic scattering

At first we applied the model to describe available experimental data on low-energy damped collision of very heavy nuclei, ^{136}Xe+^{209}Bi [21], where the DI process should dominate due to expected prevalence of the Coulomb repulsion over nuclear attraction and the impossibility of CN formation. In Fig. 6 the angular, energy and charge distributions of the Xe-like fragments are shown comparing with our calculations (histograms). In accordance with experimental conditions only the events with the total kinetic energy in the region of $260 \le E \le 546$ MeV and with the scattering angles in the region of $40^o \le \theta_{\rm c.m.} \le 100^o$ were accumulated. The total cross section corresponding to all these events is about 2200 mb (experimental estimation is 2100 mb [21]). Due to the rather high excitation energy sequential fission of the primary heavy fragments may occur in this reaction (mainly those heavier than Bi). In the experiment the yield of the heavy fragments was found to be about 30% less comparing with Xe-like fragments. Our calculation gives 354 mb for the cross section of sequential-fission, which is quite comparable with experimental data. Mass distribution of the fission fragments is shown in Fig. 6(c) by the dotted histogram. Note that it is a contamination with sequential fission products of heavy primary fragments leading to the bump around Z=40 in the experimental charge distribution in Fig. 6(c).

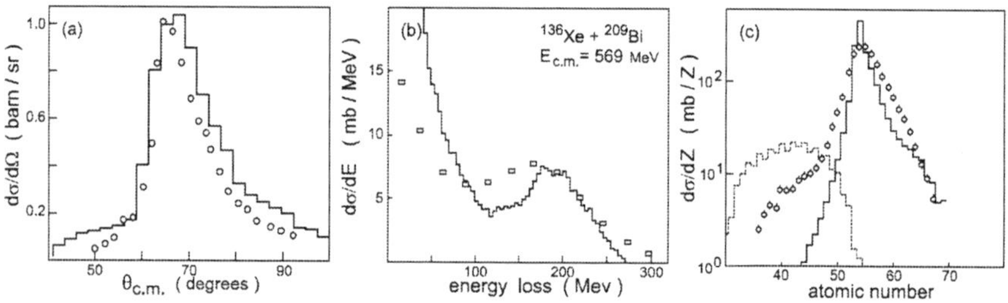

Fig. 6. Angular (a), energy-loss (b) and charge (c) distributions of the Xe-like fragments obtained in the $^{136}Xe+^{209}Bi$ reaction at $E_{c.m.} = 568$ MeV. Experimental data are taken from Ref. [21]. Histograms are the theoretical predictions. The low-Z wing of the experimental charge distribution is due to incompletely removed events of sequential fission of the heavy fragment [21]. Dotted histogram in (c) indicates the calculated total yield of sequential fission fragments.

The interaction time is one of the most important characteristics of nuclear reactions, though it cannot be measured directly. The total reaction time distribution, $\frac{d\sigma}{dlog(\tau)}$ (τ denotes the time between the moments of contact of two nuclei and re-separation of the fragments), is shown in Fig. 7 for the studied reaction. In most of the damped collisions ($E_{loss} > 35$ MeV) the interaction time is rather short ($\sim 10^{-21}$ s). These fast events correspond to collisions with intermediate impact parameters. Nevertheless, a large amount of kinetic energy is dissipated here very fast at relatively low mass transfer.

However, in some cases, in spite of an absence of attractive potential pocket the system may hold in contact rather long. During this time it moves over the multidimensional potential energy surface with almost zero kinetic energy (result of large nuclear viscosity) mainly in deformation and mass-asymmetry space. Note that it is the longest component of the time distribution (second peak in Fig. 7) which corresponds to the most dissipative collisions. Large overlap of nuclear surfaces takes place here and, as a result, significant mass rearrangement may occur. In the TKE-mass plot these events spread over a wide region of mass fragments (including symmetric splitting) with kinetic energies very close to kinetic energy of fission

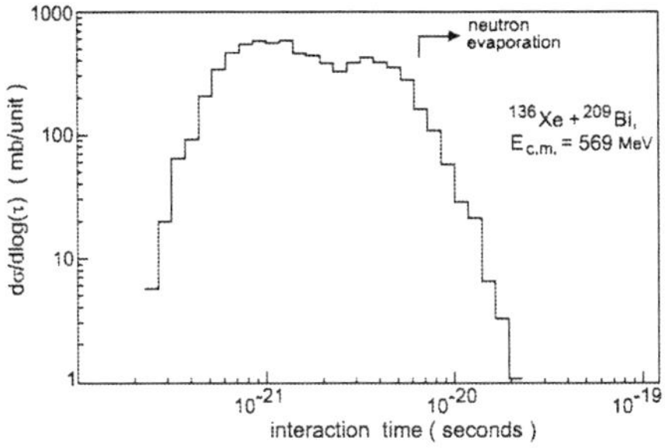

Fig. 7. Reaction time distribution for the $^{136}Xe+^{209}Bi$ collision at 569 MeV center-of-mass energy.

fragments [see pronounced bump and its tail in the energy-loss distribution in Fig. 6(b)]. Some gap between the two groups in the energy and in the time distributions can be also seen in Fig. 6(b) and Fig. 7. All these make the second group of slow events quite distinguished from the first one. These events are more similar to fission than to deep-inelastic processes. Formally, they also can be marked as quasi-fission.

5. Quasi-fission and SHE formation

Let us consider now the near-barrier $^{48}Ca+^{248}Cm$ fusion reaction (leading to the formation of a superheavy nucleus) in which the QF process plays a dominant role. The potential energy surfaces for this nuclear system are shown in Fig. 5 for two different initial orientations of the ^{248}Cm nucleus at fixed dynamic deformation, which also plays a very important role here. Our calculations show that after overcoming the Coulomb barrier the fragments become first very deformed, then the mass asymmetry gradually decreases and the system finds itself in the quasi-fission valley with one of the fragments close to the doubly magic nucleus ^{208}Pb (see deep valley at $\eta \approx 0.4$ in Fig. 5). To simulate somehow the neck formation in the QF channels and to describe properly the energy distribution of reaction fragments we assumed that the radial parameters of the formfactors of the friction forces are different in the entrance and exit channels. In the first case the contact distance was calculated as $R_{\rm cont}(\eta,\delta) = [R_1(A_1,\delta_1) + R_2(A_2,\delta_2)]$ with $r_0 = 1.16$ fm whereas for the QF channels the scission distance (up to which the friction forces keep on) was defined as $R_{\rm scission}(\eta,\delta) = (1.4/r_0)[R_1(A_1,\delta_1) + R_2(A_2,\delta_2)] + 1$ fm. The solid line in Fig. 8(b) just corresponds to the potential energy at the scission point $V(r = R_{\rm scission},\delta,\eta) + Q_{gg}(\eta)$ minimized over δ.

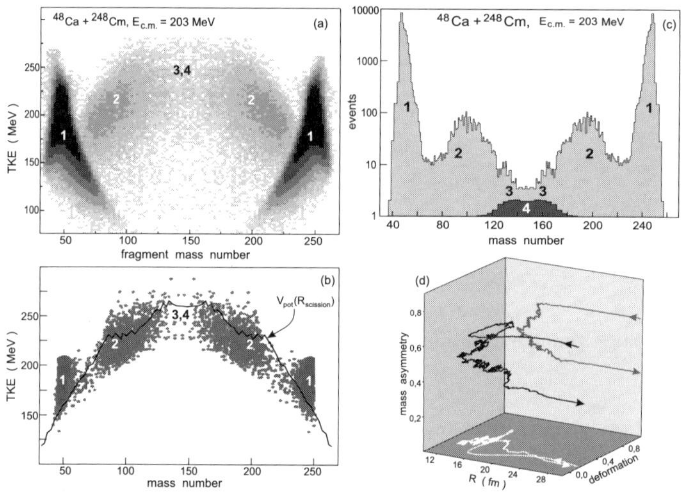

Fig. 8. Experimental (a) and calculated (b) TKE-mass distributions of reaction products in collision of ^{48}Ca+^{248}Cm at $E_{\mathrm{c.m.}} = 203$ MeV. (c) Contributions of DI (1), QF (2,3) and fusion-fission (4) processes into inclusive mass distribution. (d) One of the trajectories in collision of ^{48}Ca+^{248}Cm leading to QF channel (2).

Fig. 8 shows the experimental and calculated correlations of the total kinetic energy and the mass distributions of the primary reaction products along with inclusive mass distribution for the ^{48}Ca+^{248}Cm reaction at the near-barrier energy of $E_{\mathrm{c.m.}} = 203$ MeV. The tails of the DI component to the unphysical high energies (higher than $E_{\mathrm{c.m.}}$ at $A_1 \sim 50$ and $A_2 \sim 250$) and to very low energies with more symmetric mass combinations in Fig. 8(a) [absent in the calculations in Fig. 8(b)] are due to instrument effects. The large yield of the fragments in the region of doubly magic nucleus ^{208}Pb (and the complimentary light fragments) is the most pronounced feature of the TKE-mass distribution. These QF process ("symmetrizing" quasi-fission) is the dominant channel in reactions of such kind which protects the nuclei from fusing (formation of compound nucleus).

The probability for CN formation in this reaction was found to be very small and depended greatly on the incident energy. Due to a strong dissipation of kinetic energy just the fluctuations (random forces) define the dynamics of the system after the contact of the two nuclei. At near barrier collisions the excitation energy (temperature) of the system is rather low, the fluctuations are weak and the system chooses the most probable path to the exit channel along the quasi-fission valley (see Fig. 5). However at non-zero excitation energy there is a chance for the nuclear system to overcome the multi-dimensional inner potential barriers and find itself in the region of the CN configuration (small deformation and elongation). Within the Langevin calculations a great number of events should be tested to find this low probability. For the studied reaction, for example, only several fusion events have been found among more than 10^5 total tested events [see dark region 4 in Fig. 8(c)].

The cross section of CN formation in this reaction was found really dependent on initial orientation of the statically deformed ^{248}Cm nucleus. Having for the moment the potential energy surface only for the two limiting orientations (see Fig. 5), we performed here a simple averaging of the cross sections obtained for the tip and side configurations. Due to a lower Coulomb barrier, the tip collisions lead to larger value of the capture cross section compared with the side collisions. If we define the capture cross section as all the events in which the nuclei overcome the Coulomb barrier, come in contact and fuse or re-separate with the mass rearrangement exceeding 20 mass units (to distinguish it somehow from the DI cross section), then $\sigma_{\mathrm{cap}} \approx 45$ mb for tip collisions and only 5 mb for the side ones at the beam energy of $E_{\mathrm{c.m.}} = 203$ MeV. However, (unambiguously defined) fusion cross sections were found to be rather close for both cases (about 0.03 mb and 0.04 mb, respectively), which means that CN formation at this energy is about 10 times more probable for the side-oriented touching nuclei. This result is in a reasonable agreement with those found previously [9] and with the yield of evaporation residues in this reaction [13].

Note that a direct experimental study of an influence of static deformations of heavy nuclei on a probability of CN formation could be done by comparison of the capture and evaporation residue cross sections for two fusion reactions, ^{64}Zn+^{150}Nd and ^{70}Zn+^{144}Nd, leading to the *same* compound nucleus ^{214}Th. In the first case the nuclei, ^{64}Zn and ^{150}N, have non-zero deformations in their ground states, whereas ^{70}Zn and ^{144}N are spherical nuclei. We may expect that the excitation functions for the yields of evaporation residues will be quite different for the two reactions reflecting an influence of the orientation effects on the fusion probability.

Within our approach we estimated a possibility of SH element production in the asymmetric fusion reactions of nuclei heavier than ^{48}Ca with transuranium targets. Such reactions can be used, in principle, for a synthesis of the elements heavier then 118. Evaporation residue (EvR) cross sections for the fusion reactions ^{50}Ti+^{244}Pu, ^{50}Ti+^{243}Am, ^{54}Cr+^{248}Cm and ^{58}Fe+^{244}Pu are shown in Fig. 9. For all cases we used the fission barri-

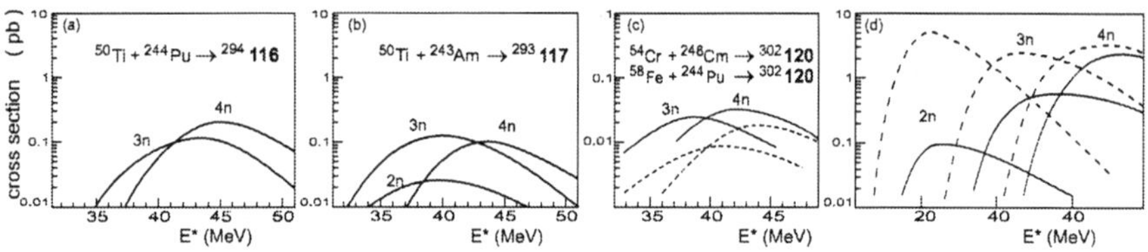

Fig. 9. Evaporation residue cross sections for the fusion reactions $^{50}Ti+^{244}Pu$ (a), $^{50}Ti+^{243}Am$ (b), $^{54}Cr+^{248}Cm$, $^{58}Fe+^{244}Pu$ (dashed curves) (c) and $^{136}Xe+^{136}Xe$ (d).

ers of CN predicted by the macro-microscopical model [6], which gives much lower fission barrier for $^{302}120$ nucleus in comparison with $^{296}116$. However, full microscopic models based on the self-consistent Hartree-Fock calculations [22] predict much higher fission barriers for the nucleus $^{302}120$ (up to 10 MeV for the Skyrme forces). This means that the 3n and 4n EvR cross sections in the $^{54}Cr+^{248}Cm$ and $^{58}Fe+^{244}Pu$ fusion reactions could be two orders of magnitudes higher as compared with those shown in Fig. 9(c).

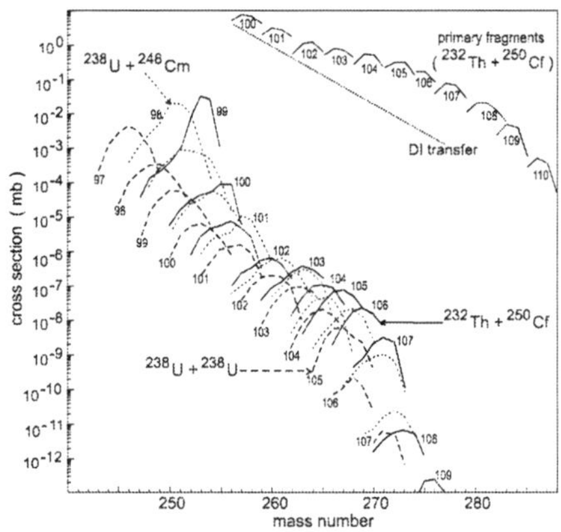

Fig. 10. Yield of superheavy nuclei in collisions of $^{238}U+^{238}U$ (dashed), $^{238}U+^{248}Cm$ (dotted) and $^{232}Th+^{250}Cf$ (solid lines) at 800 MeV center-of-mass energy. Solid curves in upper part show isotopic distribution of primary fragments in the Th+Cf reaction.

SH elements beyond 118 may be synthesized also in the fusion reactions of symmetric nuclei (fission-like fragments). However, in such reactions an uncertainty in calculation of very small cross sections for CN formation is rather large. Dashed and solid curves in Fig. 9(d) reflect this uncertainty in our estimations of the EvR cross sections in $^{136}Xe+^{136}Xe$ fusion reaction. If the experiment (planned to be performed in Dubna) will give the EvR cross sections at the level of few picobarns for this reaction then we may really dream about using neutron-rich accelerated fission fragments for production of SH elements in the region of the "island of stability" (e.g., ^{132}Sn+^{176}Yb→^{308}120).

Another possibility for a synthesis of the neutron-rich SH elements is the low-energy damped collisions of very heavy transactinide nuclei [20] (e.g., ^{238}U+^{248}Cm). Existence of rather pronounced lead valley on the potential landscape of such giant nuclear systems leads to the so-called "inverse" (anti-symmetrizing) quasi-fission process, in which one fragment transforms to the doubly magic nucleus ^{208}Pb, whereas another one transforms to complementary SH element. In spite of rather high excitation energy, this neutron-rich superheavy nucleus may survive in neutron evaporation cascade giving us an alternative way for SH element production (see Fig. 10).

References

1. J. Blocki et al., *Ann. Phys. (N.Y.)* **105**, 427 (1977).
2. R. Bass, *Nuclear Reactions with Heavy Ions* (Springer-Verlag), 1980, p. 326.
3. V.I. Zagrebaev, M.G. Itkis and Yu.Ts. Oganessian, *Yad. Fiz.* **66**, 1069 (2003).
4. U. Mosel, J. Maruhn and W. Greiner, *Phys. Lett. B* **34** 587, (1971).
5. A.I. Sierk, *Phys. Rev. C* **33**, 2039 (1986).
6. P. Möller et al., *At. Data Nucl. Data Tables* **59**, 185 (1995).
7. V.I. Zagrebaev, *Phys. Rev. C* **64**, 034606 (2001).
8. V.I. Zagrebaev, *J. Nucl. Rad. Sci.* **3**, No. 1, 13 (2002).
9. V.I. Zagrebaev, *AIP Conf. Proc.* **704**, 31 (2004).
10. V. Zagrebaev and W. Greiner, *J. Phys. G* **31**, 825 (2005).
11. W.D. Myers and W.J. Swiatecki, *Ann. Phys. (N.Y.)* **84**, 186 (1974).
12. K. Nishio, H. Ikezoe, S. Mitsuoka and J. Lu, *Phys.Rev. C* **62**, 014602 (2000).
13. Yu.Ts. Oganessian et al., *Phys. Rev. C* **70**, 064609 (2004).
14. W. Nörenberg, Phys. Lett. B **52**, 289 (1974).
15. L.G. Moretto and J.S. Sventek, Phys. Lett. B **58**, 26 (1975).
16. V.I. Zagrebaev, Phys. Rev. C **67**, 061601(R) (2003).
17. W.U. Schröder and J.R. Huizenga, *Damped Nuclear Reactions* in *Treatise on Heavy-Ion Science*, Ed. D.A. Bromley (Plenum Press, NY, 1984), V.2, p.140.
18. K.T.R. Davies, A.J. Sierk and J.R. Nix, *Phys. Rev. C* **13**, 2385 (1976).
19. H.Hofmann, Phys.Rep. **284**, 137 (1997).
20. V.I. Zagrebaev, M.G. Itkis, Yu.Ts. Oganessian and W. Greiner, *Phys. Rev. C* **73**, 031602 (2006).
21. W.W. Wilcke, J.R. Birkelund, A.D. Hoover, J.R. Huizenga, W.U. Schröder, V.E. Viola, Jr., K.L. Wolf, and A.C. Mignerey, Phys. Rev. C **22**, 128 (1980).
22. T. Bürvenich et al., Phys. Rev. C **69** (2004) 014307.

TIME-DEPENDENT POTENTIAL ENERGY FOR FUSION AND FISSION PROCESSES

A. V. KARPOV*, V. I. ZAGREBAEV, Y. ARITOMO, and M. A. NAUMENKO

Flerov Laboratory of Nuclear Reactions, JINR,
Dubna, Moscow region, Russia
**E-mail: karpov@jinr.ru*

W. GREINER

Frankfurt Institute for Advanced Studies, J. W. Goethe-Universität
Frankfurt am Main, Germany

The problem of description of low-energy nuclear dynamics and derivation of multi-dimensional potential energy surface depending on several collective degrees of freedom is discussed. Multi-dimensional adiabatic potential is constructed basing on extended version of the two-center shell model. It has correct asymptotic value and height of the Coulomb barrier in the entrance channel (fusion) and appropriate behavior in the exit one, giving required mass and energy distributions of reaction products and fission fragments. Explicit time-dependence of the driving potential was introduced in order to take into account difference of diabatic and adiabatic regimes of motion of nuclear system at above-barrier energies and also difference of nuclear shapes in fusion and fission channels (neck formation). Derived driving potential is proposed to be used for unified analysis of the processes of deep-inelastic scattering, fusion and fission at low-energy collisions of heavy ions.

Keywords: adiabatic driving potential; fusion-fission dynamics

1. Introduction

The analysis of near-barrier nucleus-nucleus collisions shows that the main reaction channels here are deep-inelastic scattering[1] and quasifission.[2,3] In particular, the quasifission processes decrease the probability of fusion of heavy nuclei appreciably. Due to this competition and substantial overlapping of all the channels of the reactions with heavy ions a unified dynamical approach for the simultaneous description of all the possible processes is needed. First of all, such an approach implies using the degrees of freedom common for all the strongly coupled channels. The most relevant degrees of freedom, in our opinion, are: elongation of the system r, dynamical deformations $\vec{\beta}$, mass asymmetry of the system η, and relative orientations of target and projectile Ω_1, Ω_2 (in the entrance channel).

Then, we need a unified potential energy which depends on the chosen collective coordinates and governs the whole process of fusion-fission. The constructed potential energy must have correct asymptotic behavior in the region of separated nuclei and be in agreement with the experimental data on fusion and fission barriers and ground state masses. Only such a potential energy allows us to perform simultaneous realistic analysis of the deep-inelastic processes, quasifission and fusion-fission. Finally, equations of motion along with the necessary initial and boundary conditions have to be formulated to perform numerical analysis of the reactions being studied.

In the present paper the unified multidimensional potential energy is developed basing on extended variant of the two-center shell model. For the fast nucleus-nucleus collisions the first non-equilibrium (diabatic) stage of the reaction is taken into account. The corresponding diabatic potential energy is calculated within the double-folding procedure.

2. Potential energy of heavy nuclear system

2.1. *Diabatic potential energy*

We use the following definition of the diabatic potential energy:

$$V_{\mathrm{diab}}(A,Z;r,\vec{\beta}_1,\Omega_1,\vec{\beta}_2,\Omega_2,\eta) = V_{12}(A_1,Z_1,A_2,Z_2;r,\vec{\beta}_1,\Omega_1,\vec{\beta}_2,\Omega_2) + M(A_1,Z_1;\vec{\beta}_1) + M(A_2,Z_2;\vec{\beta}_2) - M(A_T,Z_T;\vec{\beta}_T^{\mathrm{g.s.}}) - M(A_P,Z_P;\vec{\beta}_P^{\mathrm{g.s.}}).$$

Here V_{12} is the interaction energy of the nuclei, $M(A_{1,2},Z_{1,2})$ are the masses of future fragments, and the constant value $M(A_T,Z_T)+M(A_P,Z_P)$ (the sum of the ground state masses of target and projectile) determines zero value of the potential energy in the entrance channel at infinite distance between the nuclei. In the channels with mass rearrangement the value of V_{diab} at infinite distance equals to the Q-value of the reaction.

We apply double-folding procedure to calculate the interaction energy V_{12}. It consists in summation of the effective nucleon-nucleon interaction (see, e.g.,[6]). In this approach the effects of deformation and orientation are taken into account automatically. According to the folding procedure the interaction energy of two nuclei is given by

$$V_{12}(r;\vec{\beta}_1,\Omega_1,\vec{\beta}_2,\Omega_2) = \int_{V_1}\rho_1(\mathbf{r}_1)\int_{V_2}\rho_2(\mathbf{r}_2)v_{NN}(\mathbf{r}_{12})d^3r_1d^3r_2, \quad (1)$$

where $v_{NN}(\mathbf{r}_{12} = \mathbf{r} + \mathbf{r}_2 - \mathbf{r}_1)$ is the effective nucleon-nucleon interaction and $\rho_i(\mathbf{r}_i)$ are the density distributions of nuclear matter in the nuclei $(i = 1, 2)$. The nuclear density is usually parametrized by the Fermi-type function $\rho(\mathbf{r}) = \rho_0 \left[1 + \exp\left(\frac{r - R(\Omega_r)}{a}\right)\right]^{-1}$, where $R(\Omega_\mathbf{r})$ is the distance to the nuclear surface ($\Omega_\mathbf{r}$ are the spherical coordinates of $\mathbf{r}$), and the value ρ_0 is determined from the condition $\int \rho_i d^3\mathbf{r} = A_i$. There are two independent parameters in this formula: the diffuseness of the nuclear density a and the nuclear radius parameter r_0.

The effective nucleon-nucleon potential consists of the Coulomb and nuclear parts $v_{NN} = v_{NN}^{(N)} + v_{NN}^{(C)}$. One of the most frequently used nucleon-nucleon potential in the theory of nuclear reactions is the M3Y potential.[7–9] However, the M3Y potential leads to a very strong attraction in the region of overlapping nuclei, where due to the Pauli principle the repulsion has to appear. Therefore, we prefer to use another nucleon-nucleon potential proposed by A. B. Migdal.[10] The Migdal nucleon-nucleon potential is zero-range density-dependent potential. It has the form

$$\begin{aligned} v_{NN}^{(N)}(\mathbf{r}_1, \mathbf{r}_2) &= C \left[F_{\text{ex}} + (F_{\text{in}} - F_{\text{ex}}) \frac{\rho_1(\mathbf{r}_1) + \rho_2(\mathbf{r}_2)}{\rho_{00}}\right] \delta(\mathbf{r}_{12}) \\ &= v_{\text{eff}}(\mathbf{r}_1, \mathbf{r}_2)\delta(\mathbf{r}_{12}), \quad F_{\text{ex(in)}} = f_{\text{ex(in)}} \pm f'_{\text{ex(in)}}. \end{aligned} \tag{2}$$

Here $\rho_{00} = (\rho_{01} + \rho_{02})/2$; sign "+" in the last expression corresponds to the interaction of identical particles (proton-proton or neutron-neutron) and "-" to the proton-neutron interaction. For the fixed value of the constant C=300 MeV fm^3 the following values of the amplitudes were recommended in:[10] $f_{\text{in}} = 0.09$; $f_{\text{ex}} = -2.59$; $f'_{\text{in}} = 0.42$; $f'_{\text{ex}} = 0.54$. The potential (2) is defined by the amplitude F_{ex} for the interaction of "free" nucleons (i.e., nucleons from the tails of the nuclear density distributions, where $\rho_1 + \rho_2 \simeq 0$); by the amplitude F_{in} for the interaction of a free nucleon with a nucleon inside the nucleus ($\rho_1 + \rho_2 \simeq \rho_{00}$); and by the value $(2F_{\text{in}} - F_{\text{ex}})$ if both nucleons are inside the nuclei (double nuclear density region).

For the calculations we use equal proton and neutron densities in the nucleus center: $\rho_0^{(p)} = \rho_0^{(n)}$ and different radii of these densities. Thus, we have two free parameters: the radius of the charge distribution $R^{(p)} = r_0^{(p)} A^{1/3}$ and the diffuseness a. Parametrization of $r_0^{(p)}$ can be obtained by fitting the corresponding experimental data.[11,12] We propose the following parametrization: $r_0^{(p)}(Z) = 0.94 + 32/(Z^2 + 200)$, which can be used for the nuclei heavier than carbon. The values of the diffuseness a were fitted in order to describe the experimental fusion barriers for spherical nuclei. Using all the possible combinations of the spherical nuclei ^{16}O, ^{40}Ca, ^{48}Ca, ^{60}Ni,

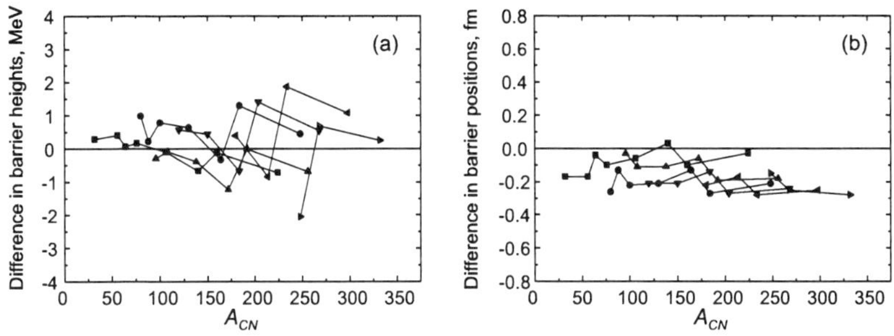

Fig. 1. Difference between the fusion barrier heights (a) and their positions (b) obtained within the folding potential with the Migdal forces and the Bass potential ("experimental data"). The calculations were performed for all the possible combinations of the spherical nuclei ^{16}O, ^{40}Ca, ^{48}Ca, ^{60}Ni, ^{90}Zr, ^{124}Sn, ^{144}Sm, ^{208}Pb.

^{90}Zr, ^{124}Sn, ^{144}Sm, ^{208}Pb we obtained the parametrization: $a(Z) = 0.734 - 150/\left(Z^2 + 500\right)$, which is recommended for the calculation of the nucleus-nucleus folding potentials for nuclei $A_{1,2} \geq 16$. The difference between the calculated and "experimental" (the Bass barriers[13]) fusion barriers is shown in Fig. 1. We reproduce the experimental data with accuracy of 2 MeV for the barrier heights and 0.3 fm for the barrier position.

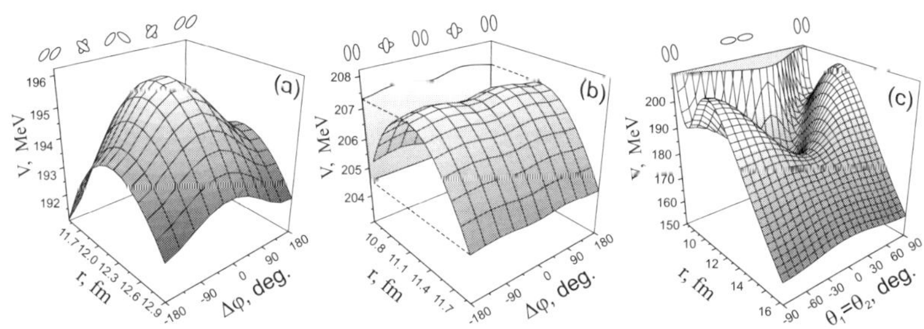

Fig. 2. Folding potential with the Migdal forces as a function of the relative distance r and various orientations of the nuclei $^{64}Zn(\beta_2^{g.s.} = 0.22)$ and $^{150}Nd(\beta_2^{g.s.} = 0.24)$. Case (a) corresponds to $\theta_1 = \theta_2 = \pi/4$, case (b) – to $\theta_1 = \theta_2 = \pi/2$, and case (c) – to $\Delta\varphi = 0$. The relative positions of the nuclei are shown schematically in the upper part of the figure.

Figure 2 shows the dependence of the folding potential with the Migdal forces on the distance between mass centers and relative orientations for the system $^{64}Zn +^{150} Nd$. Dependence on the azimuthal angle $\Delta\varphi$ is given in Fig. 2 (a) and (b). Case (c) shows the dependence on the polar angle (orientation in the reaction plane). The polar angle θ influences the diabatic potential energy significantly while the dependence on the angle $\Delta\varphi$ is very weak. In the case (a) the value of the fusion barrier changes on the value about 2 MeV and in the case (b) the change is even less (about 1 MeV). The barrier position in the cases (a) and (b) changes insignificantly too. It should be also mentioned that the diabatic double-folding potential with the Migdal forces has qualitatively correct behavior for small distances between mass centers of the interacting nuclei (see Fig. 2 (c)): the repulsive core appears in the region of overlapping nuclear densities.

2.2. *Adiabatic potential energy*

The adiabatic potential energy is defined as a difference between the mass of the whole nuclear system (the system could be either mononucleus or two separated nuclei) and the ground state masses of target and projectile: $V_{\text{adiab}}(A, Z; r, \vec{\beta}, \eta) = M(A, Z; r, \vec{\beta}, \eta) - M(A_T, Z_T; \vec{\beta}_T^{\text{g.s.}}) - M(A_P, Z_P; \vec{\beta}_P^{\text{g.s.}})$. The last two terms here provide a zero value of the adiabatic potential energy in the entrance channel for the ground state deformations of the target and projectile at infinite distance between them.

The standard macro-microscopic model based on the Strutinsky shell-correction method[14,15] is usually used for calculation of the total mass: $M(A, Z; r, \vec{\beta}, \eta) = M_{\text{mac}}(A, Z; r, \vec{\beta}, \eta) + \delta E(A, Z; r, \vec{\beta}, \eta)$. Here M_{mac} is the liquid drop mass which reproduces a smooth part of the dependence of the mass on deformation and nucleon composition. The second term δE is the microscopic shell correction which is usually calculated using the Strutinsky shell-correction method. It gives non-smooth behavior due to irregularities in shell structure. The macroscopic mass M_{mac} can be calculated in the framework of finite-range liquid-drop model[16–18] (FRLDM):

$$
\begin{aligned}
M_{\text{FRLDM}}\,(A, Z; r, \vec{\beta}, \eta) &= M_p Z + M_n N - a_v(1 - k_v I^2)A \\
+ \quad & a_s(1 - k_s I^2) B_n(r, \vec{\beta}, \eta) A^{2/3} + \frac{3}{5}\frac{e^2 Z^2}{r_0 A^{1/3}} B_C(r, \vec{\beta}, \eta) \\
+ \quad & W\left(|I| + \begin{cases} 1/A, & Z \text{ and } N \text{ equal and odd} \\ 0, & \text{otherwise} \end{cases}\right) \\
- \quad & c_a(N - Z) + a_0 A^0 + \text{ other terms.}
\end{aligned} \tag{3}
$$

Here the meanings of the terms are the following: the masses of Z protons and N neutrons; volume energy; nuclear (surface) and Coulomb energies depending on deformation via dimensionless functionals $B_n(r, \vec{\beta}, \eta)$ and $B_C(r, \vec{\beta}, \eta)$; Wigner energy; charge-asymmetry energy [$(N - Z)$-term]; and A^0-term (constant). For calculation of the shell-correction we can apply the well-known two-center shell model (TCSM) proposed in.[19,20]

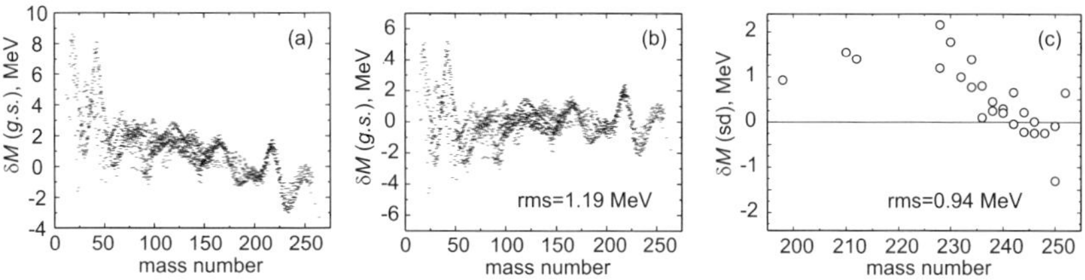

Fig. 3. Difference between the experimental and theoretical ground state masses ($\delta M = M_{\rm exp} - M_{\rm th}$): (a) with parameters recommended in;[18] (b) with parameters obtained in the present work (see Tab. 1). (c) Difference between the experimental and theoretical saddle point masses.

Figure 3 shows difference between the experimental and calculated ground state masses as a function of the mass number A. In case (a) the difference is obtained with the original values of the parameters of the macroscopic mass formula suggested by P. Möller et al.[18] We see that the dependence has a systematic slope. This slope can be corrected by additional fitting of five constants in the Weizsäcker-type formula (3). The results are shown in Fig. (b) and the values of the fitted parameters are listed in Table 1. The obtained rms error is 1.19 MeV, which is good enough for our purposes. For these calculations we restricted ourselves by ellipsoidal shapes of the nuclei. The next important characteristic of the potential energy landscape is the fission barrier which is the difference between the nuclear masses at the saddle point and ground state $B_f = M(\text{sd}) - M(\text{g.s.})$. In Fig. 3 (c) we compare the experimental ($B_f^{(\rm exp)} + M^{(\rm exp)}(\text{g.s.})$) and theoretical saddle point masses. This quantity is reproduced within 2 MeV. The saddle point deformations have been calculated in three dimensional deformation space (see section 3 for details of the degrees of freedom used).

Table 1. Parameters of macroscopic mass formula (3)

parameter	a_v (MeV)	k_v	a_0 (MeV)	c_a (MeV)	W (MeV)
work[18]	16.00126	1.92240	2.615	0.10289	30.0
present work	16.02590	1.91385	6.711	0.04998	27.276

In spite of a rather good agreement with the experimental ground state masses and fission barriers, direct application of the standard macro-microscopic approach, and in particular expression (3), to the case of highly deformed mononucleus or two separated nuclei leads to incorrect result. In

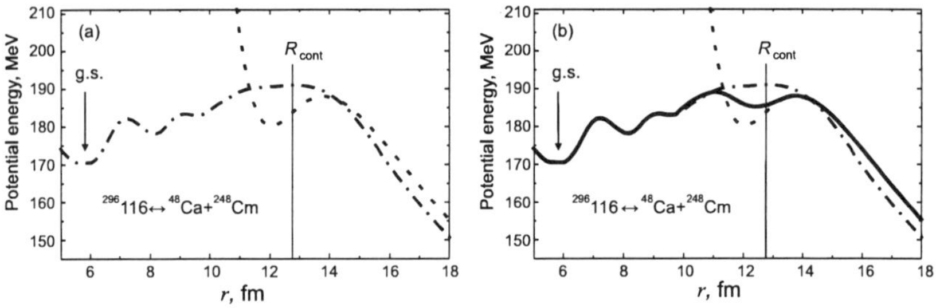

Fig. 4. The adiabatic potential energy for the system $^{296}116 \leftrightarrow ^{48}$ Ca+^{248}Cm obtained within the extended (solid curve) and standard (dash-dotted curve) version of the macro-microscopic model. The dashed curve is the diabatic potential energy calculated within the double-folding model.

Fig. 4 (a) the adiabatic potential energy calculated within the standard macro-microscopic model and the diabatic one are shown. They have to coincide in the region of well separated nuclei (see the talk of V. Zagrebaev). But in this region the standard macro-microscopic approach results in a wrong behavior of the adiabatic potential energy. In order to understand the main reason of this discrepancy we should analyze the expression for the macroscopic mass (3). We see that some of the terms in this formula are nonadditive over Z and N numbers. In fact, the only additive part in this expression is $M_pZ + M_nN - c_a(N - Z)$. In the special case of equal charge densities in the target, projectile, mononucleus, and then in reaction fragments, the volume, surface, and Coulomb terms will be also additive (but not in the general case). In the entrance channel the charge densities in the projectile and target are usually very different, i.e. $Z_P/A_P \neq Z_T/A_T$. This nonadditivity of (3) (in particular, the difference in the charge densities) results in incorrect description of transition from the ground state mass of the compound nucleus to the masses of two separated fragments. This problem with the constant and Wigner terms was pointed out in.[21] It was suggested there to take into account a deformation dependence of these terms.

In the present paper we propose to use the following procedure. It was shown above that the standard macro-microscopic model agrees well with the experimental data on the ground state masses and fission barriers. On the other hand, the double-folding model reproduces the data on the fusion barriers and the potential energy in the region of separated nuclei (in this region the diabatic and adiabatic potential energies should coincide). Thus, we propose to use the correct properties of these two potentials and to construct the adiabatic potential energy as

$$V_{\text{adiab}}(A,Z;r,\vec{\beta},\eta)=\Big\{\Big[M_{\text{FRLDM}}(A,Z;r,\vec{\beta},\eta)+\delta E_{\text{TCSM}}(A,Z;r,\vec{\beta},\eta)\Big]$$
$$-\ \Big[M_{\text{FRLDM}}(A_P,Z_P;\vec{\beta}_P^{\text{g.s.}})+\delta E_{\text{TCSM}}(A_P,Z_P;\vec{\beta}_P^{\text{g.s.}})\Big]$$
$$-\ \Big[M_{\text{FRLDM}}(A_T,Z_T;\vec{\beta}_T^{\text{g.s.}})+\delta E_{\text{TCSM}}(A_T,Z_T;\vec{\beta}_T^{\text{g.s.}})\Big]\Big\}\,B(r,\vec{\beta},\eta)$$
$$+\ V_{\text{diab}}(A,Z;r,\vec{\beta}_1,\vec{\beta}_2,\eta)\,\Big[1-B(r,\vec{\beta},\eta)\Big]\,. \tag{4}$$

The function $B(r,\vec{\beta},\eta)$ defines transition from the properties of two separated nuclei to those of the mononucleus. The function $B(r,\vec{\beta},\eta)$ is rather arbitrary. We only know that it should be unity for the ground state region of mononucleus and should tend to zero for completely separated nuclei. We use the following expression for it: $B(r,\vec{\beta},\eta)=\left[1+\exp\left(\frac{r-R_{\text{cont}}}{a_{\text{diff}}}\right)\right]^{-2}$, where $R_{\text{cont}}(\vec{\beta};A_1,A_2)$ is the distance between mass centers corresponding to the touching or scission point of the nuclei, and a_{diff} is the adjustable parameter. Using the value $a_{\text{diff}}=0.5$ fm we reproduce the fusion barriers.

We call the new procedure for the calculation of the adiabatic potential energy, defined by expression (4), the extended macro-microscopic approach. An example of the adiabatic potential energy calculated within the extended macro-microscopic approach is shown in Fig. 4 (b). This procedure leads to the correct adiabatic potential energy which reproduces the ground state properties of mononucleus properly as well as the fission and fusion barriers and the asymptotic behavior for two separated nuclei.

3. Collective dynamics of fusion-fission

The two-center parametrization[20] has been chosen for description of nuclear shapes. It has five free parameters. It is possible, consequently, to define five independent degrees of freedom determining the shape of the nucleus. We use the following set of them: r – the distance between mass centers; δ_1 and δ_2 – two ellipsoidal deformations of the nascent fragments; $\eta=(A_2-A_1)/(A_2+A_1)$ – the mass asymmetry parameter and ε – the neck parameter. This parametrization is quite flexible and gives reasonable shapes for both the fusion and fission processes. However, inclusion of all five degrees of freedom of the two-center parametrization in the dynamical equations is beyond the present computational possibilities.

In order to decrease the number of collective parameters we propose to use one unified dynamical deformation δ instead of two independent δ_1 and

δ_2. Relation between δ and δ_i is given by

$$2\delta = (\delta_1 - \delta_1^{(0)}) + (\delta_2 - \delta_2^{(0)}), \quad C_{\delta 1}(\delta_1 - \delta_1^{(0)}) = C_{\delta 2}(\delta_2 - \delta_2^{(0)}). \qquad (5)$$

The deformations $\delta_i^{(0)}$ provide a minimum of the potential energy (at fixed values of the other parameters). The first equation in (5) means that zero dynamical deformation corresponds to the bottom of the potential energy landscape. The second equation comes from the condition of equal forces of deformation between two halves of the system. We calculate these forces taking only the first term in liquid-drop expansion of the deformation energy. The quantities $C_{\delta i}$ are the stiffnesses of the potential energy with respect to the deformation δ_i. We apply the liquid drop model for the calculation of $C_{\delta i}$.

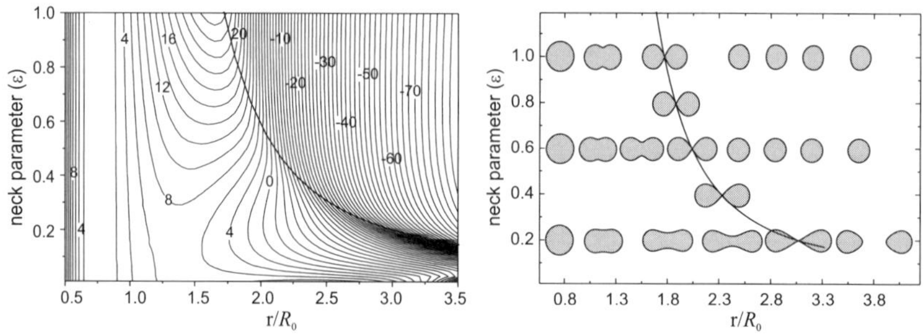

Fig. 5. The potential energy (a) and the corresponding shapes of nuclei (b) in the coordinates (r, ε) for the system ^{224}Th calculated within FRLDM[16,17] for $\eta = 0$ and $\delta_1 = \delta_2 = 0$. The potential energy is normalized to zero for the spherical compound nucleus. The thick solid curve is the scission line.

Now let us discuss the possibility of approximate consideration of the evolution of the neck parameter ε. Figure 5 shows the macroscopic potential energy in coordinates (r, ε) and the map of the respective nuclear shapes. Nuclear shapes corresponding to scission configurations in the fission channel have large distance between mass centers and a well pronounced neck. Such shapes can be described well with $\varepsilon = 1$. On the other hand, the shapes at the contact point in the fusion channel are rather compact and almost without neck. For the exit (fission) channel the value of the neck parameter should be chosen to minimize the potential energy along the fission path. The value $\varepsilon \approx 0.35$ was recommended in[22] for the fission process.

In order to construct the potential energy for the analysis of reaction at energy substantially above the Coulomb barrier (in the region above 10 MeV/nucleon) it is necessary to start from the nonequilibrium diabatic regime as an initial stage and to consider a transition to the equilibrium adiabatic one. This transition to equilibrium nucleon distribution and to adiabatic regime is rather fast. The characteristic time for the relaxation process is estimated[4,5] to be $\tau_{\rm relax} \sim 10^{-21}$ s. The value of the relaxation time can be determined from the analysis of enormous experimental data on the deep-inelastic scattering of nuclei. It is also clear that we should take into account the difference of the entrance and exit channel shapes. In order to restrict ourselves by the three-dimensional deformation space (r, η, δ) we propose to consider evolution of the neck parameter as a relaxation process with the characteristic time τ_ϵ. Finally, the potential energy for the fusion-fission process $V_{\rm fus-fis}(r, \vec{\beta}, \eta; A_P, Z_P, A_T, Z_T; \tau)$ becomes time-dependent and can be written as

$$V_{\rm fus-fis} = V_{\rm diab} \cdot \exp\left(-\frac{\tau}{\tau_{\rm relax}}\right) + \tilde{V}_{\rm adiab}(\varepsilon, \tau) \cdot \left[1 - \exp\left(-\frac{\tau}{\tau_{\rm relax}}\right)\right], \quad (6)$$

where $\tilde{V}_{\rm adiab}(\varepsilon, \tau)$ is the adibatic potential energy which also depends on time and on the neck parameter

$$\tilde{V}_{\rm adiab}(\varepsilon, \tau) = V_{\rm adiab}(\varepsilon = 1) \cdot \exp\left(-\frac{\tau}{\tau_\varepsilon}\right) + V_{\rm adiab}(\varepsilon_{\rm out}) \cdot \left[1 - \exp\left(-\frac{\tau}{\tau_\varepsilon}\right)\right], \quad (7)$$

where the first term corresponds to the entrance channel ($\varepsilon = 1$) and the second one – to the exit channel ($\varepsilon = \varepsilon_{\rm out}$). The characteristic time τ_ε is parameter of the model and should be extracted from the comparison of the experimental data with theorctically calculated.

4. Conclusions

Potential energy is a fundamental characteristic determining the statical and dynamical properties of heavy nuclear system at low energies. The unified potential energy for the simultaneous analysis of the deep-inelastic, quasifission and fusion-fission processes is proposed in this paper. The results are summarized in Fig. 6. The initial stage of the nucleus-nucleus collision is governed by the diabatic potential energy (see Fig. 6 (a)). The double-folding procedure with the density-dependent Migdal nucleon-nucleon interaction is suggested to be used for the calculation of the diabatic potential energy. It reproduces with a good accuracy the experimental fusion barriers for nuclei heavier than carbon.

We propose to use empirical time-dependent potential energy in order to take into account transition from the nonequilibrium diabatic stage of

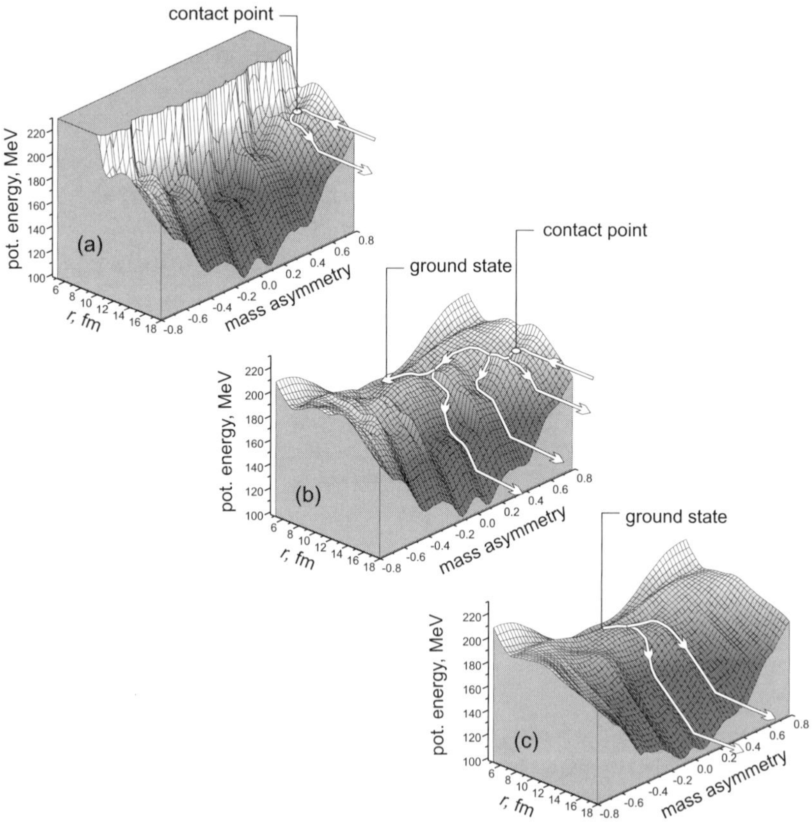

Fig. 6. Time evolution of the potential energy for the system $^{296}116 \leftrightarrow ^{48}$ Ca+^{248}Cm at zero dynamical deformation $\delta = 0$. (a) The diabatic potential energy calculated using the double-folding procedure (the first stage). The entrance-channel (b) and fission channel (c) adiabatic potential energies obtained within the extended macro-microscopic approach. The white arrows show schematically the most probable reaction channels: deep-inelastic scattering (a); deep-inelastic scattering, quasifission, and fusion (b); and multimodal fission (c).

the reaction to equilibrium adiabatic one. It allows us to analyze nucleus-nucleus collisions at above-barrier energies. The transition is treated as a relaxation process (6) with characteristic time $\tau_{\rm relax} \sim 10^{-21}$ s. The extended macro-microscopic approach is proposed for the calculations of the adiabatic potential energy. It gives correct asymptotic behavior of the potential energy and also reproduces the ground state masses, fusion (in the entrance channel) and fission barriers in contrast with the standard macro-microscopic approaches. An example of the entrance-channel adiabatic potential energy is shown in Fig. 6 (b).

Time-evolution of the neck parameter is taken into account in a phenomenological way (7). It allows us to consider very elongated nuclear configurations in the exit (fission) channel of the reaction. The corresponding fission-channel adiabatic potential energy is shown in Fig. 6 (c).

Calculation of the proposed driving potential for any nuclear system can be done at the web-server[23] with a free access.

One of us (A. V. K.) is grateful to the INTAS for financial support of the present researches (Grant No. INTAS 05-109-5058).

References

1. V. V. Volkov, *Nuclear Reactions of High-Inelastic Transfers* (Energoizdat, Moscow, 1982) [in Russian].
2. M. G. Itkis, et al., *Nucl. Phys. A* **734**, 136 (2004).
3. J. Péter, et al., *Nucl. Phys. A* **279**, 110 (2004).
4. G. F. Bertsch, *Z. Phys. A* **289**, 103 (1978); W. Cassing, W. Nörenberg, *Nucl. Phys. A* **401**, 467 (1983).
5. A. Diaz-Torres, *Phys. Rev. C* **69**, 021603 (2004); A. Diaz-Torres, W. Scheid, *Nucl. Phys. A* **757**, 373 (2005).
6. G. R. Satchler, W. G. Love, *Phys. Rep.* **55**, 183 (1979).
7. G. Bertsch, J. Borysowicz, H. McManus, W. G. Love, *Nucl. Phys. A* **284**, 399 (1977).
8. M. Lacombe, B. Loiseau, J. M. Richard, R. Vinh Mau, J. Côté, P. Pirés, R. de Tourreil, *Phys. Rev. C* **21**, 861 (1980).
9. N. Anantaraman, H. Toki, G. F. Bertsch, *Nucl. Phys. A* **398**, 269 (1983).
10. A. B. Migdal, *Theory of Finite Fermi Systems and Applications to Atomic Nuclei* (Wiley Interscience, New York, 1967).
11. E. G. Nadjakov, K. P. Marinova, Y. P. Gangrsky, *At. Data Nucl. Data Tables* **56**, 133 (1994).
12. I. Angeli, *Acta Phys. Hung. A: Heavy Ion Physics* **8**, 23 (1998).
13. R. Bass, *Nuclear Reactions with Heavy Ions* (Springer-Verlag, 1980), 326 p.
14. V. M. Strutinsky, *Nucl. Phys. A* **95**, 420 (1967); V. M. Strutinsky, *Nucl. Phys. A* **22**, 1 (1968).
15. M. Brack, et al., *Rev. Mod. Phys.* **44**, 320 (1972).
16. H. J. Krappe, J. R. Nix, A. J. Sierk, *Phys. Rev. C* **20**, 992 (1979).
17. A. J. Sierk, *Phys. Rev. C* **33**, 2039 (1986).
18. P. Möller, J. R. Nix, W. D. Myers, W. J. Swiatecki, *At. Data Nucl. Data Tables* **59**, 185 (1995).
19. P. A. Cherdantsev, V. E. Marshalkin, *Bull. Acad. Sci. USSR* **30**, 341 (1966).
20. J. Maruhn, W. Greiner, *Z. Phys. A* **251**, 431 (1972).
21. P. Möller, J. R. Nix, W. J. Swiatecki, *Nucl. Phys. A* **492**, 349 (1989); P. Möller, A. J. Sierk, A. Iwamoto, *Phys. Rev. Lett.* **92**, 072501 (2004).
22. S. Yamaji, H. Hofmann, R. Samhammer, *Nucl. Phys. A* **475**, 487 (1988).
23. NRV codes for driving potentials, *http://nrv.jinr.ru/nrv*.

ADVANCES IN THE UNDERSTANDING OF STRUCTURE AND PRODUCTION MECHANISMS FOR SUPERHEAVY ELEMENTS

Walter GREINER[1,*] and Valery ZAGREBAEV[2]

[1] *FIAS, J.W. Goethe-Universität, Frankfurt, Germany,*
[2] *Flerov Laboratory of Nuclear Reaction, JINR, Dubna, Moscow Region, Russia*
** E-mail: greiner@fias.uni-frankfurt.de*

The talk is aimed to discussion of the problems around production and study of superheavy elements. Different nuclear reactions leading to formation of superheavy nuclei are analyzed. Dynamics of heavy-ion low energy collisions is studied within the realistic model based on multi-dimensional Langevin equations. Interplay of strongly coupled deep inelastic scattering, quasi-fission and fusion-fission processes is discussed. Collisions of very heavy nuclei ($^{238}U+^{238}U$, $^{232}Th+^{250}Cf$ and $^{238}U+^{248}Cm$) are investigated as an alternative way for production of superheavy elements with increasing neutron number. Large charge and mass transfer was found in these reactions due to the inverse (anti-symmetrizing) quasi-fission process leading to formation of surviving superheavy long-lived neutron-rich nuclei. Lifetime of the composite system consisting of two touching nuclei is studied with the objective to find time delays suitable for the observation of spontaneous positron emission from super-strong electric field.

Keywords: superheavy nuclei, giant quasi-atoms.

1. Introduction

The interest in the synthesis of super-heavy nuclei has lately grown due to new experimental results demonstrating the possibility of producing and investigating the nuclei in the region of the so-called "island of stability". At the same time superheavy (SH) nuclei obtained in "cold" fusion reactions with Pb or Bi target [1] are along the proton drip line and very neutron-deficient with a short half-life. In fusion of actinides with ^{48}Ca more neutron-rich SH nuclei are produced [2] with much longer half-life. But they are still far from the center of the predicted "island of stability" formed by the neutron shell around N=184. Unfortunately a small gap between the superheavy nuclei produced in ^{48}Ca-induced fusion reactions and those which were obtained in the "cold" fusion reactions is still remain (see Fig. 1) which should be filled to get a unified nuclear map.

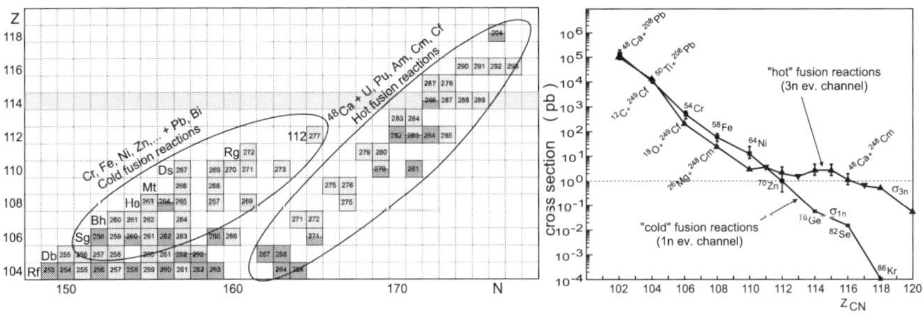

Fig. 1. Superheavy nuclei produced in "cold" and "hot" fusion reactions. By light and dark gray colors the nuclei are marked experienced aplha-decay and spontaneous fission, correspondingly.

In the "cold" fusion, the cross sections of SH nuclei formation decrease very fast with increasing charge of the projectile and become less than 1 pb for Z>112 (see Fig. 1). Heaviest transactinide, Cf, which can be used as a target in the second method, leads to the SH nucleus with Z=118 being fused with ^{48}Ca. Using the next nearest elements instead of ^{48}Ca (e.g., ^{50}Ti, ^{54}Cr, etc.) in fusion reactions with actinides is expected less encouraging, though experiments of such kind are planned to be performed. In this connection other ways to the production of SH elements in the region of the "island of stability" should be searched for.

In principle, superheavy nuclei may be produced in explosion of supernova [4]. If the half-life of these nuclei is comparable with the age of the Earth they could be searched for in nature. However, it is the heightened stability of these nuclei (rare decay) which may hinder from their discovery. To identify these more or less stable superheavy elements supersensitive mass separators should be used. Chemical methods of separation also could be useful here.

About twenty years ago transfer reactions of heavy ions with ^{248}Cm target have been evaluated for their usefulness in producing unknown neutron-rich actinide nuclides [5, 6, 7]. The cross sections were found to decrease very rapidly with increasing atomic number of surviving target-like fragments. However, Fm and Md neutron-rich isotopes have been produced at the level of 0.1 μb. Theoretical estimations for production of primary superheavy fragments in the damped U+U collision have been also performed at this time within the semiphenomenological diffusion model [8]. In spite of obtained high probabilities for the yields of superheavy primary fragments (more than 10^{-2} mb for Z=120), the cross sections for production of heavy nuclei with low excitation energies were estimated to be rather small: $\sigma_{CN}(Z = 114, E^* = 30\,\text{MeV}) \sim 10^{-6}$ mb for U+Cm collision at 7.5 Mev/nucleon beam energy. The authors concluded, however, that "fluctuations and shell effects not taken into account may conciderably increase the formation probabilities". Such is indeed the case (see below).

Renewed interest to collisions of transactinide nuclei is conditioned by the necessity to clarify much better than before the dynamics of heavy nuclear systems at low excitation energies and by a search for new ways for production of neutron rich superheavy (SH) nuclei and isotopes. SH elements obtained in "cold" fusion reactions with Pb or Bi target are situated along the proton drip line being very neutron-deficient with a short half-life. In fusion of actinides with ^{48}Ca more neutron-rich SH nuclei are produced with much longer half-life. But they are still far from the center of the predicted "island of stability" formed by the neutron shell N=184. In the "cold" fusion, the cross sections for formation of SH nuclei decrease very fast with increasing charge of the projectile and become less than 1 pb for Z$\geq$112. On the other hand, the heaviest transactinide, Cf, which can be used as a target in the second method, being fused with ^{48}Ca leads to the SH nucleus with Z=118. Using the next nearest elements instead of ^{48}Ca (e.g., ^{50}Ti, ^{54}Cr, etc.) in fusion reactions with actinides is expected less encouraging, though experiments of such kind are planned to be performed. In this connection other ways to the production of SH elements in the region of the "island of stability" should be searched for.

Recently a new model has been proposed [9] for *simultaneous* description of all these strongly coupled processes: deep inelastic (DI) scattering, quasi-fission (QF), fusion, and regular fission. In this paper we apply this model for analysis of low-energy dynamics of heavy nuclear systems formed in nucleus-nucleus collisions at the energies around the Coulomb barrier. Among others there is the purpose to find an influence of the shell structure of the driving potential (in particular, deep valley caused by the double shell closure Z=82 and N=126) on formation of compound nucleus (CN) in mass asymmetric collisions and on nucleon rearrangement between primary fragments in more symmetric collisions of actinide nuclei. In the first case, discharge of the system into the lead valley (normal or symmetrizing quasi-fission) is the main reaction channel, which decreases significantly the probability of CN formation. In collisions of heavy transactinide nuclei

(U+Cm, etc.), we expect that the existence of this valley may noticeably increase the yield of surviving neutron-rich superheavy nuclei complementary to the projectile-like fragments (PLF) around lead ("inverse" or anti-symmetrizing quasi-fission reaction mechanism).

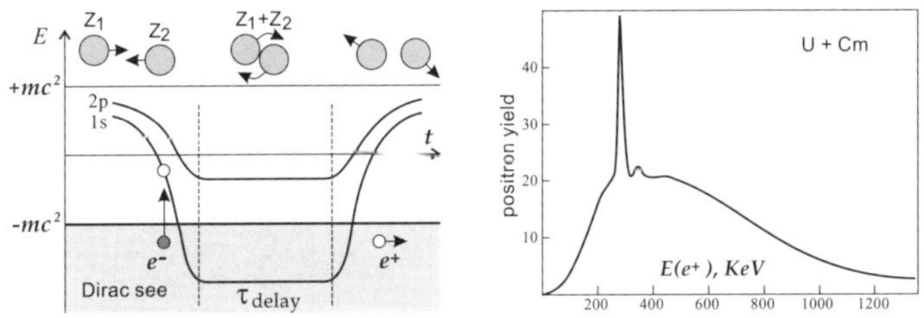

Fig. 2. Schematic figure of spontaneous decay of the vacuum and spectrum of the positrons formed in supercritical electric field ($Z_1 + Z_2 > 173$).

Direct time analysis of the collision process allows us to estimate also the lifetime of the composite system consisting of two touching heavy nuclei with total charge Z>180. Such "long-living" configurations (if they exist) may lead to spontaneous positron emission from super-strong electric fields of giant quasi-atoms by a static QED process (transition from neutral to charged QED vacuum) [10, 11], see schematic Fig. 2.

2. Nuclear shells

Quantum effects leading to the shell structure of heavy nuclei play a crucial role both in stability of these nuclei and in production of them in fusion reactions. The fission barriers of superheavy nuclei (protecting them from spontaneous fission and, thus, providing their existence) are determined completely by the shell structure. Studies of the shell structure of superheavy nuclei in the framework of the meson field theory and the Skyrme-Hartree-Fock approach show that the magic shells in the superheavy region are very isotopic dependent [12] (see Fig. 3). According to these investigations Z=120 being a magic proton number seems to be as probable as Z=114. Estimated fission barriers for nuclei with Z=120 are rather high (see Fig. 4) though depend strongly on a chosen set of the forces [13].

Interaction dynamics of two heavy nuclei at low (near-barrier) energies is defined mainly by the adiabatic potential energy, which can be calculated, for example, within the two-center shell model [14]. An example of such calculation is shown in Fig. 5 for the nuclear system consisting of

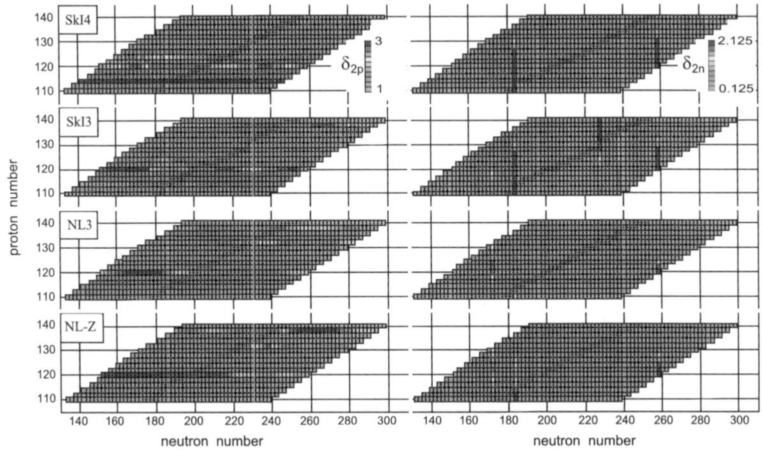

Fig. 3. Proton (left column) and neutron (right column) gaps in the $N - Z$ plane calculated within the self-consistent Hartree-Fock approach with the forces as indicated [12]. The forces with parameter set SkI4 predict both Z=114 and Z=120 as a magic numbers while the other sets predict only Z=120.

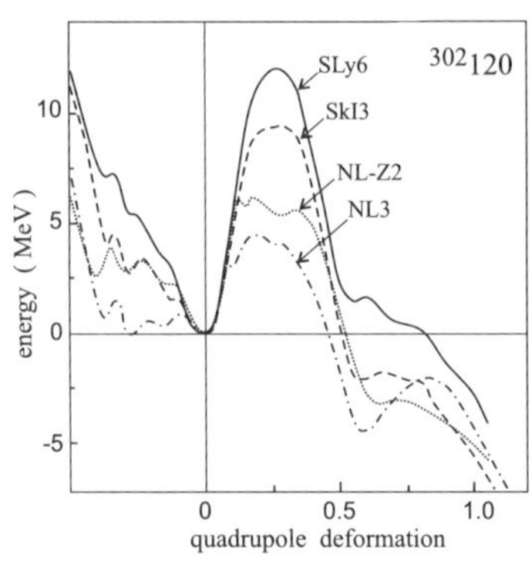

Fig. 4. Fission barriers for the nucleus $^{302}120$ calculated within the self-consistent Hartree-Fock approach [13].

108 protons and 156 neutrons. Formation of such heavy nuclear systems in fusion reactions as well as fission and quasi-fission of these systems are regulated by the deep valleys on the potential energy surface (see Fig. 5) also caused by the shell effects.

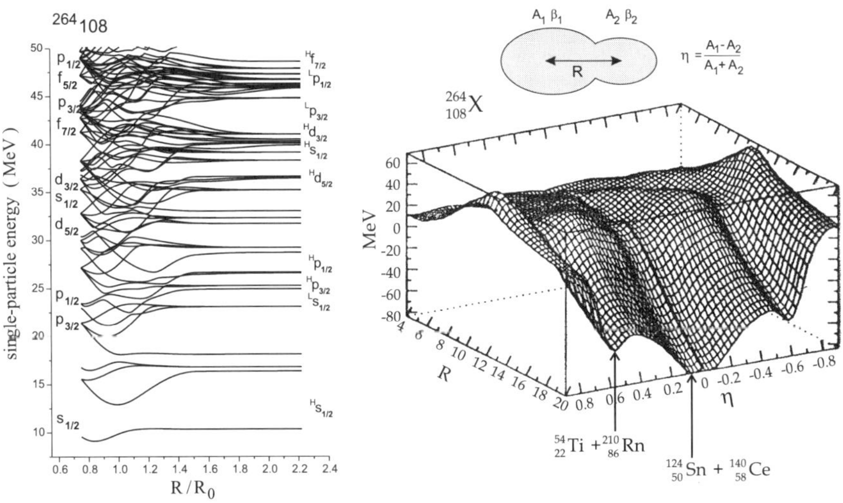

Fig. 5. Two-center single particle energy levels (left panel) and adiabatic potential energy surface for the nuclear system $^{264}108$.

3. Adiabatic dynamics of heavy nuclear system

At incident energies around the Coulomb barrier in the entrance channel the fusion probability is about 10^{-3} for mass asymmetric reactions induced by ^{48}Ca and much less for more symmetric combinations used in the "cold synthesis". DI scattering and QF are the main reaction channels here, whereas the fusion probability [formation CN] is extremely small. To estimate such a small quantity for CN formation probability, first of all, one needs to be able to describe well the main reaction channels, namely DI and QF. Moreover, the quasi-fission processes are very often indistinguishable from the deep-inelastic scattering and from regular fission, which is the main decay channel of excited heavy compound nucleus.

To describe properly and simultaneously the strongly coupled DI, QF and fusion-fission processes of low-energy heavy-ion collisions we have to choose, first, the unified set of degrees of freedom playing the principal role both at approaching stage and at the stage of separation of reaction fragments. Second, we have to determine the unified potential energy surface (depending on all the degrees of freedom) which regulates all the processes. Finally, the corresponding equations of motion should be formulated to perform numerical analysis of the studied reactions. In contrast with other models, we take into consideration all the degrees of freedom necessary for description of all the reaction stages. Thus, we need not to split artificially the whole reaction into several stages. Moreover, in that case unambiguously defined initial conditions are easily formulated at large distance, where only the Coulomb interaction and zero-vibrations of the nuclei determine the motion. The distance between the nuclear centers R (corresponding to the elongation of a mono-nucleus), dynamic spheroidal-type surface deformations β_1 and β_2, mutual in-plane orientations of deformed nuclei φ_1 and φ_2, and mass asymmetry $\eta = \frac{A_1 - A_2}{A_1 + A_2}$ are probably the relevant degrees of freedom in fusion-fission dynamics.

The two-center shell model [14] seems to be most appropriate for calculation of the adiabatic potential energy surface. A choice of dynamic equations for the considered degrees of freedom is not so evident. The main problem here is a proper description of nucleon transfer and change of the mass asymmetry which is a discrete variable by its nature. The corresponding inertia parameter μ_η, being calculated within the Werner-Wheeler approach, becomes infinite at contact (scission) point and for separated nuclei. In Ref. [9] the inertialess Langevin type equation for the mass asymmetry has been derived from the corresponding master equation for the distribution function. Finally we use a set of 13 coupled Langevin type equations for 7 degrees of freedom (relative distance, rotation and dynamic deformations of the nuclei and mass asymmetry) which are solved numerically.

The cross sections for all the processes are calculated in a simple and natural way. A large number of events (trajectories) are tested for a given impact parameter. Those events, in which the nuclear system overcame the fission barrier from the outside and entered the region of small deformations and elongations, are treated as fusion (CN formation). The other events correspond to quasi-elastic, DI and QF processes. Subsequent decay of the excited CN ($C \to B + xn + N\gamma$) is described then within the statistical model using an explicit expression for survival probability, which directly takes into account the Maxwell-Boltzmann energy distribution of evaporated neutrons [15]. The double differential cross-sections are calculated as follows

$$\frac{d^2\sigma_\eta}{d\Omega dE}(E,\theta) = \int_0^\infty b db \frac{\Delta N_\eta(b,E,\theta)}{N_{\rm tot}(b)} \frac{1}{\sin(\theta)\Delta\theta\Delta E}. \tag{1}$$

Here $\Delta N_\eta(b,E,\theta)$ is the number of events at a given impact parameter b in which the system enters into the channel η (definite mass asymmetry value) with kinetic energy in the region $(E, E+\Delta E)$ and center-of-mass outgoing angle in the region $(\theta, \theta+\Delta\theta)$, $N_{\rm tot}(b)$ is the total number of simulated events for a given value of impact parameter. In collisions of deformed nuclei an averaging over initial orientations is performed. Expression (1) describes the mass, energy and angular distributions of the *primary* fragments formed in the binary reaction (both in DI and in QF processes). Subsequent de-excitation cascades of these fragments via emission of light particles and gamma-rays in competition with fission were taken into account explicitly for each event within the statistical model leading to the *final* mass and energy distributions of the reaction fragments. The model allows us to perform also a time analysis of the studied reactions. Each tested event is characterized by the reaction time τ_{int}, which is calculated as a difference between re-separation (scission) and contact times.

4. Deep inelastic scattering of heavy nuclei

We analyzed first the collision of very heavy nuclei, ^{136}Xe+^{209}Bi at energies $E_{c.m.}$ = 568 MeV and 861 MeV [16, 17], where the DI process should dominate due to expected prevalence of the Coulomb repulsion over nuclear attraction and the impossibility of CN formation. In that case the reaction mechanism depends mainly on the nucleus-nucleus potential at contact distance (which determines the grazing angle), on the friction forces at this region (which determine the energy loss) and on nucleon transfer rate at contact.

For the nucleus-nucleus interaction we used the proximity potential [18] with $r_0 = 1.16$ fm. For separated nuclei we chose the friction forces with $\gamma_R^0 = 40 \cdot 10^{-22}$ Mev s fm^{-2}, $\rho_F = 2$ fm and $a_F = 0.6$ fm and equal tangential and radial friction strengthes $\gamma_t^0 = \gamma_R^0$ as it was recommended in [19]. In spite of a short penetration of nuclei into each other, the reaction cross sections were found to be sensitive to the value of nuclear viscosity of the mononucleus μ_0 mainly due to the large dynamic deformations of the reaction fragments. The values of $\mu_0 = 1 \cdot 10^{-22}$ and $3 \cdot 10^{-22}$ Mev s fm^{-3} have been used to describe properly the reaction cross sections at the center-of-mass beam energies of 568 and 861 MeV respectively. These values are larger than those found for low excited fissile nuclei [20]. It evidently indicates on a temperature dependence of nuclear viscosity. The nucleon transfer rate was fixed at $\lambda_0 = 0.1 \cdot 10^{22}$ s^{-1}. This rather small value was found to be sufficient to reproduce the mass distributions of reaction products at both energies.

The adiabatic potential energy surface of this nuclear system is shown on the left panel of Fig. 6 in the space of elongation and mass asymmetry at zero dynamic deformations. The colliding nuclei are very compact with almost closed shells and the potential energy has only one deep valley (just in the entrance channel, $\eta \sim 0.21$) giving rather simple mass distribution of the reaction fragments. In that case the reaction mechanism depend mainly on the nucleus-nucleus potential at contact distance, on the friction forces at this region (which determine the energy loss) and on nucleon transfer rate at contact. Note, that there is a well pronounced plateau at contact configuration in the region of zero mass asymmetry (see Fig. 6, right panel). It becomes even lower with increasing the deformations and corresponds to formation of the nuclear system consisting of strongly deformed touching fragments ^{172}Er+^{173}Tm (see Fig. 6), which means that a significant mass rearrangement may occur here leading to additional time delay of the reaction.

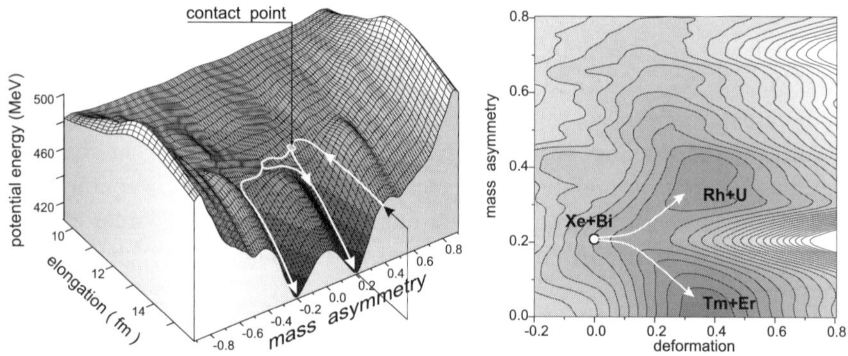

Fig. 6. Driving potential for the nuclear system formed in ^{136}Xe+^{209}Bi collision at fixed deformations (left) and at contact configuration (right). The solid lines with arrows show schematically (without fluctuations) most probable trajectories.

On the right panel of Fig. 6 the landscape of the potential energy is shown at contact configuration depending on mass asymmetry and deformation of the fragments. As can be seen, after contact and before re-separation the nuclei aim to become more deformed. Moreover, beside a regular diffusion (caused by the fluctuations), the final mass distribution is determined also by the two well marked driving paths leading the system to more and to less symmetric configurations. They are not identical and this leads to the asymmetric mass distribution of the primary fragments, see Fig. 7(c).

In Fig. 7 the angular, energy and charge distributions of the Xe-like fragments are shown comparing with our calculations (histograms). In accordance with experimental conditions at the incident energy $E_{\rm c.m.} = 861$ MeV only the events with the energy loss higher than 50 Mev and with the scattering angles in the region of $18^o \leq \theta_{\rm c.m.} \leq 128^o$ were accumulated.

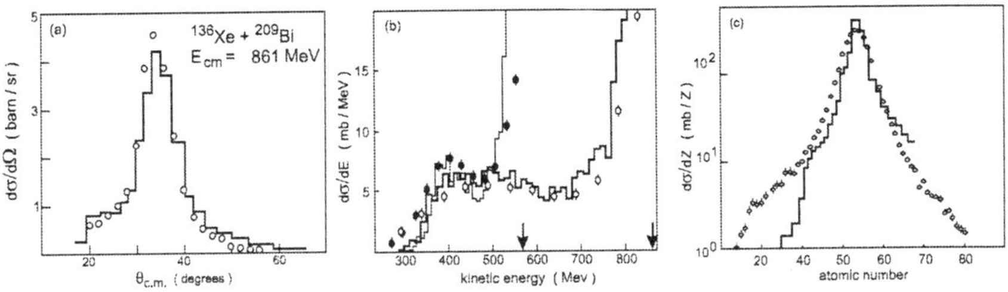

Fig. 7. Angular (a), energy (b) and charge (c) distributions of the Xe-like fragments obtained in the ^{136}Xe+^{209}Bi reaction at $E_{\rm c.m.} = 861$ MeV. Experimental data are from [17]. Histograms are the theoretical predictions. The energy distribution is also shown in (b) for $E_{\rm c.m.} = 568$ MeV [16] (filled circles and thin histogram). The arrows in (b) indicate the corresponding beam energies.

The minimal values of kinetic energy of the reaction fragments ($\sim$ 300 MeV) are almost independent on the beam energies. This indicates that these events correspond to the same re-separation distance and resemble the fission process, though the CN is not formed here. Some underestimation of the low-Z shoulder in Fig. 7(c) could be due to the contribution of sequential fission of highly excited reaction fragments not accounted in the model.

At the second step we analyzed the reaction ^{86}Kr+^{166}Er at $E_{\rm c.m.} = 464$ MeV [21], in which the nuclear attractive forces may lead, in principle, to the formation of a mono-nucleus and of CN. The adiabatic potential energy surface, QF and fusion-fission (FF) processes should in this case play a more important role. For the analysis of this reaction we used the same value of the nucleon transfer rate and the same friction forces as in the previous case. For the nuclear viscosity we choose the value $\mu_0 = 2 \cdot 10^{-22}$ Mev s fm^{-3} because of intermediate values of excitation energies available here as compared with the two previous reactions.

The interaction time is one of the most important characteristics of nuclear reactions, though it cannot be measured directly. It depends strongly on the reaction channel. The time distribution of all the ^{86}Kr+^{166}Er collisions at $E_{\rm c.m.} = 464$ MeV, in which the kinetic energy loss is higher than 35 MeV, is shown in Fig. 8. The interaction time was calculated starting from $t = 0$ at $R = R_{\rm max} = 40$ fm up to the moment of scission into two fragments ($R > R_{\rm scission}, p_R > 0$) or up to CN formation. The approaching time (path from $R_{\rm max}$ to $R_{\rm contact}$) in the entrance channel is very short ($4 \div 5 \cdot 10^{-22}$ s depending on the impact parameter) and may be ignored here. All the events are divided relatively onto the three groups:

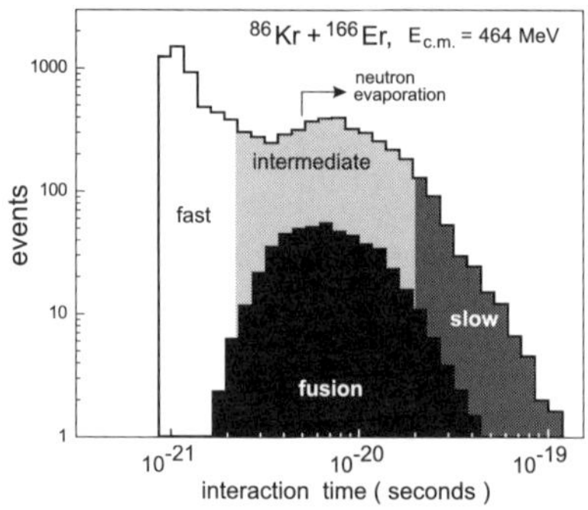

Fig. 8. Time distribution of all the simulated events for ^{86}Kr+^{166}Er collisions at $E_{\rm c.m.} = 464$ MeV, in which the energy loss was found higher than 35 MeV (totally 10^5 events). Conditionally fast ($< 2 \cdot 10^{-21}$ s), intermediate and slow ($> 2 \cdot 10^{-20}$ s) collisions are marked by the different colors (white, light gray and dark gray, respectively). The black area corresponds to CN formation (estimated cross section is 120 mb), and the arrow shows the interaction time, after which the neutron evaporation may occur.

fast ($\tau_{\rm int} < 20 \cdot 10^{-22}$ s), intermediate, and slow ($\tau_{\rm int} > 200 \cdot 10^{-22}$ s).

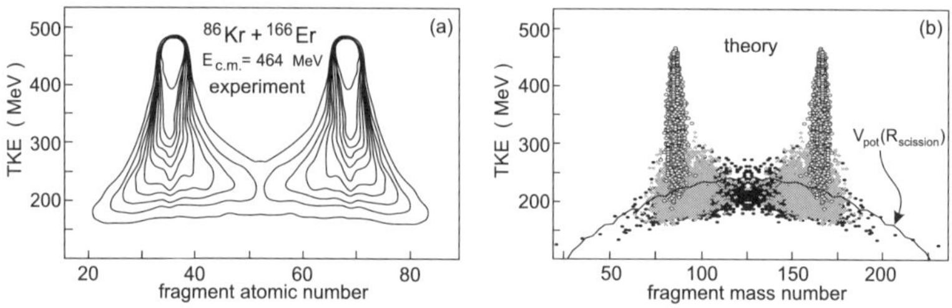

Fig. 9. (a) TKE-charge distribution of the ^{86}Kr+^{166}Er reaction products at $E_{\rm c.m.} = 464$ MeV [21]. (b) Calculated TKE-mass distribution of the primary fragments. Open, gray and black circles correspond to the fast ($< 2 \cdot 10^{-21}$ s), intermediate and long ($> 2 \cdot 10^{-20}$ s) events (overlapping each other on the plot).

A two-dimensional plot of the energy-mass distribution of the primary fragments formed in the ^{86}Kr+^{166}Er reaction at $E_{\rm c.m.} = 464$ MeV is shown in Fig. 9. Inclusive angular, charge and energy distributions of these fragments (with energy losses more than 35 MeV) are shown in Fig. 10. Rather good agreement with experimental data of all the calculated DI reaction properties can be seen, which was never obtained before in dynamic calculations. Underestimation of the yield of low-Z fragments [Fig. 10(c)] could again be due to the contribution of sequential fission of highly excited re-

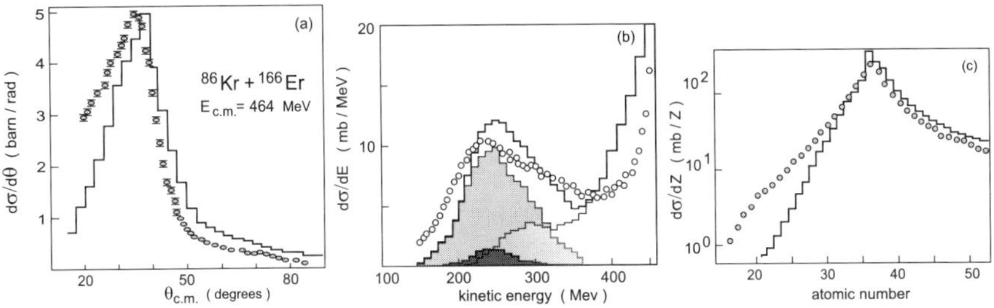

Fig. 10. Angular (a), energy (b) and charge (c) distributions of the ^{86}Kr+^{166}Er reaction products at $E_{c.m.} = 464$ MeV. Experimental data (points) are from [21]. Overlapping white, light and dark gray areas in (b) show the contributions of the fast, intermediate and slow events, respectively [see Fig. 8 and Fig. 9(b)].

action participants not accounted in the model at the moment.

In most of the damped collisions ($E_{\rm loss} > 35$ MeV) the interaction time is rather short (several units of 10^{-21} s). These fast events correspond to grazing collisions with intermediate impact parameters. They are shown by the white areas in Figs. 8 and 10(b) and by the open circles in two-dimensional TKE-mass plot [Fig. 9(b)]. Note that a large amount of kinetic energy is dissipated here very fast at relatively low mass transfer (more than 200 MeV during several units of 10^{-21} s).

The other events correspond to much slower collisions with large overlap of nuclear surfaces and significant mass rearrangement. In the TKE-mass plot these events spread over a wide region of mass fragments (including symmetric splitting) with kinetic energies very close to kinetic energy of fission fragments. The solid line in Fig. 9(b) correspond to potential energy at scission point $V(r = R_{\rm scission}, \beta, \alpha) + Q_{gg}(\alpha)$ minimized over β. Scission point is calculated here as $R_{\rm scission}(\alpha, \beta) = (1.4/r_0)[R_1(A_1, \beta_1) + R_2(A_2, \beta_2)] + 1$ fm, $Q_{gg}(\alpha) = B(A_1) + B(A_2) - B(^{86}Kr) - B(^{166}Er)$ and $B(A)$ is the binding energy of a nucleus A. Some gap between the two groups in the time and energy distributions can also be seen in Fig. 8 and Fig. 10(b). All these make the second group of slow events quite distinguished from the first one. These events are more similar to fission than to deep-inelastic processes. Formally, they also can be marked as quasi-fission.

5. Low-energy collisions of transactinide nuclei

Reasonable agreement of our calculations with experimental data on low-energy DI and QF reactions induced by heavy ions stimulated us to study the reaction dynamics of very heavy transactinide nuclei. The purpose was to find an influence of the shell structure of the driving potential (in particular, deep valley caused by the double shell closure Z=82 and N=126) on nucleon rearrangement between primary fragments. In Fig. 11 the potential energies are shown depending on mass rearrangement at contact configuration of the nuclear systems formed in ^{48}Ca+^{248}Cm and ^{232}Th+^{250}Cf collisions. The lead valley evidently reveals itself in both cases (for ^{48}Ca+^{248}Cm system there is also a tin valley). In the first case (^{48}Ca+^{248}Cm), discharge of the system into the lead valley (normal or symmetrizing quasi-fission) is the main reaction channel, which decreases significantly the probability of CN formation. In collisions of heavy nuclei (Th+Cf, U+Cm and so on) we expect that the existence of this valley may noticeably increase the yield of surviving neutron-rich superheavy nuclei complementary to the projectile-like fragments around ^{208}Pb ("inverse" or anti-symmetrizing quasi-fission process).

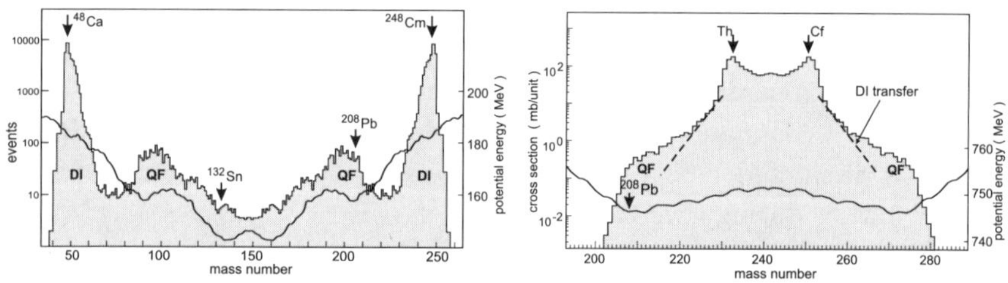

Fig. 11. Potential energy at contact "nose-to-nose" configuration and mass distribution of primary fragments for the two nuclear systems formed in ^{48}Ca+^{248}Cm (left) and ^{232}Th+^{250}Cf (right) collisions.

Direct time analysis of the reaction dynamics allows us to estimate also the lifetime of the composite system consisting of two touching heavy nuclei with total charge Z>180. Such "long-living" configurations may lead to spontaneous positron emission from super-strong electric field of giant quasi-atoms by a static QED process (transition from neutral to charged QED vacuum) [10]. About twenty years ago an extended search for this fundamental process was carried out and narrow line structures in the positron spectra were first reported at GSI. Unfortunately these results were not confirmed later, neither at ANL, nor in the last experiments performed at GSI. These negative finding, however, were contradicted by Jack Greenberg (private communication and supervised thesis at Wright Nuclear Structure Laboratory, Yale university). Thus the situation remains unclear, while the experimental efforts in this field have ended. We hope that new experiments and new analysis, performed according to the results of our dynamical model, may shed additional light on this problem and also answer the principal question: are there some reaction features (triggers) testifying a long reaction delays? If they are, new experiments should be planned to detect the spontaneous positrons in the specific reaction channels.

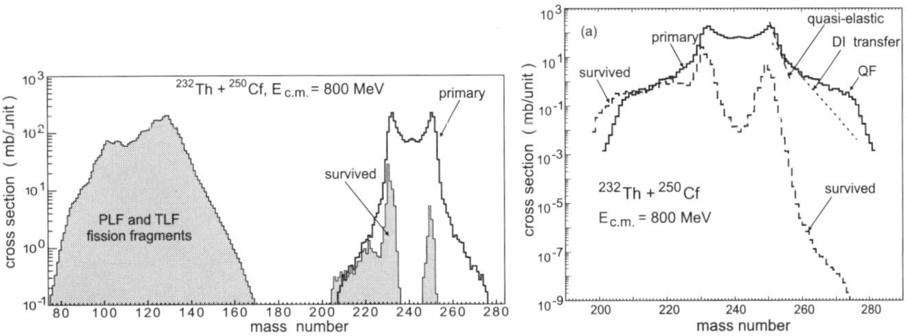

Fig. 12. Mass distributions of primary (solid histogram), surviving and sequential fission fragments (hatched areas) in the ^{232}Th+^{250}Cf collision at 800 MeV center-of-mass energy. On the right the result of longer calculation is shown.

Using the same parameters of nuclear viscosity and nucleon transfer rate as for the system Xe+Bi we calculated the yield of primary and surviving fragments formed in the ^{232}Th+^{250}Cf collision at 800 MeV center-of mass energy. Low fission barriers of the colliding nuclei and of most of the reaction products jointly with rather high excitation energies of them in the exit channel will lead to very low yield of surviving heavy fragments. Indeed, sequential fission of the projectile-like and target-like fragments dominate in these collisions, see Fig. 12. At first sight, there is no chances to get surviving superheavy nuclei in such reactions. However, as mentioned above, the yield of the primary fragments will increase due to the QF effect (lead valley) as compared to the gradual monotonic decrease typical for damped mass transfer reactions. Secondly, with increasing neutron number the fission barriers increase on average (also there is the closed sub-shell at N=162). Thus we may expect a non-negligible yield (at the level of 1 pb) of surviving superheavy neutron rich nuclei produced in these reactions [22].

Result of much longer calculations is shown on the right panel of Fig. 12. The pronounced shoulder can be seen in the mass distribution of the primary fragments near the mass number A=208 (274). It is explained by the existence of a valley in the potential energy surface [see Fig. 11(b)], which corresponds to the formation of doubly magic nucleus

^{208}Pb ($\eta = 0.137$). The emerging of the nuclear system into this valley resembles the well-known quasi-fission process and may be called "inverse (or anti-symmetrizing) quasi-fission" (the final mass asymmetry is larger than the initial one). For $\eta > 0.137$ (one fragment becomes lighter than lead) the potential energy sharply increases and the mass distribution of the primary fragments decreases rapidly at A<208 (A>274).

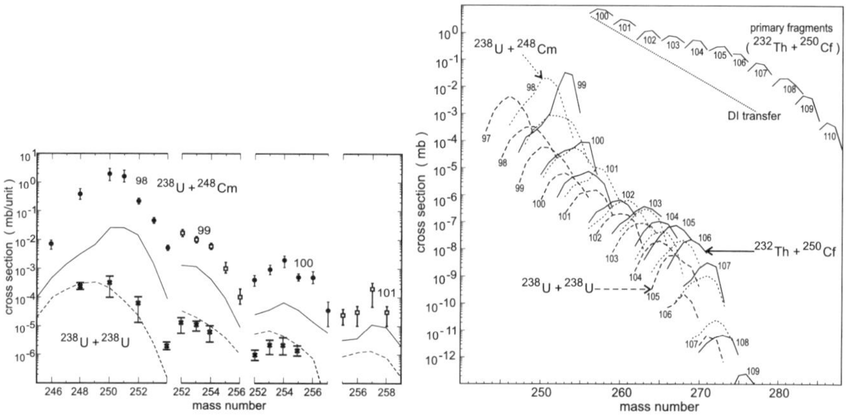

Fig. 13. (Left panel) Experimental and calculated yields of the elements 98÷101 in the reactions ^{238}U+^{238}U (crosses) [5] and ^{238}U+^{248}Cm (circles and squares) [6]. (Right panel) Predicted yields of superheavy nuclei in collisions of ^{238}U+^{238}U (dashed), ^{238}U+^{248}Cm (dotted) and ^{232}Th+^{250}Cf (solid lines) at 800 MeV center-of-mass energy. Solid curves in upper part show isotopic distributions of primary fragments in the Th+Cf reaction.

In Fig. 13 the available experimental data on the yield of SH nuclei in collisions of ^{238}U+^{238}U [5] and ^{238}U+^{248}Cm [6] are compared with our calculations. The estimated isotopic yields of survived SH nuclei in the ^{232}Th+^{250}Cf, ^{238}U+^{238}U and ^{238}U+^{248}Cm collisions at 800 MeV center-of-mass energy are shown on the right panel of Fig. 13. Thus, as we can see, there is a real chance for production of the long-lived neutron-rich SH nuclei in such reactions. As the first step, chemical identification and study of the nuclei up to $^{274}_{107}$Bh produced in the reaction ^{232}Th+^{250}Cf may be performed.

The time analysis of the reactions studied shows that in spite of absence of an attractive potential pocket the system consisting of two very heavy nuclei may hold in contact rather long in some cases. During this time the giant nuclear system moves over the multidimensional potential energy surface with almost zero kinetic energy (result of large nuclear viscosity), see

Fig. 14. The total reaction time distribution, $\frac{d\sigma}{dlog(\tau)}$ (τ denotes the time after the contact of two nuclei), is shown in Fig. 15 for the $^{238}U+^{248}Cm$ collision. The dynamic deformations are mainly responsible here for the time delay of the nucleus-nucleus collision. Ignoring the dynamic deformations in the equations of motion significantly decreases the reaction time, see Fig. 15(a). With increase of the energy loss and mass transfer the reaction time becomes longer and its distribution becomes more narrow.

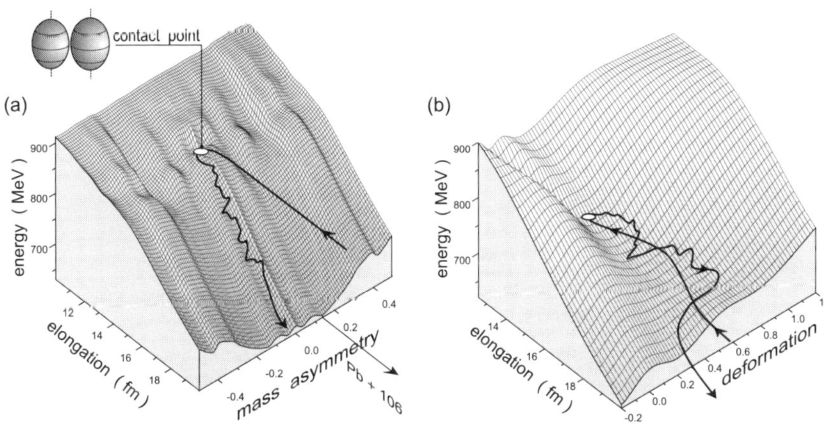

Fig. 14. Potential energy surface for the nuclear system formed by $^{232}Th+^{250}Cf$ as a function of R and α ($\beta = 0.22$) (a) and as a function of R and β ($\alpha = 0.037$) (b). Typical trajectories are shown by the thick curves with arrows.

As mentioned earlier, the lifetime of a giant composite system more than 10^{-20} s is quite enough to expect positron line structure emerging on top of the dynamical positron spectrum due to spontaneous e^+e^- production from the supercritical electric fields as a fundamental QED process ("decay of the vacuum") [10]. The absolute cross section for long events is found to be maximal just at the beam energy ensuring the two nuclei to be in contact, see Fig. 15(c). The same energy is also optimal for the production of the most neutron-rich SH nuclei. Of course, there are some uncertainties in the used parameters, mostly in the value of nuclear viscosity. However we found only a linear dependence of the reaction time on the strength of nuclear viscosity, which means that the obtained reaction time distribution is rather reliable, see logarithmic scale on both axes in Fig. 15(a).

Formation of the background positrons in these reactions forces one to find some additional trigger for the longest events. Such long events

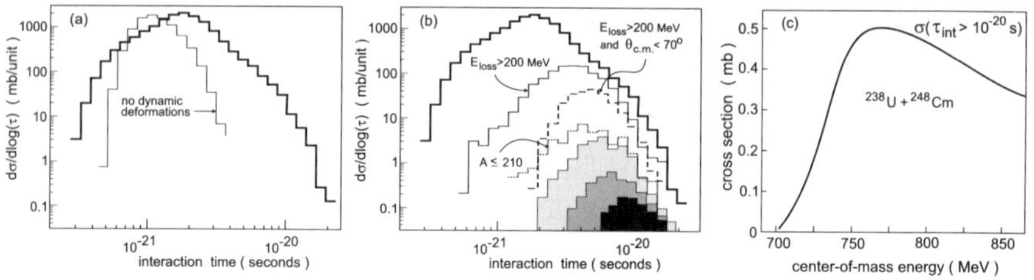

Fig. 15. Reaction time distributions for the ^{238}U+^{248}Cm collision at 800 MeV center-of-mass energy. Thick solid histograms correspond to all events with energy loss more than 30 MeV. (a) Thin solid histogram shows the effect of switching-off dynamic deformations. (b) Thin solid, dashed and dotted histograms show reaction time distributions in the channels with formation of primary fragments with $E_{loss} > 200$ MeV, $E_{loss} > 200$ MeV and $\theta_{c.m.} < 70^o$ and $A \leq 210$, correspondingly. Hatched areas show time distributions of events with formation of the primary fragments with $A \leq 220$ (light gray), $A \leq 210$ (gray), $A \leq 204$ (dark) having $E_{loss} > 200$ MeV and $\theta_{c.m.} < 70^o$. (c) Cross section for events with interaction time longer than 10^{-20} s.

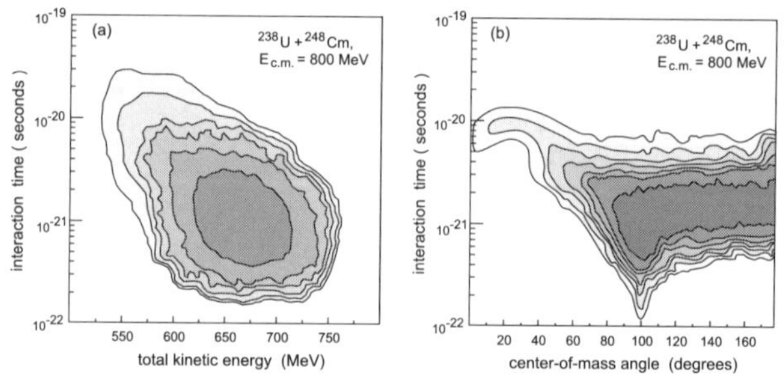

Fig. 16. Energy-time (a) and angular-time (b) distributions of primary fragments in the ^{238}U+^{248}Cm collision at 800 MeV ($E_{loss} > 15$ MeV).

correspond to the most damped collisions with formation of mostly excited primary fragments decaying by fission, see Figs. 16(a). However there is also a chance for production of the primary fragments in the region of doubly magic nucleus ^{208}Pb, which could survive against fission due to nucleon evaporation. The number of the longest events depends weakly on impact parameter up to some critical value. On the other hand, in the angular distribution of all the excited primary fragments (strongly peaked at the center-of-mass angle slightly larger than 90^0) there is the rapidly decreasing tail at small angles, see Fig. 16(b). Time distribution for the most damped events ($E_{loss} > 150$ MeV), in which a large mass transfer occurs and primary fragments scatter in forward angles ($\theta_{c.m.} < 70^o$), is rather narrow and really shifted to longer time delay, see hatched areas in

Fig. 15. For the considered case of $^{238}U+^{248}Cm$ collision at 800 MeV center-of-mass energy, the detection of the surviving nuclei in the lead region at the laboratory angles of about 25° and at the low-energy border of their spectrum (around 1000 MeV for Pb) could be a real trigger for longest reaction time.

6. Conclusion

For near-barrier collisions of heavy ions it is very important to perform a combined (unified) analysis of all strongly coupled channels: deep-inelastic scattering, quasi-fission, fusion and regular fission. This ambitious goal has now become possible. A unified set of dynamic Langevin type equations is proposed for the *simultaneous* description of DI and fusion-fission processes. For the first time, the whole evolution of the heavy nuclear system can be traced starting from the approaching stage and ending in DI, QF, and/or fusion-fission channels. Good agreement of our calculations with experimental data gives us hope to obtain rather accurate predictions of the probabilities for superheavy element formation and clarify much better than before the mechanisms of quasi-fission and fusion-fission processes. The determination of such fundamental characteristics of nuclear dynamics as the nuclear viscosity and the nucleon transfer rate is now possible. The production of long-lived neutron-rich SH nuclei in the region of the "island of stability" in collisions of transuranium ions seems to be quite possible due to a large mass rearrangement in the inverse (anti-symmetrized) quasi-fission process caused by the Z=82 and N=126 nuclear shells. A search for spontaneous positron emission from a supercritical electric field of long-living giant quasi-atoms formed in these reactions is also quite promising.

References

1. S. Hofmann and G. Münzenberg, *Rev. Mod. Phys.* **72**, 733 (2000).
2. Yu.Ts. Oganessian, V.K. Utyonkov, Yu.V. Lobanov, F.Sh. Abdullin, A.N. Polyakov, I.V. Shirokovsky, Yu.S. Tsyganov, G.G. Gulbekian, S.L. Bogomolov, B.N. Gikal, A.N. Mezentsev, S. Iliev, V.G. Subbotin, A.M. Sukhov, A.A. Voinov, G.V. Buklanov, K. Subotic, V.I. Zagrebaev, M.G. Itkis, J.B. Patin, K.J. Moody, J.F. Wild, M.A. Stoyer, N.J. Stoyer, D.A. Shaughnessy, J.M. Kenneally, P.A. Wilk, R.W. Lougheed, R.I. Il'kaev, and S.P. Vesnovskii, *Phys. Rev.* **C70**, 064609 (2004).
3. V.I. Zagrebaev, M.G. Itkis, Yu.Ts. Oganessian, *Yad. Fiz.*, **66**, 1069 (2003).
4. A.S. Botvina, I.N. Mishustin, *Phys. Lett. B* **584**, 233 (2004); I.N. Mishustin, *Proc. ISHIP Conf.*, Frankfurt, April 3-6, 2006.

5. M. Schädel, J.V. Kratz, H. Ahrens, W.Brüchle, G. Franz, H. Gäggeler, I. Warnecke, G. Wirth, G. Herrmann, N. Trautmann, and M. Weis, *Phys. Rev. Lett.* **41**, 469 (1978).
6. M. Schädel, W. Brüchle, H. Gäggeler, J.V. Kratz, K. Sümmerer, G. Wirth, G. Herrmann, R. Stakemann, G. Tittel, N. Trautmann, J.M. Nitschke, E.K. Hulet, R.W. Lougheed, R.L. Hahn, and R.L. Ferguson, Phys. Rev. Lett. **48**, 852 (1982).
7. K.J. Moody, D. Lee, R.B. Welch, K.E. Gregorich, G.T. Seaborg, R.W. Lougheed, and E.K. Hulet, *Phys. Rev. C* **33**, 1315 (1986).
8. C. Riedel, W. Nörenberg, *Z. Phys. A* **290**, 385 (1979).
9. V. Zagrebaev and W. Greiner, *J. Phys. G* **G31**, 825 (2005).
10. J. Reinhard, U. Müller and W. Greiner, *Z. Phys. A* **303**, 173 (1981).
11. W. Greiner (Editor), *Quantum Electrodynamics of Strong Fields*, (Plenum Press, New York and London, 1983); W. Greiner, B. Müller and J. Rafelski, *QED of Strong Fields* (Springer, Berlin and New York, 2nd edition, 1985)
12. K. Rutz, M. Bender, T. Bürvenich, T. Schilling P.-G, Reinhard J. Maruhn, W. Greiner, *Phys. Rev. C* **56**, 238 (1997).
13. T. Bürvenich, M. Bender, J. Maruhn, P.-G, Reinhard, *Phys. Rev. C* **69**, 014307 (2004).
14. J. Maruhn and W. Greiner, *Z. Phys.* **251**, 431 (1972).
15. V.I. Zagrebaev, Y. Aritomo, M.G. Itkis, Yu.Ts. Oganessian, M. Ohta, *Phys. Rev. C* **65**, 014607 (2002).
16. W.W. Wilcke, J.R. Birkelund, A.D. Hoover, J.R. Huizenga, W.U. Schröder, V.E. Viola, Jr., K.L. Wolf, and A.C. Mignerey, *Phys. Rev. C* **22**, 128 (1980).
17. H.J. Wollersheim, W.W. Wilcke, J.R. Birkelund, J.R. Huizenga, W.U. Schröder, H. Freiesleben, and D. Hilscher, *Phys. Rev. C* **24**, 2114 (1981).
18. J. Blocki, J. Randrup, W.J. Swiatecki, and C.F. Tsang, *Ann. Phys. (N.Y.)* **105**, 427 (1977).
19. H.H. Deubler and K. Dietrich, *Nucl. Phys. A* **277**, 493 (1977).
20. K.T.R. Davies, R.A. Managan, J.R. Nix and A.J. Sierk, *Phys. Rev. C* **16**, 1890 (1977).
21. A. Gobbi, U. Lynen, A. Olmi, G. Rudolf, and H. Sann, in *Proceedings of Int. School of Phys. "Enrico Fermi", Course LXXVII*, Varenna, 1979 (North-Holl., 1981), p. 1.
22. V.I. Zagrebaev, Yu.Ts. Oganessian, M.I. Itkis and Walter Greiner, *Phys. Rev. C* **73**, 031602(R) (2006).

FISSION BARRIERS OF HEAVIEST NUCLEI

A. SOBICZEWSKI*, M. KOWAL and L. SHVEDOV

Soltan Institute for Nuclear Studies
ul. Hoża 69, PL-00-681 Warsaw, Poland
**E-mail: adam.sobiczewski@fuw.edu.pl*

Recent macroscopic-microscopic studies on the static fission-barrier height B_f^{st} of heaviest nuclei, done in our Warsaw group, are shortly reviewed. The studies have been motivated by the importance of this quantity in calculations of cross sections for synthesis of these nuclei. Large deformation spaces, including as high multipolarities of deformation as λ=8, are used in the analysis of B_f^{st}. Effects of various kinds of deformations, included into these spaces, on the potential energy of a nucleus are illustrated. In particular, the importance of non-axial shapes for this energy is demonstrated. They may reduce B_f^{st} by up to more than 2 MeV.

1. Introduction

Fission barriers of heavy nuclei are intensively studied recently by a number of groups (e.g., [1–7])), in particular by our group in Warsaw (e.g., [8–11])). The main scope is the calculation of the heights B_f^{st} of the static fission barriers of heaviest nuclei. The motivation for this is the importance of the height B_f^{st} in the calculations of cross sections σ for the synthesis of these nuclei (e.g., [12,13]). This height is a decisive quantity in the competition between neutron evaporation and fission of a compound nucleus in the process of its cooling. A large sensitivity of σ to B_f^{st} stresses a need for accurate calculations of B_f^{st}. For example, a change of B_f^{st} by 1 MeV may result in a change of σ by about one order of magnitude or even more [14]. The basic role, in reaching this accuracy, is played by the deformation space admitted in the calculations of B_f^{st}.

The objective of this paper is to give a short review of recent results of the studies of the potential energy (and, in particular, of the barrier heights B_f^{st}) done in our Warsaw group with the use of large deformation spaces.

2. Theoretical model

A macroscopic-microscopic approach is used to describe the potential energy of a nucleus. The Yukawa-plus-exponential model [15] is taken for the macroscopic part of the energy and the Strutinski shell correction, based on the Woods-Saxon single-particle potential, is used for its microscopic part. Details of the approach are specified in [16].

Especially important in the calculations is the deformation space admitted in them. Generally, a 10-dimensional deformation space is used in our studies. In particular, it includes the general hexadecapole space (if one assumes the reflexion symmetry of a nucleus with respect to all three planes of the intrinsic coordinate system [17]), not considered in earlier studies. The space is specified by the following expression for the nuclear radius $R(\vartheta,\varphi)$ (in the intrinsic frame of reference) in terms of spherical harmonics $Y_{\lambda\mu}$:

$$\begin{aligned} R(\vartheta,\varphi) = R_0\,\{1\, + & \quad \beta_2\left[\cos\gamma_2 \mathrm{Y}_{20} + \sin\gamma_2 \mathrm{Y}_{22}^{(+)}\right] \\ + & \frac{1}{\sqrt{12}}\beta_4\Big[(\sqrt{7}\cos\delta_4 + \sqrt{5}\sin\delta_4\cos\gamma_4)\mathrm{Y}_{40} \\ & \qquad -\sqrt{12}\sin\delta_4\sin\gamma_4 \mathrm{Y}_{42}^{(+)} \\ & \qquad + (\sqrt{5}\cos\delta_4 - \sqrt{7}\sin\delta_4\cos\gamma_4)\mathrm{Y}_{44}^{(+)}\Big] \\ + & \quad \beta_6 \mathrm{Y}_{60} + \beta_8 \mathrm{Y}_{80} \\ + & \quad \beta_3 \mathrm{Y}_{30} + \beta_5 \mathrm{Y}_{50} + \beta_7 \mathrm{Y}_{70}\}\,, \end{aligned} \tag{1}$$

where γ_2 is the Bohr quadrupole non-axiality parameter, δ_4 and γ_4 are the hexadecapole non-axiality parameters [17], and the dependence of R_0 on the deformation parameters is determined by the volume-conservation condition. The functions $\mathrm{Y}_{\lambda\mu}^{(+)}$ are defined as:

$$\mathrm{Y}_{\lambda\mu}^{(+)} = \frac{1}{\sqrt{2}}\left[\mathrm{Y}_{\lambda\mu} + (-1)^{\mu}\mathrm{Y}_{\lambda-\mu}\right], \qquad \text{for} \quad \mu \neq 0. \tag{2}$$

The regions of variation of the deformation parameters are

$$\beta_\lambda \geq 0, \quad (\lambda = 2, 3, ..., 8), \tag{3}$$

$$0^\circ \leq \gamma_2 \leq 60^\circ, \tag{4}$$

$$0^\circ \leq \delta_4 \leq 180^\circ, \quad 0^\circ \leq \gamma_4 \leq 60^\circ. \tag{5}$$

In our studies, the deformation parameters β_3, β_5, β_7 are only used to show that the potential energy of the studied nuclei is not influenced by the reflexion-asymmetric shapes at both the equilibrium and the saddle-point configurations.

The behavior of the energy in the remaining 7-dimensional space has been studied in details. To avoid, however, too big calculations, the analysis is divided into two steps. In the first one, it is done in the most important 5- or 6-dimensional space and then the influence of the remaining two or one dimensions is checked in a separate calculation. For example, the energy is calculated in the 5-dimensional space $\{\beta_2, \gamma_2, \beta_4, \delta_4, \beta_6\}$. In this space, the equilibrium and the saddle point (and the energy corresponding to them) are found. Then, at these points, the energy is corrected by minimization of it in the 2-dimensional space $\{\gamma_4, \beta_8\}$. Such division is certainly an approximation, as it assumes that the equilibrium and the saddle points are not changed by the minimization of the energy in the γ_4 and β_8 degrees of freedom. We check, however, that, when the division of the large space into two smaller ones is done properly, the approximation is quite good and that the minimization in the second space ($\{\gamma_4, \beta_8\}$ in our example) leads to only a small correction of the energy, in particular of $B_{\rm f}^{\rm st}$.

To illustrate the numerical size of such calculation, let us specify some details. The potential energy is calculated on the following grid points (numbers in parentheses indicate the step length with which the calculation is performed for a given variable):

$$\begin{aligned} \beta_2 \cos\gamma_2 &= 0(0.05)0.65, \\ \beta_2 \sin\gamma_2 &= 0(0.075)0.375, \\ \beta_4 \cos\delta_4 &= -0.20(0.05)0.20, \\ \beta_4 \sin\delta_4 &= 0(0.075)0.225, \\ \beta_6 &= -0.12(0.06)0.12, \end{aligned} \tag{6}$$

corresponding to $14 \times 6 \times 9 \times 4 \times 5 = 15120$ points. Then, the energy is interpolated (by the standard SPLIN3 procedure of the IMSL library) to the five times denser grid in each variable. Thus, only in this 5-dimensional space, we have the values of the potential energy on a huge number of $15120 \times 5^5 = 4.725 \cdot 10^7$ points, i.e. on about 50 million points.

Minimization of the energy at the equilibrium and the saddle points, found in the above 5-dimensional space, is done on the following grid points in the remaining two degrees of freedom γ_4 and β_8:

$$\begin{aligned} \gamma_4 &= 0^\circ(20^\circ)60^\circ, \\ \beta_8 &= -0.12(0.06)0.12. \end{aligned} \tag{7}$$

This calculation is very small with respect to the previous one, done in the 5-dimensional space.

3. Results

We are going to illustrate here the results for the barrier heights B_f^{st} obtained in the two main cases of axially symmetric and axially asymmetric shapes of a nucleus.

3.1. *Axially symmetric shapes*

Figure 1, taken from [9], shows an example of the ground-state static fission barrier for the superheavy nucleus $^{278}112$ (this is the compound nucleus obtained in the reaction which has lead to the discovery of the element 112 [18]). One can see that a rather high barrier is obtained for this very heavy nucleus, which is entirely created by the effects on the energy of the shell structure of this nucleus. Without this structure (see macroscopic part of the energy, $E_{\rm macr}$), no barrier is obtained. The largest shell correction to the macroscopic part of the energy is obtained at the (deformed) equilibrium point (about 6 MeV), smaller (about 1.8 MeV) at the first, and the smallest (about 0.5 MeV) at the second saddle point. Significant shell corrections at the saddle points are worth to be noticed, as these corrections are quite often neglected in various estimates of the static fission barriers of superheavy nuclei.

The height of the barrier is defined as the difference between the potential energy at the highest saddle point and the ground-state energy. The latter is the potential energy at the equilibrium point, increased by the zero-point energy in the fission degree of freedom, for which 0.7 MeV is taken [19]. Thus, as a matter of fact, we are only interested in the two values of the potential energy: at the equilibrium point β_λ^0 and the highest saddle point β_λ^s. To find, however, these points, knowledge of the energy in a large deformation region is needed.

Figure 2, taken from [20], shows a contour map of the potential energy of the nucleus ^{250}Cf projected on the plane (β_2, β_4). This means that at each point (β_2, β_4), the energy, which is minimal in the β_6 and β_8 degrees of freedom, is taken. (The energy is normalized in such a way that its macroscopic part is zero at the spherical shape of a nucleus). The saddle point is obtained at the deformation $(\beta_2^s, \beta_4^s, \beta_6^s, \beta_8^s) = (0.432, 0.084, 0.015, 0.005)$ and the equilibrium point at $(\beta_2^0, \beta_4^0, \beta_6^0, \beta_8^0) = (0.247, 0.029, -0.046, 0.002)$. The respective energies are 3.5 MeV and -4.7 MeV. Thus, the barrier height is 3.5−(−4.7+0.7) MeV = 7.5 MeV, because the zero-point energy, E_{zp} = 0.7 MeV, is taken [19].

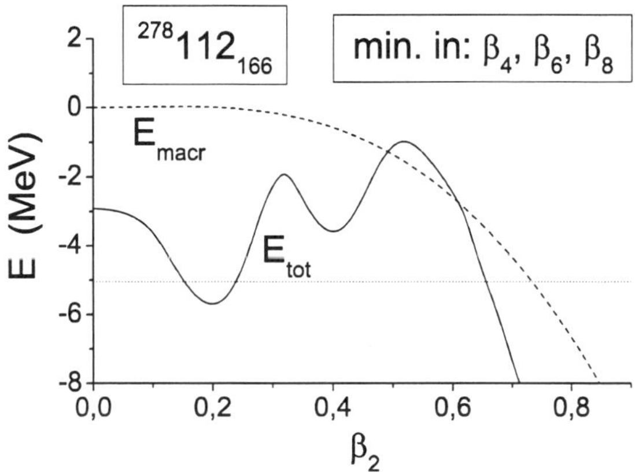

Fig. 1. Static spontaneous-fission barrier calculated for the nucleus $^{278}112$ in two cases: when only the macroscopic ($E_{\rm macr}$) and when the total ($E_{\rm tot}$) energy of it is considered [9].

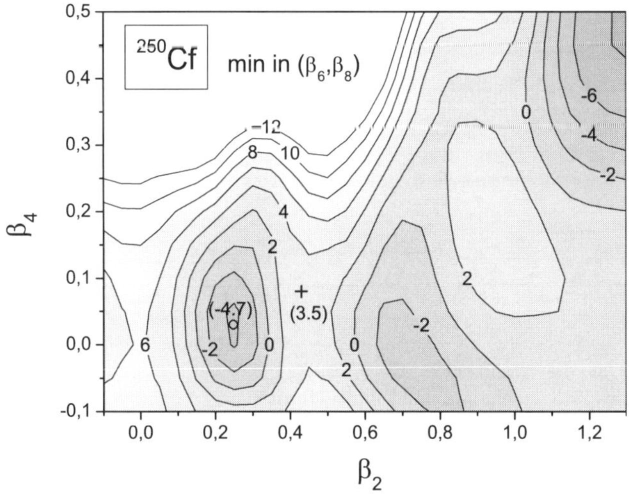

Fig. 2. Contour map of the potential energy of the nucleus ^{250}Cf. Numbers at the contour lines specify the value of the energy in MeV. Position of the equilibrium point is marked by the symbol "o" and of the saddle point by the symbol "+". Numbers in the parentheses give the values of the energy at these points [20].

To see the role of higher multipolarity deformations in the barrier height B_f^{st}, we calculate this quantity in 1-, 2-, 3- and 4-dimensional deformation spaces. As we use only even-multipolarity deformations (to describe thin barriers of very heavy nuclei), this is the calculation of B_f^{st} as a function of the maximal multipolarity $\lambda_{max} = 2,4,6$ and 8 taken into account. Figure 3 shows the dependence of the potential energy at the equilibrium, E_{min}, and at the saddle, E_s, points, as well as of B_f^{st}, on λ_{max}. Two nuclei are taken for the illustration. One (^{250}Cf) [20] which is deformed in its ground state, and the other (^{294}116) [21] which is spherical in this state. One can see that, in the deformed nucleus, E_{min} decreases more strongly than E_s with increasing λ_{max}, resulting in the increase of the barrier height B_f^{st} with the increase of λ_{max}. For the spherical nucleus, E_s is decreasing, while E_{min} is constant when λ_{max} increases, resulting in the decrease of B_f^{st} with the increase of λ_{max}. This difference, between a deformed and a spherical nucleus, in the behavior of B_f^{st} as a function of the dimension of the deformation space, in which B_f^{st} is analyzed, is worth to be noticed.

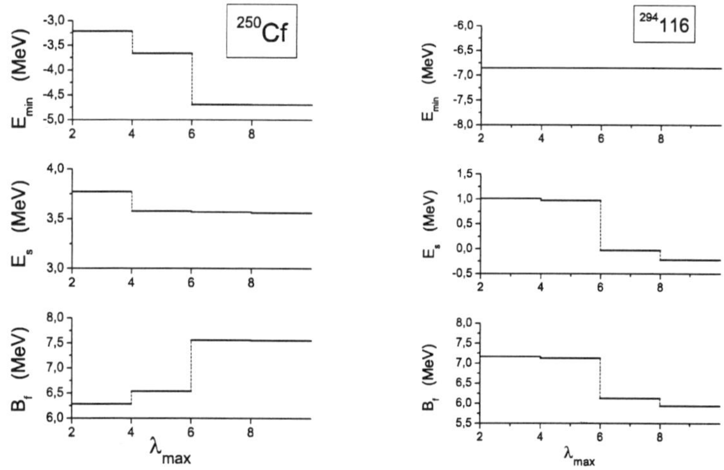

Fig. 3. Dependence of the potential energy of the nucleus ^{250}Cf (left part) and ^{294}116 (right part) at the equilibrium, E_{min}, and at the saddle point, E_s, and also of the barrier height, B_f^{st}, on the maximal multipolarity λ_{max} of the deformation taken in the analysis [20,21].

The values of B_f^{st}, calculated by a macro-micro method for many superheavy nuclei with the atomic number Z=106-120, have been given in [8]. Other calculations, done by another (Extended Thomas-Fermi plus Strutinski Integral) method, but still with the assumption of the axial symmetry of nuclear shapes, have been done in [1,2].

3.2. *Axially asymmetric shapes*

Figure 4 [22] shows a contour map of the potential energy of ^{250}Cf when non-axial deformation is taken into account. The energy is obtained in the 2-dimensional quadrupole-deformation space $\{\beta_2, \gamma_2\}$. One can see that, in the case of axial symmetry (γ_2=0°), the saddle point (denoted by the

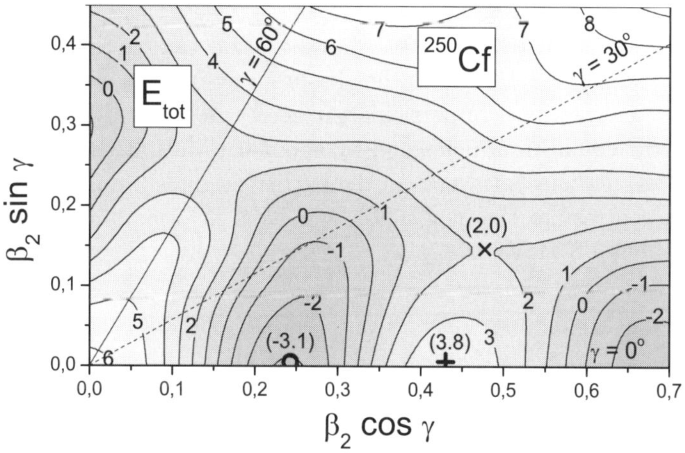

Fig. 4. Contour map of the potential energy of ^{250}Cf calculated in the 2-dimensional deformation space $\{\beta_2, \gamma_2\}$ (γ_2 is denoted by γ in the figure). Position of the saddle point is denoted by "+", when the axial symmetry of the nucleus is assumed, and by "×", when the non-axiality is taken into account. Position of the equilibrium point is denoted by "o". Numbers in the parentheses give the values of the energy at these points [22].

symbol "+") has the energy 3.8 MeV, while the non-axiality shifts it to the point denoted by the symbol "×" and decreases its energy to 2.0 MeV, i.e. by a large value of 1.8 MeV. As the energy at the equilibrium point is not changed by the non-axial deformation γ_2, the barrier height B_f^{st} is decreased by γ_2 by the same amount as the saddle-point energy, i.e. by 1.8 MeV. Only after the inclusion of this decrease, the calculated barrier height: $B_f^{st} = 7.5 - 1.8 = 5.7$ MeV, becomes close to the measured value: (5.6 ± 0.3) MeV [23]. It is worth mentioning that in the case of ^{250}Cf, practically the whole decrease of B_f^{st} comes from the quadrupole non-axiallity, as will be illustrated later. The value 7.5 MeV, obtained with the use of the space of the axially symmetric deformations $\{\beta_\lambda\}$, λ=2,4,6,8, is taken from Fig. 2:

B_f^{st}(sym)=[3.5-(-4.7+0.7)]MeV=7.5 MeV, where 0.7 MeV is the ground-state zero-point energy in the fission degree of freedom [19], as already stated above in the description of Fig. 2.

For some nuclei, the decrease of B_f^{st}, due to the quadrupole non-axial shapes of a nucleus, may be even larger. This is seen in Fig. 5 [24], where the saddle-point energy and the barrier height B_f^{st} are decreased by γ_2 by 2.3 MeV.

The effect of the hexadecapole deformation on the potential energy is relatively small for the nucleus ^{250}Cf, as might be expected on the basis of Fig. 3. This is better illustrated in Fig. 6 [25], where this effect is shown in a large region of the deformations β_2 and γ_2. One can see that this effect is smaller than about 1.1 MeV (in the absolute value) in the whole considered region of deformations. In particular, it decreases the energy by about 0.5 MeV at the equilibrium point and by about 0.4 MeV at the saddle point. As a result, it changes (increases) the barrier height B_f^{st} only by about 0.1 MeV, in contrast to the quadrupole non-axial deformation, which lowers B_f^{st} by about 1.8 MeV [22].

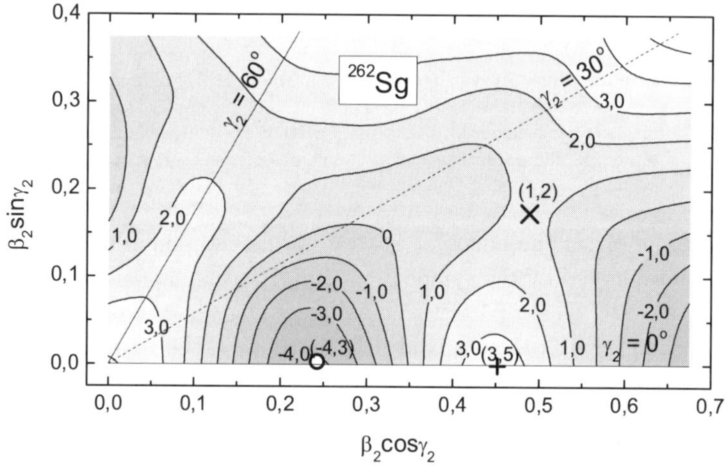

Fig. 5. Same as in Fig. 4, but for the nucleus ^{262}Sg (Z=106) [24].

The above effect is much larger for the nucleus ^{262}Sg, as can be seen in

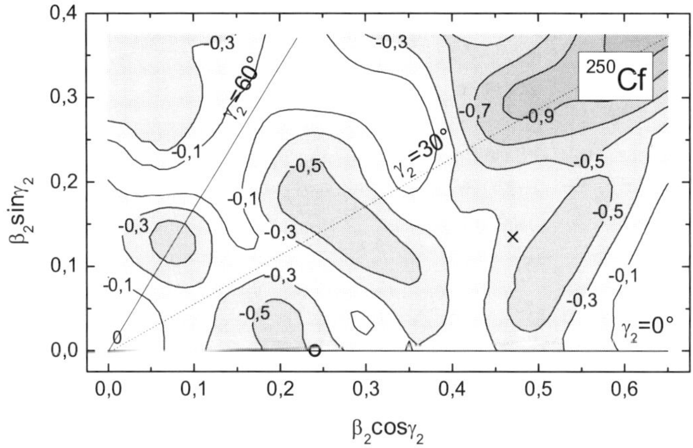

Fig. 6. The effect on the potential energy of the total hexadecapole deformation: $E(\beta_2, \gamma_2; \beta_4^{\min}, \delta_4^{\min}, \gamma_4^{\min}) - E(\beta_2, \gamma_2; \beta_4 = 0)$, calculated for the nucleus ^{250}Cf [25].

Fig. 7 [24]. In particular, the saddle-point mass (and also the barrier height B_f^{st}) is decreased by about 1.5 MeV by the hexadecapole deformations of this nucleus.

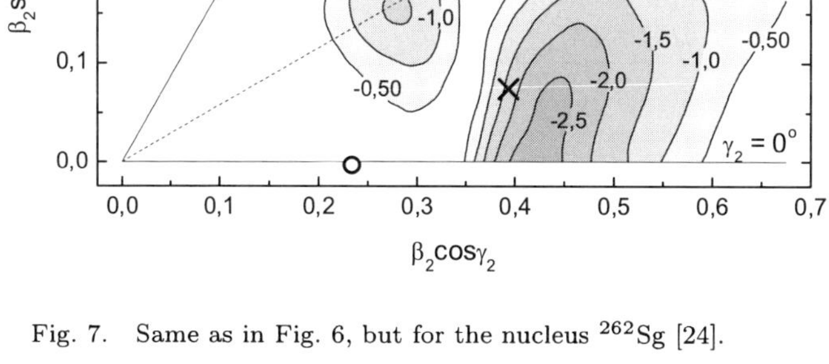

Fig. 7. Same as in Fig. 6, but for the nucleus ^{262}Sg [24].

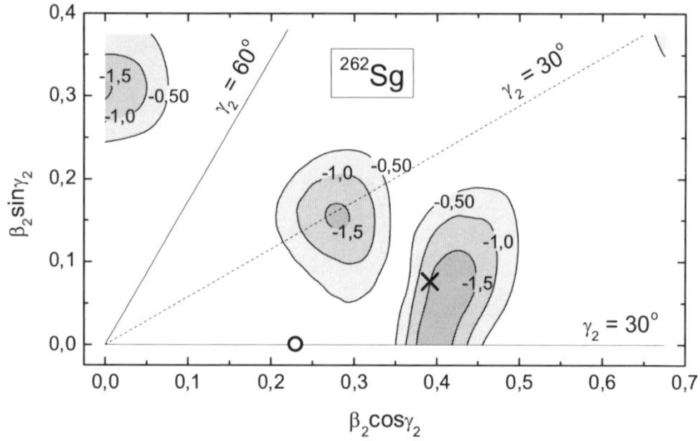

Fig. 8. Same as in Fig. 7, but for the difference: $E(\beta_2, \gamma_2; \beta_4^{\rm min}, \delta_4^{\rm min}, \gamma_4^{\rm min}) - E(\beta_2, \gamma_2; \beta_4^{\rm min}, \delta_4 = 0, \gamma_4^{\rm min})$, i.e. for the effect on energy of the hexadecapole non-axial deformation of ^{262}Sg described by the parameter δ_4 [24].

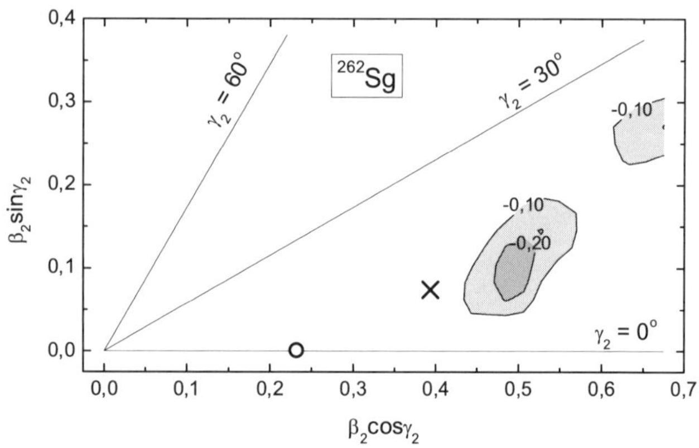

Fig. 9. Same as in Fig. 7, but for the difference: $E(\beta_2, \gamma_2; \beta_4^{\rm min}, \delta_4^{\rm min}, \gamma_4^{\rm min}) - E(\beta_2, \gamma_2; \beta_4^{\rm min}, \delta_4^{\rm min}, \gamma_4 = 0)$, i.e. for the effect on energy of the hexadecapole non-axial deformation of ^{262}Sg described by the parameter γ_4 [24].

It is interesting to see how significant are the contributions of the two non-axial hexadecapole deformations δ_4 and γ_4 to this large effect of all hexadecapole shapes described by the parameters β_4, δ_4 and γ_4. Figure 8 [24] shows the contribution of δ_4 to this effect. One can see that this contribution is large, up to about 2 MeV in the considered region of deformations. In particular, this deformation decreases the saddle-point mass by about 1.5 MeV, which amounts to about the total effect of the hexadecapole deformations on this mass.

The effect of the deformation γ_4 on the potential energy of the nucleus ^{262}Sg is illustrated in Fig. 9 [24]. It is seen that this effect is small, less than about 0.3 MeV in the whole considered region of deformations.

In conclusion, one can say that the dependence of the potential energy (and in particular of the fission-barrier height B_f^{st}) of a heavy nucleus on its deformation is a very individual property of each nucleus, because of its individual shell structure. Due to this, one should be careful in drawing general conclusions from considered examples. Still, the results illustrated in this paper show that non-axial deformations of a heavy nucleus may be large and should not be disregarded in calculations of its potential energy and, in particular, of its barrier height B_f^{st}. These deformations may very strongly modify the height, making the result obtained in the case of axial symmetry completely unrealistic.

Acknowledgements

Support by the Polish State Committee for Scientific Research, grant no. 1 P03B 042 30, and the Polish-JINR (Dubna) Cooperation Programme is gratefully acknowledged.

References

1. A. Mamdouh, J.M. Pearson, M. Rayet and F. Tondeur, *Nucl. Phys. A* **644**, 389 (1998).
2. A. Mamdouh, J.M. Pearson, M. Rayet and F. Tondeur, *Nucl. Phys. A* **679**, 337 (2001).
3. P. Möller, A.J. Sierk and A. Iwamoto, *Phys. Rev. Lett.* **92**, 072501 (2004).
4. T. Bürvenich, M. Bender, J.A. Maruhn and P.-G. Reinhard, *Phys. Rev. C* **69**, 014307 (2004).
5. J. Dudek, K. Mazurek and B. Nerlo-Pomorska, *Acta Phys. Pol. B* **35**, 1263 (2004).
6. A. Staszczak, J. Dobaczewski and W. Nazarewicz, *Int. J. Mod. Phys. E* **14**, 395 (2005).

7. A. Baran, Z. Łojewski, K. Sieja and M. Kowal, *Phys, Rev. C* **72**, 044310 (2005).
8. I. Muntian, Z. Patyk and A. Sobiczewski, *Acta Phys. Pol. B* **34**, 2141 (2003).
9. A. Sobiczewski and I. Muntian, *Nucl. Phys. A* **734**, 176 (2004).
10. I. Muntian, O. Parkhomenko and A. Sobiczewski, Proc. Intern. Tours Symp. on Nuclear Physics V, Tours (France) 2003 , ed. by. M. Arnould et al. (AIP Conf. Proc., vol. 704, New York, 2004) p. 41.
11. A. Sobiczewski and M. Kowal, *Phys. Scr. T* **125**, 68 (2006).
12. W.J. Świątecki, K. Siwek-Wilczyńska and J. Wilczyński, *Acta Phys. Pol. B* **34**, 2049 (2003).
13. V.V. Volkov, *Fiz. Element. Chastits i At. Yadra* **35**, 797 (2004).
14. M.G. Itkis, Yu.Ts. Oganessian and V.I. Zagrebaev, *Phys. Rev. C* **65**, 044602 (2002).
15. H.J. Krappe, J.R. Nix and A.J. Sierk, *Phys. Rev. C* **20**, 992 (1979).
16. I. Muntian, Z. Patyk and A. Sobiczewski, *Acta Phys. Pol. B* **32**, 691 (2001).
17. S.G. Rohoziński and A. Sobiczewski, *Acta Phys. Pol. B* **12**, 1001 (1981).
18. S. Hofmann *et al.*, *Z. Phys. A* **354**, 229 (1996).
19. R. Smolańczuk, J. Skalski and A. Sobiczewski, *Phys. Rev. C* **52**, 1871 (1995).
20. A. Sobiczewski and I. Muntian, *Int. J. Mod. Phys. E* **14**, 409 (2005).
21. I. Muntian and A. Sobiczewski, *Int. J. Mod. Phys. E* **14**, 417 (2005).
22. I. Muntian and A. Sobiczewski, *Acta Phys. Pol. B* **36**, 1359 (2005).
23. S. Bjørnholm and J.E. Lynn, *Rev. Mod. Phys.* **52**, 725 (1980).
24. A. Sobiczewski, M. Kowal and L. Shvedov, *Acta Phys. Pol. B* **38** (2007) in press.
25. L. Shvedov and A. Sobiczewski, *Acta Phys. Pol. B* **38** (2007) in press.

POSSIBILITY OF SYNTHESIZING DOUBLY MAGIC SUPERHEAVY NUCLEI

Y. ARITOMO

Flerov Laboratory of Nuclear Reactions, JINR, Dubna, Moscow Region 141980, Russia
**E-mail: aritomo24@muj.biglobe.ne.jp*

M. OHTA

Department of Physics, Konan University, 8-9-1 Okamoto, Kobe, Japan

F. HANAPPE and T. MATERNA

Universite Libre de Bruxelles, 1050 Bruxelles, Belgium

L. STUTTGÉ and O. DORVAUX

Institut de Recherches Subatomiques, F-67037 Strasbourg Cedex, France

Fusion-fission process in superheavy mass region is investigated on the basis of the fluctuation-dissipation dynamics. We analyze the trajectory in three-dimensional coordinate space with the Langevin equation. To investigate the fusion-fission process more precisely, we propose to take into account the pre-scission neutron multiplicity in connection with fission fragments. For the survival process, we apply the dynamical model instead of the statistical model. The possibility of synthesizing a doubly magic superheavy nucleus, $^{298}114_{184}$, is investigated taking into account the temperature dependence of the potential energy owing to neutron emission.

Keywords: superheavy elements, fluctuation-dissipation dynamics, fusion-fission process, quasi-fission process, neutron emission

1. Introduction

The existence of the "Island of Stability" in the superheavy mass region has been predicted using the nuclear shell model.[1] Recent experiments on the synthesis of superheavy elements have been focused on the investigation of this region and seem to support this hypothesis. Recently, the production of new superheavy elements has been reported by Dubna (Z=113,114,115,116 and 118) using hot fusion reaction[2] and GSI ($Z = 110, 111, 112$) using cold fusion reaction.[3] Using the facility in RIKEN, nuclei with Z=110,111,112 and 113 elements have also been produced by cold fusion reaction with high accuracy.[4]

In theoretical approaches, the clarification of the reaction mechanism in the superheavy mass region and the prediction of the favorable conditions (projectile-target combination and beam energy values) to synthesis new elements have advanced step by step in the last few years. However, we are still seeking suitable models for calculating the fusion and the survival probabilities, and investigating the unknown parameters involved in them. The fusion-fission mechanism in the superheavy-mass region must be elucidated; in particular, a more accurate estimation of fusion probability should be established.

Here, we discuss two topics separately; one is the fusion process and another is the survival process. For fusion process, we have analyzed the trajectory in three-dimensional coordinate space with the Langevin equation on the basis of the fluctuation-dissipation model.[5] To investigate the fusion process more precisely, we propose to take into account the pre-scission neutron multiplicity in connection with fission fragments. For survival process, the statistical model does not work correctly for highly excited compound nucleus in superheavy mass region, which does not have the fission barrier. We apply the dynamical model instead of the statistical model. The possibility of synthesizing a doubly magic superheavy nucleus, $^{298}114_{184}$, is investigated taking into account the temperature dependence of the potential energy owing to neutron emission.

In section 2, we discuss the fusion process taking into account the neutron emission. The survival process and possibility of synthesizing a doubly magic superheavy nucleus, $^{298}114_{184}$, are presented in section 3.

2. Fusion process

Generally, on the base of analysis of the experimental mass, kinetic energy, and angular distributions of the reaction fragments, the whole reaction process in heavy nucleus collisions is classified into the fusion-fission, quasi-fission and deep inelastic collision processes and so on. Generally, mass symmetric fission fragments have been considered to originate from a compound nucleus. Therefore, in the experiment, the fusion-fission cross section has been derived by counting mass symmetric fission events.[6]

Using the Langevin equation, which enables to calculate the time development of a trajectory of a colliding system in a shape parameter space, we attempted to distinguish different dynamical paths by analyzing the trajectories and mass distribution of fission fragments. We identified the fusion-fission (FF) path, the quasi-fission (QF) path and the deep inelastic collision (DIC) path. In addition, in the reaction ^{48}Ca+^{244}Pu, we found a new path, called the deep quasi-fission (DQF) path, whose trajectory does not reach the spherical compound nucleus but goes to the mass symmetric fission area.[5]

The conclusion is that the analysis of only the mass distribution of fission fragments is insufficient to identify the reaction process described between the FF and DQF paths. To estimate the fusion probability accurately, the precise identification of the FF path is very important. On the basis of experimental results, it has also been suggested that the reaction process may be classified by the measurement of the pre-scission neutron multiplicity, which correlates with the mass distribution of fission fragments.[6–8] This means that each process is expected to have its own characteristic reaction time, which is related to the pre-scission neutron multiplicity.

Hence, to identify the reaction process more precisely, we undertake to extend our model discussed in reference,[5] by taking into account the effect of neutron emission along the dynamical path. We introduce the effect of neutron emission into the three-dimensional Langevin calculation code, or we combine the fluctuation and dissipation dynamical model with a statistical model. Here, we show theoretically the usefulness of the methods of classifying the dynamical process on the basis of the pre-scission neutron multiplicity correlated with the mass distribution of fission fragments.

2.1. *Model*

Using the same procedure as described in reference[5] to investigate dynamically the fusion-fission process, we use the fluctuation-dissipation model and employ the Langevin equation. We adopt the three-dimensional nuclear deformation space given by two-center parameterization.[9,10] The three collective parameters involved in the Langevin equation are as follows: z_0 (distance between two potential centers), δ (deformation of fragments) and α (mass asymmetry of the colliding nuclei); $\alpha = (A_1 - A_2)/(A_1 + A_2)$, where A_1 and A_2 denote the mass numbers of the target and the projectile, respectively.

The multidimensional Langevin equation is given as

$$\frac{dq_i}{dt} = \left(m^{-1}\right)_{ij} p_j,$$
$$\frac{dp_i}{dt} = -\frac{\partial V}{dq_i} - \frac{1}{2}\frac{\partial}{\partial q_i}\left(m^{-1}\right)_{jk} p_j p_k - \gamma_{ij}\left(m^{-1}\right)_{jk} p_k + g_{ij} R_j(t), \quad (1)$$

where a summation over repeated indices is assumed. q_i denotes the deformation coordinate specified by z_0, δ and α. p_i is the conjugate momentum of q_i. V is the potential energy, and m_{ij} and γ_{ij} are the shape-dependent collective inertia parameter and dissipation tensor, respectively. A hydrodynamical inertia tensor is adopted in the Werner-Wheeler approximation for the velocity field, and the wall-and-window one-body dissipation is adopted for the dissipation tensor. The normalized random force $R_i(t)$ is assumed to be white noise, *i.e.*, $\langle R_i(t)\rangle=0$ and $\langle R_i(t_1)R_j(t_2)\rangle = 2\delta_{ij}\delta(t_1 - t_2)$. The strength of random force g_{ij} is given by $\gamma_{ij}T = \sum_k g_{ij}g_{jk}$, where T is the temperature of the compound nucleus calculated from the intrinsic energy of the composite system as $E_{int} = aT^2$, with a denoting the level density parameter. Using the procedure outlined by Fröbrich et al.,[11] in which the Langevin calculation and the statistical model were combined, the emission of neutrons has been coupled to the three-dimensional Langevin equation for the fusion-fission process. The detail is explained in reference.[5]

2.2. *Neutron emission during fusion-fission process*

Using our model, we analyze the experimental data in the reaction ^{58}Ni+^{208}Pb at E^*=185.9 MeV.[7] Figure 1(a) shows the experimental result for the distribution of the pre-scission neutron multiplicity ν_n in correlation with fission fragments whose mass number is greater than $\frac{A}{2} - 30$ and less than $\frac{A}{2} + 30$.

An interesting feature of the distribution of pre-scission neutron multiplicity is its shape having two components, which are located around $\nu_n = 4$ and $\nu_n = 8$. This structure may be the sign of a simultaneous coexistence of two mechanisms corresponding to different life time of the composite system but to the phenomena giving nearly the same mass fragment in the fission process. The first one, defined as the QF process, would lead to the emission of about four neutrons and for the second one, associated with the FF process via a compound nucleus, would be found around 8 neutron emission. In this case, the fission process would take enough long time to allow the emission of nearly 8 neutrons. As will be shown later, from the results of our model calculation, we can confirm that the different dynamical processes, i.e., the FF process and the QF process, are giving different pre-scission neutron multiplicities in spite of the phenomenon belonging to the similar fission mass fragment.

The distribution of the pre-scission neutron multiplicity ν_n calculated by our model is shown in Fig. 1(b). The pre-scission neutron multiplicity for the QF and the FF processes are denoted by the gray and the black lines,

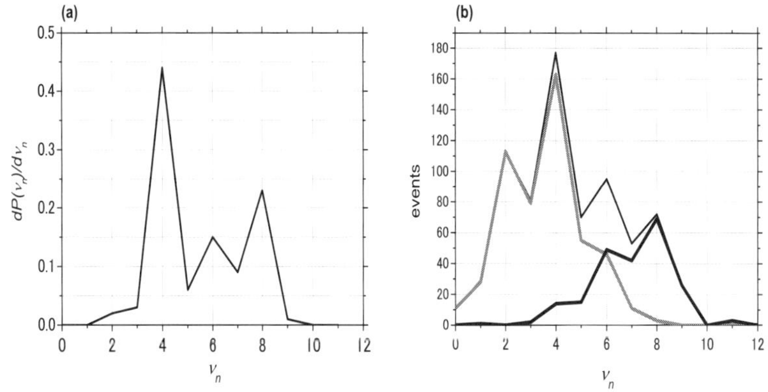

Fig. 1. (a) Experimental pre-scission neutron multiplicity associated with fission fragment measurements in the reaction ^{58}Ni+^{208}Pb at E^*=185.9 MeV.[7] (b) Calculation results. The neutron multiplicity from the QF process and the FF process are denoted by the gray and black lines, respectively. The thin line denotes the total processes.

respectively. The thin line shows the total multiplicity of each process. The FF and the QF trajectories occupy 0.256 % and 0.509 %, respectively.

We can clearly see that the two components are coming from the QF process and the FF process. It is shown that for the large neutron multiplicity it originates from the FF process, and on the other hand for small neutron multiplicity it comes from the QF process. This means that the pre-scission neutron multiplicity has a strong correlation with dynamical paths. On the pre-scission neutron multiplicity, the odd-even oscillations appear clearly, due to the odd-even effect on neutron binding energies. The initial number of neutrons involved in the reaction is even, so the probability of emitting an even number of neutrons is larger. The calculations show the similar structure observed in the experimental measurements in Fig. 1(a).

Next, we discuss the details on this calculation. Figure 2(a) show the potential energy surface of the liquid drop model for ^{266}Ds on the $z - \alpha$ ($\delta = 0$) plane, in the case of $l = 0$. This potential energy surface is calculated using the two-center shell model code.[12,13] The contour lines of the potential energy surface are drawn with step of 2 MeV. In Fig. 2(a), the position at $z = \alpha = \delta = 0$ corresponds to a spherical compound nucleus. The injection point of this system is indicated by the arrow. The top of the arrow

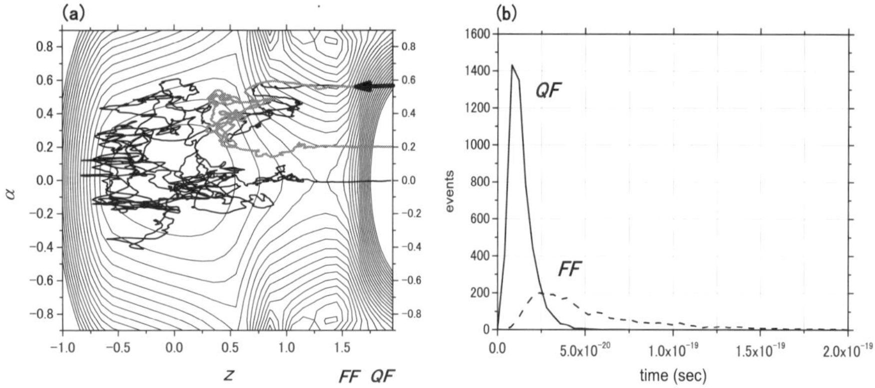

Fig. 2. (a) Sample trajectories projected onto $z - \alpha$ ($\delta = 0$) plane at $E^* = 185.9$ MeV in the reaction ^{58}Ni+^{208}Pb. The trajectories of the QF and the FF processes are denoted by gray and black lines, respectively. The potential energy surface is presented by the liquid drop model in nuclear deformation space for ^{266}Ds. The arrow denotes the injection point of the reaction. (b) The distribution of travelling time t_{trav}. The t_{trav} from the QF and FF processes are denoted by the solid and dashed lines, respectively.

corresponds to the point of contact in the system. We start the calculation of the three-dimensional Langevin equation at the point of contact, which is located at $z = 1.57, \delta = 0.0, \alpha = 0.56$. Whether the trajectory takes the FF process or the QF process, it depends on the potential landscape and the random force (or random number) in the fusion-fission process. The sample trajectories of the QF process and the FF process are also shown in Fig. 2(a). The trajectories are projected onto the $z - \alpha$ plane ($\delta = 0$). The trajectories of the QF and the FF processes are denoted by gray line and black line, respectively.

We define the travelling time t_{trav} as a time duration during which the trajectory moves from the point of contact to the scission point. Figure 2(b) shows the distribution of t_{trav}. The t_{trav} from the QF and FF processes are denoted by the solid and dashed lines, respectively. On the FF process, as we can see in Fig. 2(a), the trajectory is trapped in the pocket around the spherical region. The trajectory spends a relatively long time in the pocket and it has a large chance to emit neutrons. In average, the time duration spending in the pocket fluctuate around 7×10^{-20}sec. The time scale of the

FF process is about 3 or 4 times longer than that of the QF process.

3. Survival process

According to macroscopic-microscopic calculations,[1] there should be a magic island of stability surrounding the doubly magic superheavy nucleus containing 114 protons and 184 neutrons. Actually, if we plan to synthesize the doubly magic superheavy nucleus $^{298}114_{184}$, we must fabricate more neutron-rich compound nuclei because of the neutron emissions from excited compound nuclei. Since combinations of stable nuclei do not provide such neutron-rich nuclei, the reaction mechanism for nuclei with $Z = 114, N > 184$ has rarely been investigated until now. However, because of the characteristic properties of these nuclei, we find an unexpected reaction mechanism for enhancing the evaporation residue cross section. We report this mechanism here.

In superheavy mass region, the fission barrier of highly excited compound nucleus disappears. Therefore, Bohr-Wheeler as well as Kramers formulas are not valid. Moreover, since we must treat extremely small probabilities in the decay process of the compound nucleus, we investigate the evolution of the probability distribution $P(q,l;t)$ in the collective coordinate space with the Smoluchowski equation,[14] which is a strong friction limit of Fokker-Planck equation. We employ the one-dimensional Smoluchowski equation in the elongation degree of freedom z_0, which is expressed as follows;

$$\frac{\partial}{\partial t}P(q,l;t) = \frac{1}{\mu\beta}\frac{\partial}{\partial q}\left\{\frac{\partial V(q,l;t)}{\partial q}P(q,l;t)\right\} + \frac{T}{\mu\beta}\frac{\partial^2}{\partial q^2}P(q,l;t). \qquad (2)$$

q denotes the coordinate specified by z_0. $V(q,l;t)$ is the potential energy, and the angular momentum of the system is expressed by l. μ and β are the inertia mass and the reduced friction, respectively. For these quantities we use the same values as in references.[14] T is the temperature of the compound nucleus calculated from the excitation energy as $E^* = aT^2$ with a denoting the level density parameter of Töke and Swiatecki.[15] The temperature dependent shell correction energy is added to the macroscopic potential energy,

$$V(q,l;t) = V_{\rm DM}(q) + \frac{\hbar^2 l(l+1)}{2I(q)} + V_{\rm shell}(q)\Phi(t), \qquad (3)$$

where $I(q)$ is the moment of inertia of rigid body at coordinate q. $V_{\rm DM}$ is the potential energy of the finite range droplet model and $V_{\rm shell}$ is the shell correction energy at $T = 0$.[5]

The temperature dependence of the shell correction energy is extracted from the free energy calculated with single particle energies.[14,16] The temperature-dependent factor $\Phi(t)$ in Eq. (3) is parameterized as;

$$\Phi(t) = \exp\left(-\frac{aT^2(t)}{E_d}\right), \tag{4}$$

following the work by Ignatyuk *et al.*[17] The shell-damping energy E_d is chosen as 20 MeV. The cooling curve $T(t)$ is calculated by the statistical model code SIMDEC.[14,16] We assume that the particle emissions in the composite system are limited to neutron evaporation in the neutron-rich heavy nuclei. When the temperature decreases as a result of neutron evaporation, the potential energy $V(q, l; t)$ changes due to the restoration of shell correction energy.

The survival probability $W(E_0^*, l; t)$ is defined as the probability which is left inside the fission barrier in the decay process;

$$W(E_0^*, l; t) = \int_{\text{inside saddle}} P(q, l; t) dq. \tag{5}$$

Here, E_0^* is the initial excitation energy of the compound nucleus. For the purpose of understanding well the characteristic enhancement in the excitation function, we first discuss the evaporation residue probability of one partial wave, i.e., of $l = 10$, which is one of the dominantly contributing partial waves.[14]

The neutron separation energy depends on the neutron number. Figure 3(a) shows the neutron separation energies averaged over four successive neutron emissions $\langle B_n \rangle$ for the isotopes with $Z = 114$. We use the mass table in reference.[18] With increasing neutron number of the nucleus, the neutron separation energy becomes small. Therefore many neutrons evaporate easily from the neutron-rich compound nuclei.

Because of rapid neutron emissions, the cooling speed of the compound nucleus is very high. Figure 3(b) shows the cooling curves of $A = 292, 298$ and 304 at the initial excitation energy $E_0^* = 40$ MeV, that were derived using the statistical code SIMDEC.[14,16] In the case of $A = 304$, the excited compound nucleus cools rapidly and the fission barrier recovers at a low excitation energy.

Moreover, owing to the neutron emissions, the neutron number of the de-exciting nucleus with $A = 304$ approaches that of a nucleus with the double closed shell $Z = 114, N = 184$. Figure 4(a) shows the shell correction energies V_{shell} of isotopes with $Z = 114$.[18] V_{shell} of the $A = 304$ ($N =$

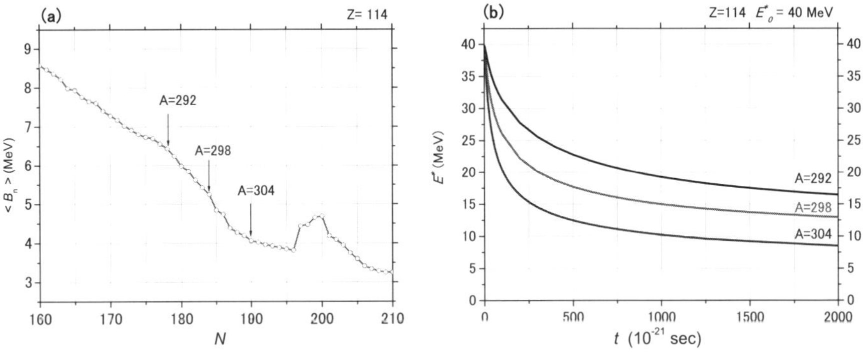

Fig. 3. (a) Neutron separation energies averaged over four successive neutron emissions $\langle B_n \rangle$ for the isotopes with $Z = 114$.[18] (b) Cooling curves of $A = 292, 298$ and 304 with $Z = 114$ at the initial excitation energy $E_0^* = 40$ MeV, that are derived by the statistical code SIMDEC.[14,16]

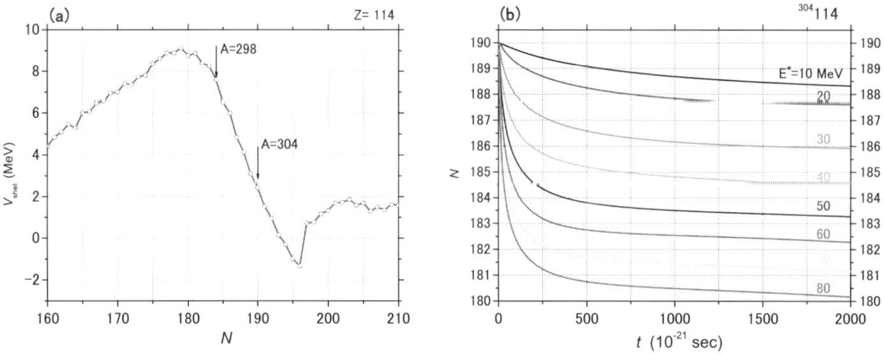

Fig. 4. (a) Shell correction energies V_{shell} of isotopes with $Z = 114$.[18] (b) Time evolution of the neutron number for the de-exciting nucleus $^{304}114_{190}$ for eight different initial excitation energies.

190) nucleus is smaller than that of the $A = 298$ ($N = 184$) nucleus. However, in the de-exciting process of the nucleus with $A = 304$ ($N = 190$),

the neutron number approaches $N = 184$ because of neutron emission. In Fig. 4(b), the time evolution of the neutron number for the compound nucleus $^{304}114_{190}$ is shown for eight different initial excitation energies, as calculated by SIMDEC.[14,16] At a high initial excitation energy, the neutron number of the compound nucleus quickly approaches $N \sim 184$, which is that of a neutron closed shell. This means the rapid appearance of a large fission barrier.

The compound nucleus with $^{304}114$ has two advantages to obtaining a high survival probability. First, because of small neutron separation energy and rapid cooling, the shell correction energy recovers quickly. Secondly, because of neutron emissions, the number of neutrons in the nucleus approaches that in the double closed shell, and a large shell correction energy is attained.

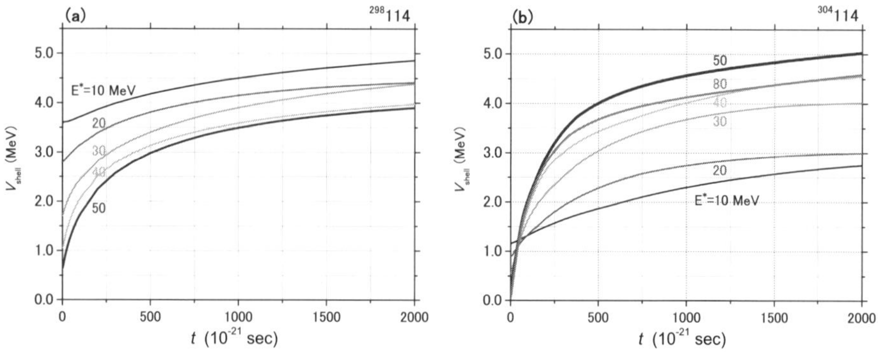

Fig. 5. Time evolution of the fission barrier height B_f for the de-exciting nuclei (a) $^{298}114$ and (b)$^{304}114$.

Generally, at a high excitation energy, the recovery of the shell correction energy is delayed. On the other hand, at a low excitation energy, the shell correction energy is established. Figure 5(a) shows the time evolution of the fission barrier height B_f for $^{298}114$. We can see that the restoration of shell correction energy is increasingly delayed with increasing excitation energy. Using the Smoluchowski equation, we calculate the survival probability in Fig. 6. With increasing excitation energy, the survival probability decreases drastically.

However, for $^{304}114$, the situation is opposite. At an excitation energy of 50 MeV, the fission barrier recovers faster than in the cases with lower excitation energies, as shown in Fig. 5(b). The reason is the double effects, that is to say, the rapid cooling and rapid approach to $N \sim 184$. The survival probability of $^{304}114$ is denoted in Fig. 6. It is very interesting that the excitation function of the survival probability has a flat region around $E^* = 20 \sim 50$ MeV. At $E^* = 50$ MeV, the survival probability of $^{304}114$ is three orders magnitude larger than that of $^{298}114$. For reference, the survival probability of $^{300}114$ is denoted in Fig. 6. These properties lead to a rather high evaporation reside cross section. As a more realistic model, we plan to take into account the emission of the charged particles from the compound nucleus.

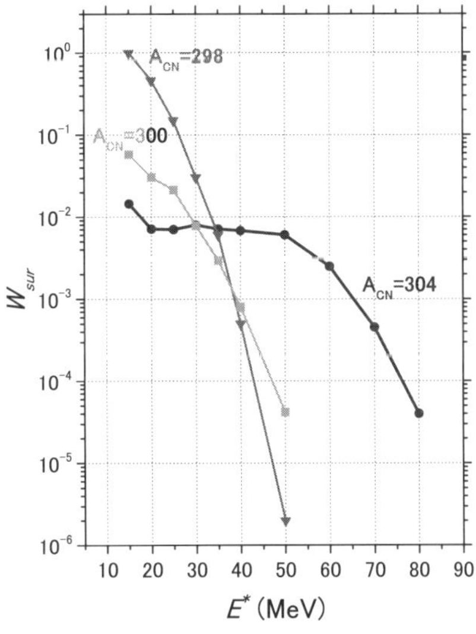

Fig. 6. Survival probabilities for $^{298}114$,^{300} 114 and $^{304}114$, which are calculated by the one-dimensional Smoluchowski equation.

Although the combinations of stable nuclei cannot yield such neutron-rich nuclei as $Z = 114$ and $N > 184$, we hope to make use of secondary beams in the future. We believe, the mechanism that we discussed here can inspire new experimental studies on the synthesis of superheavy elements.

Also, such a mechanism is very interesting and can be applied to any system that has the same properties, small neutron separation energy and slightly larger neutron number than the closed shell.

The author is grateful to Professor Yu. Ts. Oganessian, Professor M.G. Itkis, Professor V.I. Zagrebaev and Professor T. Wada for their helpful suggestions and valuable discussion throughout the present work. The authors thank Dr. S. Yamaji and his collaborators, who developed the calculation code for potential energy with two-center parameterization. This work has been in part supported by INTAS projects 03-01-6417.

References

1. W.D. Myers and W.J. Swiatecki, Nucl. Phys. **81** 1 (1966); A. Sobiczewski *et. al.*, Phys. Lett. **22** 500 (1966).
2. Yu.Ts. Oganessian et al., Nature **400** 242 (1999) ; Phys. Rev. Lett. **83** 3154 (1999); Phys. Rev. C **63** 011301(R) (2001); Phys. Rev. C **69** 021601(R) (2004).
3. S. Hofmann and G. Munzenberg, Rev. Mod. Phys. **72** 733 (2000); S. Hofmann et al., Eur. Phys. J. A **14** 147 (2002).
4. K. Morita et al., Nucl Phys, A **734** 101 (2004); Jap. Phys. Soc. J. **73** 1738 (2004); Journal of the Physical Society of Japan, **73** 2593 (2004).
5. Y. Aritomo and M. Ohta, Nucl. Phys. A **744** 3 (2004).
6. M.G. Itkis et al., Proc. of Fusion Dynamics at the Extremes (World Scientific, Singapore, 2001) p93.
7. L. Donadiile et al, Nucl. Phys. A **656** 259 (1999).
8. T. Materna et al., Nucl. Phys. **A734** 184 (2004); T. Materna et al., Prog. Theo. Phys. **154** 442 (2004).
9. J. Maruhn and W. Greiner, Z. Phys. **251** 431 (1972).
10. K. Sato, A. Iwamoto, K. Harada, S. Yamaji, and S. Yoshida, Z. Phys. A **288** 383 (1978).
11. P. Fröbrich, I.I. Gontchar and N.D. Mavlitov, Nucl. Phys. A **556** 281 (1993).
12. S. Suekane, A. Iwamoto, S. Yamaji and K. Harada, JAERI-memo, 5918 (1974).
13. A. Iwamoto, S. Yamaji, S. Suekane and K. Harada, Prog. Theor. Phys. **55** 115 (1976).
14. Y. Aritomo, T. Wada, M. Ohta and Y. Abe, Phys. Rev. C **59** 796 (1999).
15. J. Töke and W.J. Swiatecki, Nucl. Phys. A **372** 141 (1981).
16. M. Ohta, Y. Aritomo, T. Tokuda and Y. Abe, *Proc. of Tours Symp. on Nuclear Physics II* (World Scientific, Singapore, 1995) p.480.
17. A.V. Ignatyuk, G.N. Smirenkin and A.S. Tishin, Sov. J. Nucl. Phys. **21**, 255 (1975).
18. P. Möller, J.R. Nix, W.D. Myers and W.J. Swiatecki, Atomic Data and Nuclear Data Tables **59**, 185 (1995).

SYNTHESIS OF SUPERHEAVY NUCLEI IN ^{48}CA-INDUCED REACTIONS

YU.TS. OGANESSIAN, V.K. UTYONKOV, YU.V. LOBANOV, F.SH. ABDULLIN, A.N. POLYAKOV, R.N. SAGAIDAK, I.V. SHIROKOVSKY, YU.S. TSYGANOV, A.A. VOINOV, G.G. GULBEKIAN, S.L. BOGOMOLOV, B.N. GIKAL, A.N. MEZENTSEV, S. ILIEV, V.G. SUBBOTIN, A.M. SUKHOV, K. SUBOTIC, V.I. ZAGREBAEV, G.K. VOSTOKIN, AND M.G. ITKIS

Joint Institute for Nuclear Research, Dubna, Moscow reg. 141980, Russian Federation

K.J. MOODY, J.B. PATIN, D.A. SHAUGHNESSY, M.A. STOYER, N.J. STOYER, P.A. WILK, J.M. KENNEALLY, J.H. LANDRUM, J.F. WILD, AND R.W. LOUGHEED

University of California, Lawrence Livermore National Laboratory, Livermore, California 94551, USA

Thirty-four new nuclides with Z=104-116, 118 and N=161-177 have been synthesized in the complete-fusion reactions of ^{238}U, ^{237}Np, ^{242,244}Pu, ^{243}Am, ^{245,248}Cm, and ^{249}Cf targets with ^{48}Ca beams. The masses of evaporation residues were identified through measurements of the excitation functions of the xn-evaporation channels and from cross bombardments. The decay properties of the new nuclei agree with those of previously known heavy nuclei and with predictions from different theoretical models. A discussion of self-consistent interpretations of all observed decay chains originating from the parent isotopes ^{282,283}112, ^{282}113, $^{286\text{-}289}$114, ^{287,288}115, $^{290\text{-}293}$116, and ^{294}118 is presented. Decay energies and lifetimes of the neutron-rich superheavy nuclei as well as their production cross sections indicate a considerable increase in the stability of nuclei with an increasing number of neutrons, which agrees with the predictions of theoretical models concerning the decisive dependence of the structure and radioactive properties of superheavy elements on their proximity to the nuclear shells with N=184 and Z=114.

1. Introduction

The existence of a region of superheavy nuclei located beyond the domain of the heaviest known nuclei has been hypothesized for about 40 years. Calculations performed with different versions of the nuclear shell model predict a substantial enhancement of the stability of heavy nuclei when approaching the closed spherical shells at Z=114 and N=184, the next spherical shells predicted after ^{208}Pb. Superheavy nuclei that are close to the predicted magic neutron shell N=184 and are consequently relatively stable, can be synthesized in complete fusion reactions of target and projectile nuclei with significant neutron excess. In the reactions of the doubly magic ^{48}Ca projectile with isotopes of heavy actinide elements, e.g., ^{244}Pu or ^{248}Cm, the resulting compound nuclei should have excitation energies of about 30 MeV at the Coulomb barrier. Nuclear shell effects are still expected to persist in the excited nucleus, thus increasing the survival probability of the evaporation residues (ER), as compared to "hot fusion" reactions ($E^* \approx 45$-55 MeV), which were used for the synthesis of heavy isotopes of elements with atomic numbers Z=106-110. Additionally, the high mass asymmetry in the entrance channel should decrease the dynamic limitations on nuclear fusion that arise in more symmetrical "cold fusion" reactions.

In spite of the advantages of ^{48}Ca-induced reactions in comparison with hot or cold complete-fusion reactions, past attempts to synthesize new elements in the reactions of ^{48}Ca projectiles with actinide targets resulted only in upper limits on their production cross sections [1,2]. In view of the more recent experimental data on the production of the heaviest nuclides (see, e.g., [3-5]), it became obvious that the sensitivity level of the previous experiments was insufficient to detect superheavy nuclides. Our present experiments are designed to attempt the production of elements 112-116 and 118 in reactions of ^{233,238}U, ^{237}Np, ^{242,244}Pu, ^{243}Am, ^{245,248}Cm, and ^{249}Cf with ^{48}Ca at the picobarn cross-section level, thus exceeding the sensitivity of the previous experiments by at least two orders of magnitude.

According to predictions, the decay chains of superheavy nuclei that would be synthesized in ^{48}Ca-induced reactions should be terminated by spontaneous fission (SF) of previously unknown nuclides [6-8]. In addition, because of the lack of available target and projectile reaction combinations, these unknown descendant nuclei cannot be produced as primary reaction products. Thus, the method of genetic correlations to known nuclei for the identification of the parent nuclide can be applied in this region of nuclei only after an independent identification, such as the determination of the chemical properties of any one decay-chain member. In these experiments, we identified the masses of evaporation residues using the characteristic dependence of their production cross sections on the excitation energy of the compound nucleus (thus defining the number of emitted neutrons) and from cross bombardments, i.e., varying mass and/or atomic number of the projectile or target nuclei, which changes the relative yields of the xn-evaporation channels. Both of these methods were successfully used in previous experiments for the identification of unknown artificial nuclei (see [9] and Refs. therein), particularly those with short SF half-lives. Moreover, the identification of superheavy nuclei in this region is based on a comparison of experimental results with theoretical predictions and the systematics of experimental nuclear properties and reaction cross sections.

2. Experimental technique

The ^{48}Ca ion beam was accelerated by the U400 cyclotron at the Flerov Laboratory of Nuclear Reactions. The typical beam intensity at the target was 1.2 pμA. The beam energy was determined with a precision of 1 MeV by a time-of-flight technique. The 32-cm^2 rotating targets consisted of the enriched ($\geq$ 97.3%) isotopes of U to Cf deposited as oxides onto 1.5-μm Ti foils to thicknesses of about 0.34-0.40 mg cm^{-2}.

The ERs recoiling from the target were spatially separated in flight from ^{48}Ca beam ions, scattered particles and transfer-reaction products by the Dubna Gas-filled Recoil Separator. The transmission efficiency of the separator for Z=112 to 118 nuclei was estimated to be about 35-40%. Evaporation recoils passed through a time-of-flight system and were implanted in a 4×12-cm^2 semiconductor detector array with 12 vertical position-sensitive strips, located at the separator's focal plane. This detector was surrounded by eight 4×4-cm^2 side detectors without position sensitivity, forming a box of detectors open from the beam side. The position-averaged detection efficiency for full-energy α particles from the decay of implanted nuclei was 87%. The detection system was tested by registering the recoil nuclei and decays (α or SF) of known isotopes of No and Th, as well as their descendants, produced in the reactions ^{206}Pb(^{48}Ca,xn) and ^{nat}Yb(^{48}Ca,xn). Fission fragments from ^{252}No implants produced in the ^{206}Pb+^{48}Ca reaction were used for an approximate fission-energy calibration.

For detection of sequential decays of synthesized nuclides in the absence of beam-associated background, the beam was switched off automatically after a recoil was detected with an implantation energy expected for complete-fusion ERs, followed by an α-like signal with an energy expected for α decays of the parent and sometimes the daughter nuclei. Both ER and α-particle signals were required to be detected within a narrow position window in the same strip during an appropriate time interval estimated for the decays of heavy nuclei. Thus, the decays of the daughter nuclides were observed under very low-background conditions. The probability that all of the observed events are due to random detector background is very low, even for decay chains detected in beam, and negligible for those decay chains registered during the beam-off periods.

3. Experimental results and discussion

The most neutron-rich nuclei with the magic proton number 114 that is predicted by macroscopic-microscopic (MM) theory can be produced in the fusion reaction ^{244}Pu+^{48}Ca. During the years 1998-2003, we studied this reaction at different projectile energies [10]. The decay properties of nuclei observed in these experiments and their corresponding excitation functions are shown in Figs. 1 and 2, respectively. At the three lowest ^{48}Ca energies above the Coulomb barrier [11], we synthesized an isotope with E_α=9.82 MeV and $T_{1/2}$=2.6 s that underwent two consecutive α decays followed by SF. At the three higher projectile energies, the neighboring isotope was observed, with E_α=9.95 MeV and $T_{1/2}$=0.80 s; its α decay was followed by SF of the daughter nuclide with $T_{1/2}$=97 ms. Finally, a third isotope with an even greater α-particle energy and a correspondingly lower half-life was produced at the highest bombarding energy.

As seen in Fig. 2, the measured excitation functions for the three isotopes are in agreement with empirical expectations and calculations [12] for the complete-fusion reaction ^{244}Pu+^{48}Ca followed by evaporation of 3, 4 and 5 neutrons. It is reasonable to assume that the first of the aforementioned isotopes was produced in the [*x*]*n*-evaporation channel leading to ^{289}114 and the other two isotopes were produced in the [*x*+1]*n* and [*x*+2]*n* channels (^{288}114 and ^{287}114).

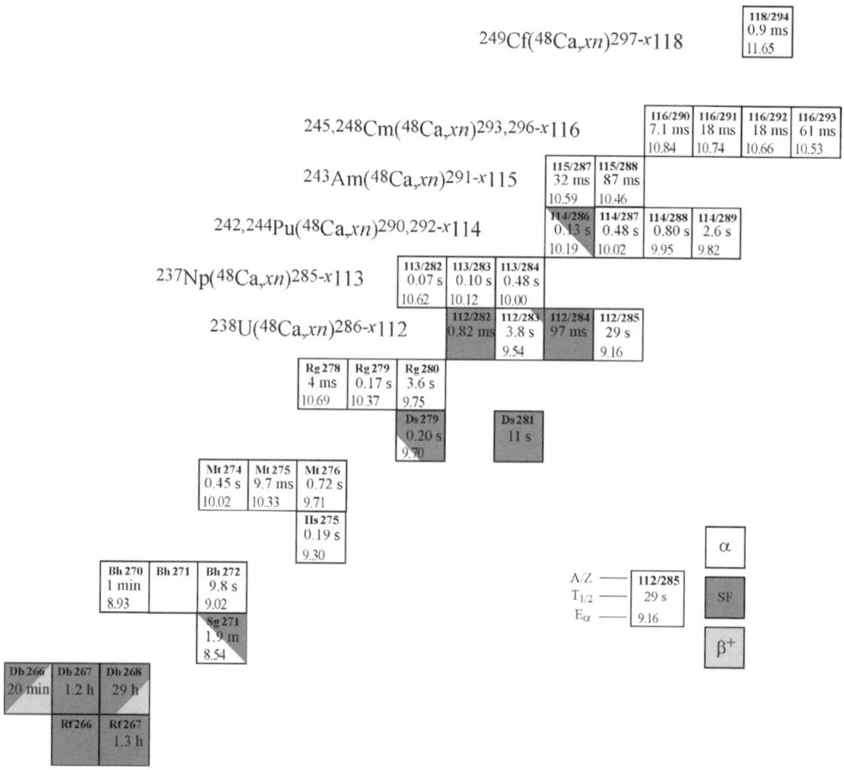

Figure 1. Half-lives, α-particle energies and decay modes of nuclei produced in corresponding reactions of ^{48}Ca ions with actinide targets.

A more solid mass identification is possible if the mass number of the target nucleus is varied. Thus, if the suggested mass assignment is correct, the isotopes $^{288}114$ and $^{287}114$ could be observed in the $2n$ and $3n$ channels of the reaction ^{242}Pu+^{48}Ca, and another even-even isotope, $^{286}114$, with a higher α-particle energy and lower lifetime could be produced via the $4n$-evaporation. Indeed, in the experiments aimed at the cross section measurement of the reaction ^{242}Pu(^{48}Ca,xn)$^{290-x}114$, we observed $^{288}114$ at the lowest ^{48}Ca energy and $^{287}114$ at the three lowest energies. The α-decay of the new isotope, $^{286}114$, which undergoes SF and α decay with equal probability (E_α=10.19 MeV, $T_{1/2}$=0.13 s), was followed by a SF nuclide with $T_{1/2}$=0.82 ms. This nuclide was produced at the two highest energies [13]. Thus, following the previous considerations, these isotopes should be the products of the $[x-1]n$, $[x]n$ and $[x+1]n$-reaction channels.

Considering the decay properties of the observed nuclei, particularly their decay modes and the number of observed α decays before a SF is encountered, one can conclude that the first (highest mass) and third of the four consecutive element-114 isotopes should possess an odd number of neutrons while the second and fourth ones should be even-N isotopes. The unpaired neutron in the odd-N isotopes increases the SF partial half life by 3-5 orders of magnitude; therefore, the value x can only be equal to 1, 3, 5 etc.

Together with cross-section measurements, another approach for the mass and atomic-number identification of unknown nuclei is the method of cross bombardments, which was widely applied in previous experiments [9]. In our

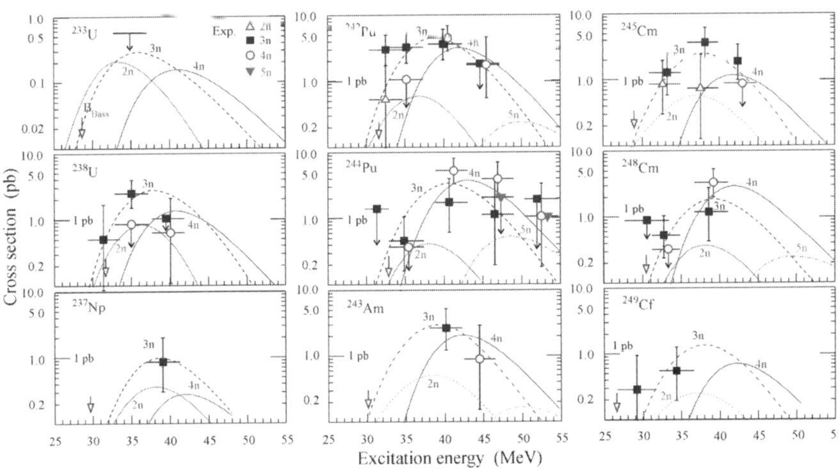

Figure 2. Excitation functions for the $2n$ to $5n$ evaporation channels from different complete-fusion reactions. The Bass barrier [11] is shown by an open arrow. Lines show the results of calculations of xn-evaporation channel [12].

case this means the production of the same isotopes of element 114 as the daughter nuclei following the α decays of heavier parent nuclei with Z=116. This method was first used in our experiment aimed at the synthesis of superheavy nuclei with Z=116 in the complete-fusion reaction ^{248}Cm+^{48}Ca [13,14] in which two isotopes were observed. All the decays following the first α emission agree well with the decay chains of the [x]n and [x+1]n channels, ^{289}114 and ^{288}114, previously observed in the ^{244}Pu+^{48}Ca reaction [10]. Thus, it was reasonable to assign the observed decays to the nuclides ^{293}116 and ^{292}116, produced via evaporation of the same number of neutrons in the reaction ^{248}Cm+^{48}Ca.

Two lighter neighboring isotopes of element 116 were produced in the reaction ^{245}Cm+^{48}Ca at three ^{48}Ca energies [10,15]. As in the previous case, the decay chains following the first α particle of the isotope observed at the two lowest projectile energies agree with those observed for the ^{287}114 parent nuclide from the reactions ^{244}Pu+^{48}Ca at the highest ^{48}Ca energy and ^{242}Pu+^{48}Ca at the three lowest energies. The decay properties of the descendant nuclei of the lighter isotope observed at three energies agree with those of ^{286}114 produced in the reaction ^{242}Pu+^{48}Ca at the two highest energies. Moreover, this lighter isotope was also observed in the reaction ^{249}Cf+^{48}Ca after the α decay of the parent element-118 nucleus [15]. Note that the granddaughter nuclei from the parent nuclides of the reaction ^{245}Cm+^{48}Ca were also produced in the direct reaction ^{238}U+^{48}Ca, ^{282}112 and ^{283}112 [13]. Therefore, the isotopes observed in the reaction ^{245}Cm+^{48}Ca should be the products of the [x-1]n and [x]n channels.

The results from all of these experiments allow us to consider a possible value for x more definitely. Assuming x=1, we would conclude that the 0n channel was observed in the reaction ^{245}Cm+^{48}Ca at excitation energies E^*=33-43 MeV. However, the γ channel was not observed even in cold fusion reactions at much lower excitation energies of the compound nuclei (see Ref. [3,4] and Refs. therein). The value x=5 would result in the conclusion that the parent isotopes of the reaction ^{244}Pu+^{48}Ca are the products of the 5-7n channels, which is unreasonable based both on the compound nucleus excitation energies and the competition with de-excitation by fission. Thus, in our consideration of the xn-evaporation channels, we conclude that the only reasonable value for x is x=3.

Now one can discuss the possibility of other reaction channels accompanied by evaporation or emission of light charged particles. The reactions with odd-Z target nuclei are important for consideration of the pxn channel. In the reaction ^{243}Am+^{48}Ca, two different isotopes ^{287,288}115 were synthesized [16]. In both cases, the parent isotopes underwent five consecutive α decays followed by SF. The isotope ^{282}113 produced recently in the reaction ^{237}Np+^{48}Ca (two decay chains) demonstrates comparable behavior. Comparison of the decay properties of nuclei produced in the reactions with even-Z and odd-Z target nuclei indicates that all synthesized nuclides cannot originate from the pxn channel because hindrance against SF of nuclei possessing an odd number of protons increases their SF stability by orders of magnitude. Indeed, the three isotopes of element 113, $^{282-284}113$, undergo four consecutive α decays whereas the neighboring isotopes, even-even $^{286}114$ and $^{282,284}112$ or even-odd ^{279,281}Ds, decay by SF.

Comparison of the decay properties of previously known heavy nuclides and those of the new superheavy nuclei also supports their assignment to the products of the xn-reaction channels. This can be seen in Figure 3 where the dependence of T_α on Q_α for known even-even nuclei with Z=100-110 are shown along with the data for nuclides produced in ^{48}Ca-induced reactions. The lines are drawn in accordance with the formula by Viola and Seaborg with parameters fit to the T_α values of 65 even-even nuclei with Z>82 and N>126 [13]. The measured $T_{1/2}$ *vs.* Q_α values for all superheavy nuclei with Z=112-118, including $^{283,284}113$ and $^{287,288}115$, are in agreement with values expected for *allowed* α decays of isotopes of the corresponding elements. Thus, the assumption that these nuclei were produced in reactions accompanied by emission of charged particles (αxn etc.) would demand a change of assignment of all 15 $T_{1/2}$ *vs.* Q_α values to lower Z-values. The correlation between the experimental data and the empirical systematics for the heaviest nuclei with Z=112-118 indicates rather low hindrance factors, if any, for α decay. For the lighter isotopes of elements 106-113, the difference between measured and calculated T_α values results in hindrance factors of 3-10 which is consistent with values that can be extracted for the deformed nuclei located near the neutron shell N=162 (see, e.g., [3,4]).

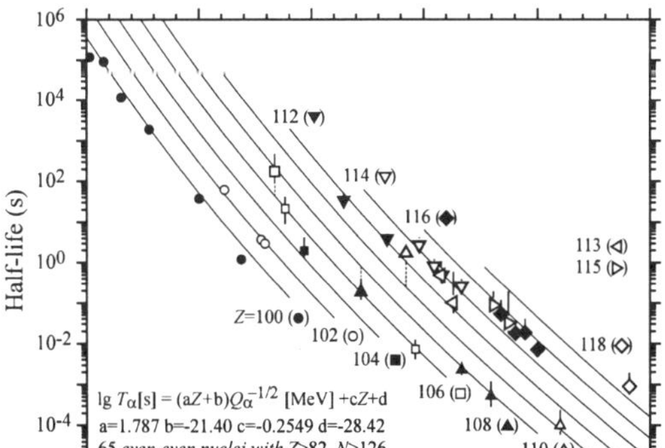

Figure 3. The dependence of T_α *vs.* Q_α for known heavy even-even isotopes with Z=100-110 and nuclei shown in Fig. 1. The lines are drawn according to the formula by Viola and Seaborg. The experimental data are taken from Refs. [3,10,13-17] and Refs. therein.

One can postulate that in this region of nuclei, a noticeable transition from spherical to deformed shapes occurs at Z=111-112, resulting in the complication of the level structures of these nuclei and in an increased probability of α transitions going through excited states. This assumption is in agreement with MM calculations [6,7]. The deformation parameter β_2 was calculated to be 0.072-0.138 for $^{288}115$, $^{284}113$ or $^{291}116$, $^{287}114$, and $^{283}112$ (see Fig. 4). As the decay chain recedes from the predicted shell closure at Z=114, the deformation parameter β_2 increases to 0.197-0.247 for the descendant nuclei ^{280}Rg-^{272}Bh or ^{279}Ds-^{271}Sg.

The experimental α-decay energies of the newly synthesized isotopes as well as previously known nuclei are plotted in Fig. 4. The α-decay energies attributed to isotopes of Mt and Bh coincide well with theoretical values [6,7]. The same is true for the last nuclei in the decay chain $^{275}\mathrm{Hs} \xrightarrow{\alpha} {}^{271}\mathrm{Sg} \xrightarrow{\alpha}$.

For the isotopes $^{278\text{-}280}$Rg and $^{282\text{-}284}113$ the difference between theoretical and experimental Q_α values is 0.5-0.9 MeV. Part of this energy can be accounted for by γ-ray emission from excited levels populated during α decay. For the even-Z nuclei as well, the agreement between theory and experiment becomes somewhat worse as one moves from the deformed nuclei in the vicinity of neutron shells N=152 and N=162 to the more neutron-rich nuclides with $N\geq169$. In this region, experimentally measured values of Q_α are less than the values calculated from the model by ≤0.5 MeV. While the predicted Q_α values for the heaviest nuclei observed in our experiments are systematically larger than the experimental data as a whole, the trends of the predictions are in good agreement for the 27 nuclides with Z=106-118 and N=163-177, especially considering that the theoretical predictions of the MM model match the experimental data over a broad, previously unexplored region of nuclides. One should note that the predictions of other models for even-Z and odd-Z nuclei within the Skyrme-Hartree-Fock-Bogoliubov and the relativistic mean field theories also compare

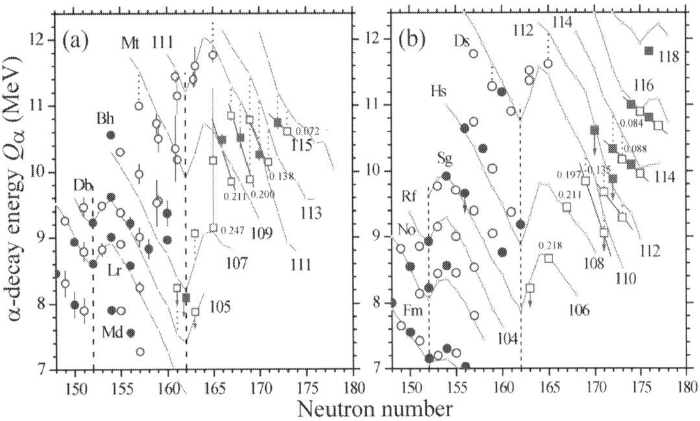

Figure 4. (a) α-decay energy *vs.* neutron number for isotopes of odd-Z elements (solid symbols – odd-even isotopes, open symbols – odd-odd isotopes [17], squares – data from [10,13-16] and present work). Solid lines show the theoretical Q_α values [6,7] for odd Z=103-115 elements. (b) The same for even-Z elements.

well with the experimental results (see [13-16] and Refs. therein). These models predict the same spherical neutron shell at N=184 but different proton shells, Z=114 (MM) and Z=120, 124 or 126 (SHFB, RMF), yet all describe the experimental data equally well.

All of the decay chains presented in Fig. 1 end in spontaneous fission. In cases when the detector array registered both fission fragments, their sum energy E_{tot} could be used to evaluate the total kinetic energy (TKE) of the fission fragments. The value of TKE for all of the SF isotopes obtained from the ^{48}Ca-induced reactions is plotted in Fig. 5, together with the previously known data for isotopes with $Z \geq 96$ [18]. One can observe that, with the transition $Z \geq 110$, the TKE increases with increasing Z in agreement with the previously established dependence of TKE vs. $Z^2/A^{1/3}$ typical for the asymmetric fission of lighter nuclei. We speculate that this is due to the influence of the spherical shells at Z=50 and N=82 on the formation of the light fission fragment in the scission of superheavy nuclei. In the spontaneous fission of the lighter nuclides ^{267,268}Db, ^{271}Sg, and ^{267}Rf, the same effect apparently results in symmetric fission with high kinetic energy, characteristic of the close scission configuration that takes place in the symmetric fission of the heavy isotopes of Fm, Md, and No.

The calculations of SF half-lives are more complicated and correspondingly far less definite than those for α decay. Nevertheless, reasonable agreement between experiment and theory can be seen for terminal spontaneously fissioning isotopes in the decay chains of superheavy nuclei (see Fig. 6). The measured partial SF half-lives for three even-even isotopes ^{282,284}112 and ^{286}114 are lower than calculated values [8] by less than two orders of magnitude. The same can be found for the isotopes ^{267}Rf, ^{271}Sg, and ^{279,281}Ds if one assumes some hindrance factor, say 3 orders of magnitude, for SF of even-odd nuclei.

Therefore, we conclude that the superheavy nuclei produced in the ^{48}Ca-induced reactions [10,13-16]

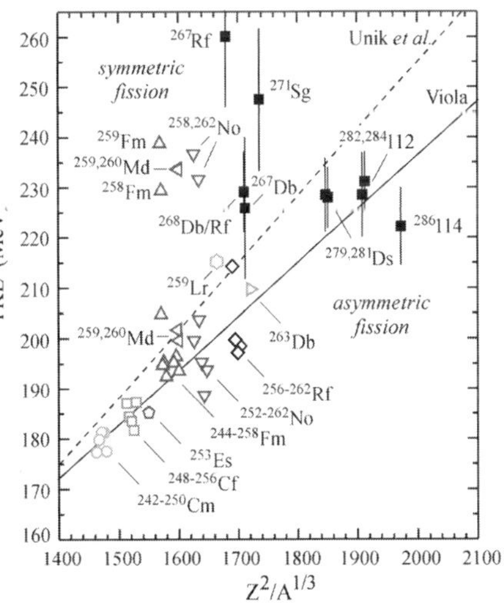

Figure 5. Experimental values of TKE *vs.* $Z^2/A^{1/3}$ (previously measured data from [18] and Refs. therein) – open symbols; experimental data from the present work - solid squares. The lines are the linear fit to the data, excluding the mass-symmetric fissioners [19].

are the result of complete fusion followed by evaporation of two to five neutrons from the excited compound nuclei.

Independent confirmation of these results was recently obtained in experiments aimed at the determination of the chemical properties of isotopes of Db and element 112. This method allows the identification of the atomic number of a nucleus, and was used in the first identification and characterization of many of the artificial elements heavier than uranium [20].

According to experimental data from the gas-filled separator, the 3*n*-evaporation channel of the reaction ^{243}Am+^{48}Ca resulted in an isotope of element 115, ^{288}115, which underwent five consecutive α decays followed by a SF nuclide with a half-life of T_{SF}=16 h. This long lifetime allowed us to perform chemical separations in order to verify the atomic number of the final nuclide in the ^{288}115 decay chain, ^{268}Db (*Z*=105). In this chemistry experiment, using the same production *reaction*, a SF activity was observed with the identical *cross section*, *half-life*, *decay mode* (15 SF events) and *total kinetic energy* [16] as the ^{268}Db produced in the experiment with the gas-filled separator. This nuclide was found to be chemically consistent with the 4th or 5th group of the Periodic Table. Using a more refined chemical procedure that allowed not only a +4/+5 group separation, but also an intra-group separation, five more SF events were observed during a subsequent experiment with similar decay properties in the Ta-like fraction [21]. Since all of the consecutive α decays and the SF are strongly correlated with each other and the order of occurrence of the nuclei in the decay chains has been determined, the identification of the atomic number of the final nucleus in the chain originating from ^{288}115 (^{268}Db) independently supports the synthesis of the previously unknown elements 115 and 113.

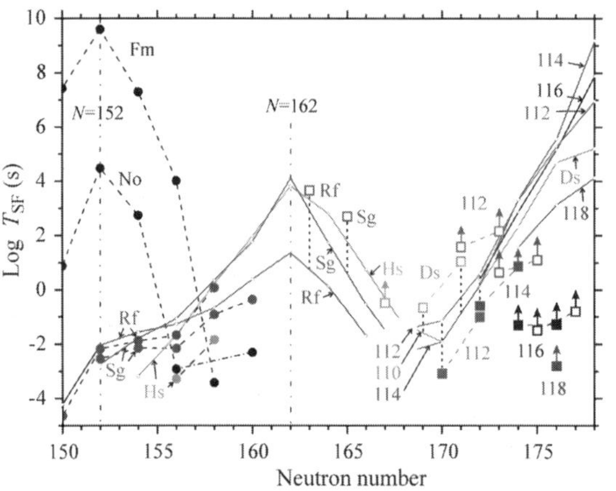

Figure 6. Common logarithm of partial spontaneous fission half-life *vs.* neutron number for isotopes of even-*Z* elements with *Z*≥100 (circles – even-even isotopes [17]). Data at *N*>162 are from [10,13-15] and the present work (open squares – odd-*N* isotopes; solid squares – even-*N* isotopes). Solid lines show the theoretical T_{SF} values [8] for even-even *Z*=104-118 isotopes.

In another experiment, the decay properties of the previously observed isotopes

$^{283}112 \xrightarrow{\alpha} {}^{279}Ds \xrightarrow{SF}$ [10,13,15] were confirmed and the Hg-like behavior of element 112 was established [22]. The nuclei were produced in the reaction $^{242}Pu(^{48}Ca,3n)^{287}114$ and were thermalized in a He/Ar gas volume where the parent isotope $^{287}114$ ($T_{1/2}$=0.48 s) decayed to $^{283}112$ ($T_{1/2}$=3.8 s), which was subsequently transported via a gas-jet to a set of detector pairs covered on one side by a Au layer. The detector assembly was operated under a temperature gradient. Two decay chains of $^{283}112$ were observed. The decay properties of the parent and daughter nuclei were measured and found to be in agreement with those determined in our previous experiments. The position of the two atoms on the detector assembly points to Hg-like condensation behavior of element 112. All of the even-Z nuclei, including $^{283}112$ and ^{279}Ds, were produced in various cross bombardments [10,13-15]. Therefore, confirmation of the decay properties of $^{283}112$ and the determination of the chemical properties of element 112 simultaneously signify the independent identification of all of the other even-Z nuclei observed in ^{48}Ca-induced reactions.

4. Conclusions

The existence of enhanced stability in the region of the superheavy nuclei has been validated through recent experiments. Decay energies and lifetimes of 34 new nuclides with Z=104-118 and N=161-177 that have been synthesized in the complete-fusion reactions of ^{238}U, ^{237}Np, $^{242,244}Pu$, ^{243}Am, $^{245,248}Cm$, and ^{249}Cf targets with ^{48}Ca beams indicate a considerable increase of the stability of superheavy nuclei with an increasing number of neutrons. The comparison of the decay properties of the isotopes $^{278}113$ and $^{277}112$ produced in the cold fusion reactions $^{209}Bi,^{208}Pb$ ($^{70}Zn,1n$) [3,4] with those of $^{284}113$ and $^{285}112$ reveals a decrease in α-decay energy by 1.6 and 2.1 MeV, and a corresponding increase in half-lives by a factors of 250 and 4×10^4, respectively, due to their closer proximity to the region of spherical nuclei. The isotopes ^{274}Rg ($T_{1/2}$=3.1 ms) [4] and ^{280}Rg ($T_{1/2}$=3.6 s) [16], or ^{273}Ds ($T_{1/2}$=0.2 ms) [3,4] and ^{281}Ds ($T_{1/2}$=11 s) [13] demonstrate comparable behavior. As a whole, the results of these experiments agree with the predictions of theoretical models concerning the properties of superheavy nuclei in the vicinity of closed nuclear shells.

Acknowledgments

This work was performed with the support of the Russian Ministry of Atomic Energy and grant of RFBR No. 04-02-17186. Much of the support for the LLNL authors was provided through the U.S. DOE under Contract No. W-7405-Eng-48. These studies were performed in the framework of the Russian

Federation/U.S. Joint Coordinating Committee for Research on Fundamental Properties of Matter.

References

1. G.N. Flerov and G.M. Ter-Akopian, *Rep. Prog. Phys.* **46**, 817 (1983).
2. P. Armbruster *et al.*, *Phys. Rev. Lett.* **54**, 406 (1985).
3. S. Hofmann and G. Münzenberg, *Rev. Mod. Phys.* **72**, 733 (2000).
4. K. Morita *et al.*, *J. Phys. Soc. Jpn.* **73**, 2593 (2004); *Proceedings of the International Symposium on Exotic Nuclei (EXON2004), Peterhof, Russia, 2004*, (World Scientific, Singapore, 2005), p. 188.
5. Yu.A. Lazarev *et al.*, *Phys. Rev. Lett.* **75**, 1903 (1995).
6. I. Muntian, Z. Patyk and A. Sobiczewski, *Phys. At. Nucl.* **66**, 1015 (2003).
7. I. Muntian *et al.*, *Acta Phys. Pol.* B **34**, 2073 (2003).
8. R. Smolańczuk, J. Skalski and A. Sobiczewski, *Phys. Rev.* C **52**, 1871 (1995); R. Smolańczuk, *Phys. Rev.* C **56**, 812 (1997).
9. R.C. Barber *et al.*, *Prog. Part. Nucl. Phys.* **29**, 453 (1992).
10. Yu.Ts. Oganessian *et al.*, *Phys. Rev.* C **62**, 041604(R) (2000); C **69**, 054607 (2004).
11. R. Bass, *Proceedings of the Symposium on Deep Inelastic and Fusion Reactions with Heavy Ions, West Berlin, 1979*, Lecture Notes in Physics, Vol. 117 (Springer-Verlag, Berlin, 1980), p. 281.
12. V.I. Zagrebaev, M.G. Itkis and Yu.Ts. Oganessian, *Phys. At. Nucl.* **66**, 1033 (2003); V.I. Zagrebaev, *Nucl. Phys.* **A734**, 164c (2004).
13. Yu.Ts. Oganessian *et al.*, *Phys. Rev.* C **70**, 064609 (2004).
14. Yu.Ts. Oganessian *et al.*, *Phys. Rev.* C **63**, 011301(R) (2001).
15. Yu.Ts. Oganessian *et al.*, *Phys. Rev.* C **74**, 044602 (2006).
16. Yu.Ts. Oganessian *et al.*, *Phys. Rev.* C **69**, 021601(R) (2004); C **72**, 034611 (2005); S.N. Dmitriev *et al.*, *Mendeleev Commun.* 1 (2005).
17. Evaluated Nuclear Structure Data File (ENSDF), Experimental Unevaluated Nuclear Data List (XUNDL). http://www.nndc.bnl.gov/ensdf.
18. D.C. Hoffman and M.R. Lane, *Radiochim. Acta* **70/71**, 135 (1995).
19. V.E. Viola, Jr., *Nucl. Data Tables* A **1**, 391 (1996); J.P. Unik *et al.*, *Proceedings of the Third International Atomic Energy Symposium on the Physics and Chemistry of Fission, Rochester, 1973,* (IAEA, Vienna, 1974) Vol. II, p. 19.
20. E.K. Hyde, I. Perlman and G.T. Seaborg, *The Nuclear Properties of the Heavy Elements, Detailed Radioactive Properties* (Prentice-Hall, Englewood Cliffs, New Jersey, 1964).
21. N.J. Stoyer *et al.*, *Proceedings of the IX International Conference on Nucleus-Nucleus Collisions (NN2006), Brazil, 2006,* to be published.
22. R. Eichler *et al.*, in Ref. [21], to be published.

Production of neutron-rich nuclei in the nucleus-nucleus collisions around the Fermi energy

M. Veselsky

Institute of Physics, Slovak Academy of Sciences, Bratislava

An overview of the recent progress in production of neutron-rich nuclei in the nucleus-nucleus collisions around the Fermi energy is presented and the possibilities to produce the very neutron-rich nuclei in the region of mid-heavy to heavy nuclei is examined. Possible scenarios for the new generation of rare nuclear beam facilities such as Eurisol are discussed. Isoscaling is investigated as a possible tool to predict the production rates of exotic species in reactions induced by both stable and radioactive beams.

Keywords: Neutron-rich nuclei; nucleus-nucleus collisions; Fermi-energy domain

1. Introduction

Nucleus-nucleus collisions in the Fermi energy domain exhibit a large variety of contributing reaction mechanisms and reaction products (see e.g. ref. [1]) and offer the principal possibility to produce mid-heavy to heavy neutron-rich nuclei in very peripheral collisions. In the reactions of massive heavy ions such as ^{124}Sn+^{124}Sn[2] and ^{86}Kr+^{124}Sn,[3] an enhancement was observed over the yields expected in cold fragmentation which is at present the method of choice to produce neutron-rich nuclei. In this case the neutron-rich nuclei are produced in damped symmetric nucleus-nucleus collisions with intense nucleon exchange leading to the large width of isotopic distributions. Further enhancement of yields of n-rich nuclei was observed in the reaction ^{86}Kr+^{64}Ni[4] in the very peripheral collisions, thus pointing to the possible importance of neutron and proton density profiles at the projectile and target surfaces. In this contribution, a review of various aspects of the nucleus-nucleus collisions in the Fermi energy domain will be presented, specifically concerning the possibilities to produce neutron-rich nuclei in both peripheral and central collisions, the scenarios for the use of such collisions at the new generation of rare nuclear beam (RNB) facilities, and the role of isoscaling as a possible prediction tool for the yields of very neutron-rich nuclei in reactions with unstable beams.

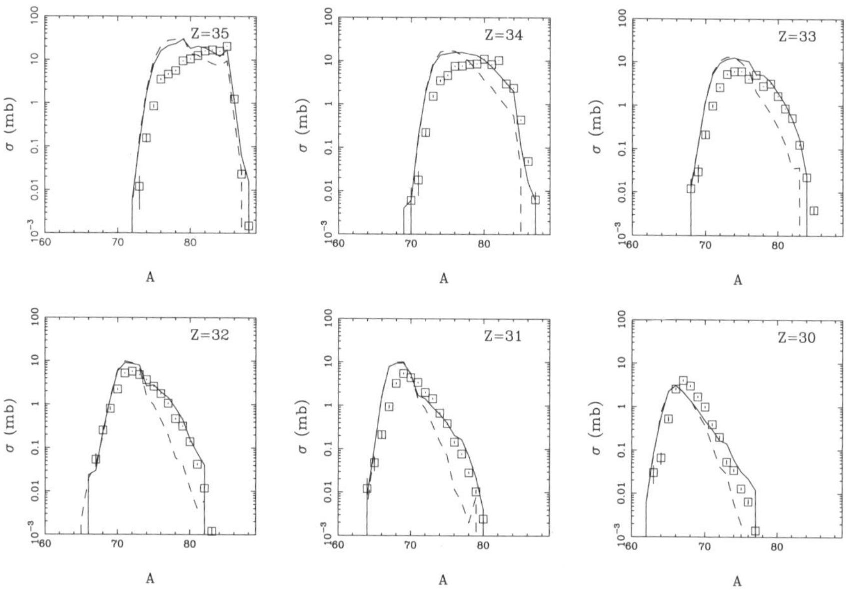

Fig. 1. Experimental mass distributions (symbols) of elements with Z = 30 - 35 observed in the reaction ^{86}Kr+^{64}Ni at 25 AMeV[4] compared to the results of the modified and standard DIT calculations (solid and dashed line, respectively) combined with the de-excitation code SMM.[5]

2. Effect of nuclear periphery on nucleon transfer in peripheral collisions

A comparison of experimental heavy residue cross sections from the reactions ^{86}Kr+^{64}Ni,^{112,124}Sn at projectile energy 25 AMeV[4] with the model of deep-inelastic transfer (DIT)[6] is carried out in ref. [7], where the model of deep-inelastic transfer was supplemented with a phenomenological correction introducing the effect of shell structure on nuclear periphery. A modified expression for nucleon transfer probabilities is used at non-overlapping projectile-target configurations, thus introducing a dependence on isospin asymmetry at the nuclear periphery. The experimental yields of neutron-rich nuclei close to the projectile are reproduced better and the trend deviating from the bulk isospin equilibration is explained. For the neutron-rich products further from the projectile, originating from hot quasiprojectiles, the statistical multifragmentation model reproduces the mass distributions better than the model of sequential binary decay. In the reaction with proton-rich target ^{112}Sn the nucleon exchange appears to depend on isospin asymmetry of nuclear periphery only when the surface separation is larger than 0.8 fm due to the stronger Coulomb interaction at more compact dinuclear configuration.

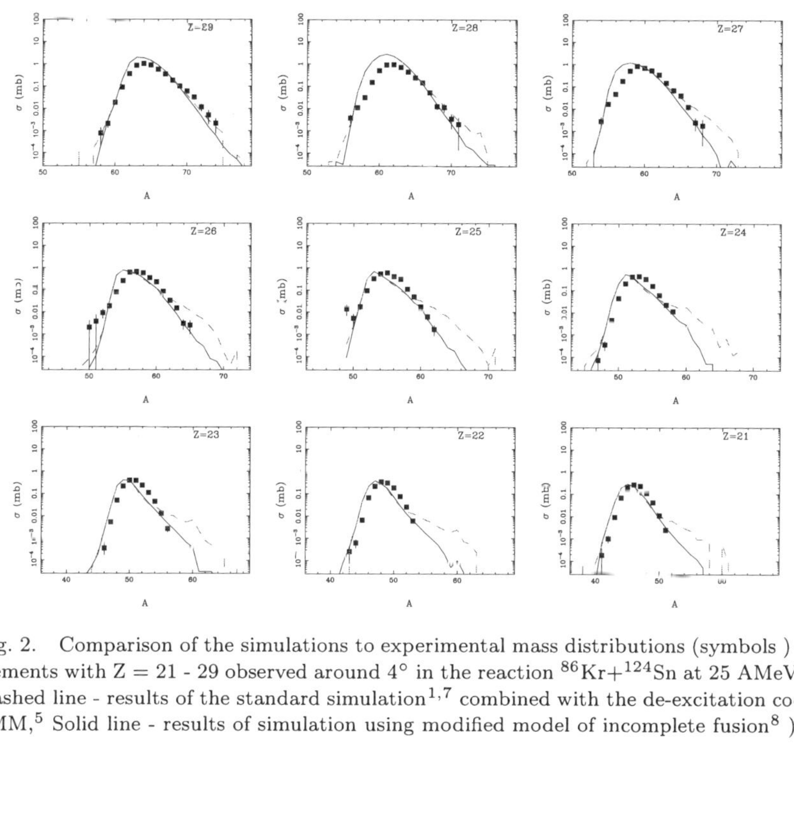

Fig. 2. Comparison of the simulations to experimental mass distributions (symbols) of elements with Z = 21 - 29 observed around 4° in the reaction ^{86}Kr+^{124}Sn at 25 AMeV.[3] Dashed line - results of the standard simulation[1,7] combined with the de-excitation code SMM,[5] Solid line - results of simulation using modified model of incomplete fusion[8]).

3. Production of cold fragments in nucleus-nucleus collisions in the Fermi-energy domain

The reaction mechanism of the nucleus-nucleus collisions at projectile energies around the Fermi energy was investigated[8] with emphasis on the production of fragmentation-like residues. The results of simulations were compared to experimental mass distributions of elements with Z = 21 - 29 observed around 4° in the reaction ^{86}Kr+^{124,112}Sn at 25 AMeV. The model of incomplete fusion[1] was modified and a component of excitation energy of the cold fragment dependent on isospin asymmetry was introduced. The modifications in the model of incomplete fusion appear consistent with both overall model framework and available experimental data (see Figs. 2, 3).

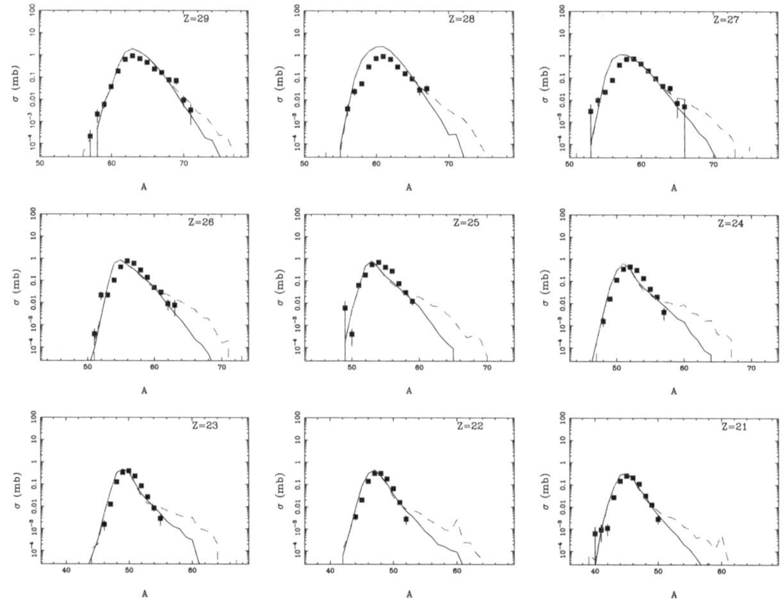

Fig. 3. Comparison of the simulations[8] to experimental mass distributions (symbols) of elements with Z = 21 - 29 observed around 4° in the reaction ^{86}Kr+^{112}Sn at 25 AMeV.[3] Solid, dashed lines - as in Fig. 2.

4. EURISOL

Projects for future secondary- beam facilities aim at exploiting the very technological limits of the production of rare nuclide beams. Such a goal requires careful work on the optimization of the production techniques. That is an interdisciplinary task, as it involves the fields of accelerator technology, nuclear reactions, extraction techniques, target handling and others. EURISOL[9] is the European project for constructing a secondary-beam facility based on the ISOL approach, which should provide the highest beam intensities and which will give access to the most exotic nuclides within the technological limits. Guided by the experience with the long-term operation of the ISOLDE facility, a 1 GeV proton beam has been chosen as the baseline option for the driver accelerator. Using different target material, this option allows for producing a large number of isotopes of many elements. However, this option might not be the optimum solution for all nuclides of interest. The report[10] investigates the benefit of extended capabilities of the driver accelerator is considered in connection with a quantitative discussion of nuclear-reaction aspects and the technical limitations of the ISOL method.

A number of approaches for the production of radioactive beams have been examined[10] with regard to their potential to be used as possible additional options for the EURISOL project in order to enhance the production of specific nuclear species with respect to the baseline 1-GeV proton case. The spallation of suitable target material by 1 GeV protons and the fission induced by secondary neutrons in a uranium target provide overall high intensities for secondary beams almost all over the chart of the nuclides. Still, there are cases where the 1 GeV proton beam together with the restrictions of the ISOL method in view of suitable target material do not allow optimizing the reaction parameters. In addition, low extraction efficiencies for certain elements let gaps in available beams. Some of these problems can be overcome by providing heavier projectiles and/or higher beam energies: A 2 GeV 3He2+ beam, introducing more energy into the reaction, and the fragmentation of heavy projectiles, overcoming the limited choice of target materials, allow bridging gaps in the nuclide production. The higher energy introduced also enhances the production of neutron-rich intermediate-mass fragments emitted from actinide targets. In addition, the characteristics of specific nuclear reactions can be exploited to obtain a benefit in the production of nuclei in specific regions of the chart of the nuclides. Thus, deuterons can be used to produce secondary neutrons of higher energy to extend the production of fission fragments, or heavy-ion reactions at Fermi energy help to exploit isospin diffusion in deep-inelastic or incomplete-fusion reaction for the production of very neutron-rich species. Such approaches require extended capabilities of the driver accelerator. Specifically for heavy-ion reactions at Fermi energy, it is possible to consider their use as a regime for production of the very neutron-rich nuclei in the reactions of secondary unstable beams.

5. Production of neutron-rich nuclei around N=82

One of the most promising ways to produce extremely neutron-rich nuclei around the neutron shell N=82 is fragmentation of a secondary beam of

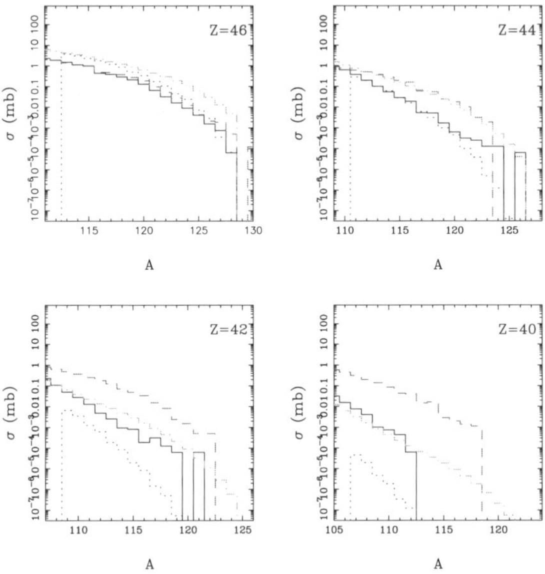

Fig. 4. Comparison of production cross sections for reaction $^{132}Sn+^{238}U$ at 28 AMeV using standard (dashed lines) and modified simulation[8] (solid lines) with fragmentation cross sections of ^{132}Sn beam with Be target using COFRA[11] (dotted lines) and EPAX[12] (dash-dotted lines).

^{132}Sn. Nevertheless, based on the results of the previous section, one can in principle consider also the reaction in the Fermi-energy domain at energies below 50 AMeV. The comparison of production cross sections for the reaction $^{132}Sn+^{238}U$ at 28 AMeV with fragmentation cross section of ^{132}Sn beam with Be target is provided in Fig. 4. For the reaction $^{132}Sn+^{238}U$ the modified DIT code[7] was used for peripheral collisions together with original model of incomplete fusion[1] (dashed lines) and its modification presented in ref. [8] (solid lines) for central collisions, while for the fragmentation of ^{132}Sn beam the codes COFRA[11] (dotted lines) and EPAX[12] (dash-dotted lines) were used. The production cross sections calculated using both the original and modified model of incomplete fusion for Z=46 are comparable with results of EPAX and COFRA, while for elements with lower atomic numbers the reaction $^{132}Sn+^{238}U$ leads, according to both the original and modified model of incomplete fusion, to still more favorable cross sections exceeding both COFRA and even EPAX cross sections.

The in-target yields calculated using the production cross sections from Fig. 4 are shown in Fig. 5. For the reaction $^{132}Sn+^{238}U$ at 28 AMeV a target

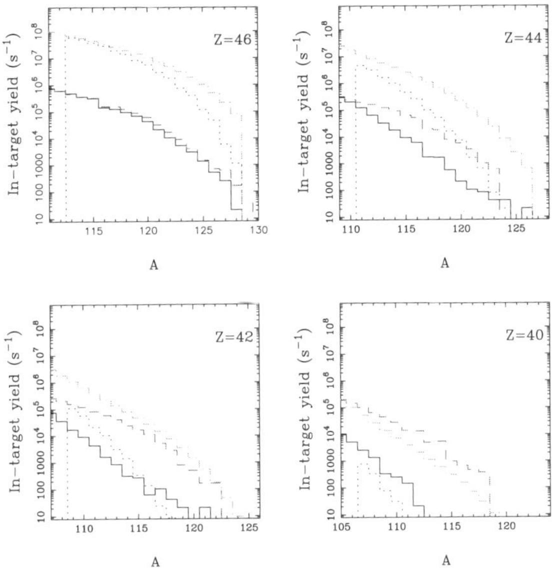

Fig. 5. The in-target yields (for the intensity of ^{132}Sn beam 10^{12} s^{-1}) calculated using the production cross sections from Fig. 4. Meaning of lines is analogous to Fig. 4.

thickness 40 mg/cm^2 was assumed while for fragmentation regime an initial beam energy of 100 AMeV was chosen. The intensity of ^{132}Sn secondary beam of 10^{12} s^{-1} was adopted from Eurisol RTD Report.[9] Due to larger target thickness, the in-target yield for fragmentation option calculated using both COFRA and EPAX dominate for elements Z=44 and above, for lighter nuclei nevertheless the larger production cross sections in the Fermi-energy domain lead also to larger in-target yields despite relatively thin target and for Z=40 the in-target yield calculated using the modified model of incomplete fusion exceeds the COFRA value (and the EPAX value is exceeded by original model of incomplete fusion). However, the angular distribution of reaction products at 28 AMeV would require a large-acceptance separator with angular coverage up to 10 degrees and a highly efficient gas-cell in order to form a secondary beam.

6. Isoscaling

The yield ratios from two nuclear reactions which differ only in isospin asymmetry can exhibit an exponential scaling with neutron and proton numbers. Such behavior was observed experimentally in multifragmentation data from collisions of high energy light particles with massive target nuclei[13] and from collisions between mass symmetric projectiles and targets at intermediate energies[14] and it is called isotopic scaling or isoscaling[14] (the slope parameters $\alpha, \beta, \alpha', \beta'$ being referred to as isoscaling parameters). An isoscaling behavior was also reported in studies of heavy residues[15] and in fission data.[16] The values of the isoscaling parameters were related by several authors to various physical quantities such as the symmetry energy,[13,14] the level of isospin equilibration[15] and the transport coefficients.[16] Isoscaling can be also envisioned as a powerful tool for prediction of the production rates of exotic nuclei, such as e.g. the extremely neutron-rich nuclei, in the reactions induced by the beams of the exotic unstable nuclei delivered by the new generation of RNB facilities such as EURISOL.

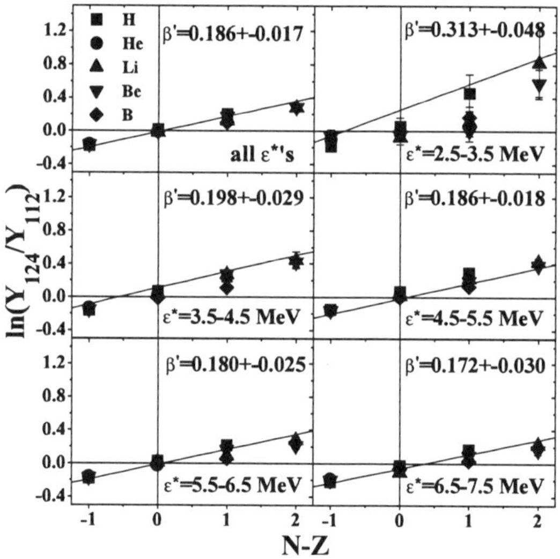

Fig. 6. Simulated isoscaling plots (symbols) and fits to experimental data (lines) from the statistical decay of hot quasi-projectiles in the reactions ^{28}Si+^{124,112}Sn at incident energies of 50 AMeV.[17] The upper left panel corresponds to inclusive data while the other panels correspond to the five excitation energy bins.

As demonstrated by the simulations in the recent study,[17] the increasing width of initial isotopic distributions (represented by Gaussians) at the dynamical stage and the corresponding decrease of the initial (dynamical) isoscaling slope is reflected by significant modification of the final isoscaling slope after de-excitation. For narrow initial distributions, the isoscaling slope assumes the limiting value fully determined by the details of the de-excitation stage. For wide initial distributions, the isoscaling slope for hot fragments approaches the slope of initial isoscaling plots and it is thus fully determined by the initial stage. This correspondence is modified by secondary emission and the isoscaling slopes for final cold fragments are larger possibly due to a corresponding decrease of the temperature during secondary emissions. It is noteworthy that the width of initial Gaussian distributions induces a decrease of the isoscaling parameters comparable to the values, reported in the literature,[13,14] and explained as an effect of a decreasing symmetry energy, according to liquid-drop based formula that relates the symmetry energy coefficient directly to the isoscaling parameter. However, the effect of the dynamical stage and specifically of the width of the initial distributions was not considered in the analysis and the estimates are based on simulation for individual initial nuclei, which appears to be an over-simplified approach.

In Fig. 6 are presented simulated isoscaling data (symbols) from statistical decay of hot quasi-projectiles produced in the reactions $^{28}Si+^{124,112}Sn$ at projectile energy 50 AMeV. The isoscaling plots are presented not only for the inclusive data (upper left panel) but also for five bins of excitation energy. The isoscaling slope in the simulations depends on the excitation energy almost identically as in the experimental data, represented by the solid lines. The DIT+SMM simulation fully reproduces experimental isoscaling behavior for the observed sample of the peripheral damped nucleus-nucleus collisions.

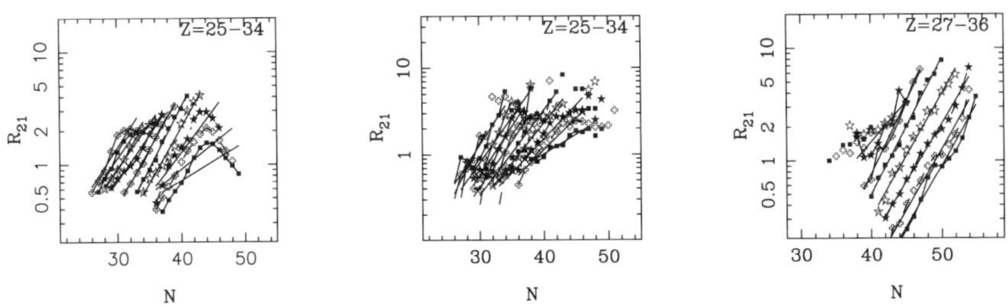

Fig. 7. Isoscaling plots for the reactions of $^{86}Kr+^{124,112}Sn$ at an incident energy of 25 AMeV.[17] Left panel - simulated data for final fragments, middle panel - experimental data,[3] right panel - simulated data after dynamical stage. The lines represent exponential fits.

The left panel of Fig. 7 shows isoscaling plots corresponding to the simulations of the reactions of ^{86}Kr (25AMeV) with ^{124,112}Sn.[17] The used simulation is the same as in ref. [8] where it allowed to reproduce experimental cross sections for production of the neutron-rich heavy residues after de-excitation of the projectile-like nuclei produced in the dynamical stage of the collision. As a comparison, in the middle panel the experimental isoscaling plots are shown. For nuclei with Z=25-30 the simulation and experiment lead to a similar behavior with constant slopes and consistent values of the isoscaling parameters. For heavier nuclei with N>44, the simulation leads to a reverse trend of the yield ratios toward unity, possibly signaling the onset of a reaction mechanism independent of the N/Z of the target, possibly quasi-elastic (direct) few-nucleon transfer taking place in very peripheral collisions. The experimental isoscaling behavior for these nuclei shows signs of a similar reverted trend, the transition is not as regular as in the simulation. A decrease of the slope of exponential ("isoscaling") fits is shown by the lines in the left panel of Fig. 7, despite the very poor quality of such fits. Both the experimental and simulated data suggest a mixing of two components: one component very sensitive to the N/Z of the target, possibly due to an intense nucleon exchange; a second component, insensitive to the N/Z of the target, possibly quasi-elastic few-nucleon exchange.

This situation is demonstrated in the right panel of Fig. 7 where simulated isoscaling plots are shown for the dynamical stage prior to de-excitation. The isotopes with Z = 30 - 36 exhibit regular isoscaling behavior, except for a structure around N = 50 corresponding to elements close to the projectile charge, which can be identified with quasi-elastic processes. Despite minor effect on isoscaling plots, these points represent a significant portion of the reaction cross section. The discrepancy of the final simulated and experimental isoscaling behavior, corresponding mostly to the residues from quasi-elastic collisions, can be possibly attributed to an underestimated probability for the emission of complex fragments below multifragmentation threshold in the SMM.

A further possibility to explore the nucleus-nucleus collisions at the Fermi energy is to use a fissile primary beam, which would undergo a deep-inelastic collision with the target followed by subsequent fission of the quasi projectile. The fragment yield ratios were investigated in the fission of ^{238,233}U targets induced by 14 MeV neutrons.[16] The isoscaling behavior was typically observed for isotopic chains ranging from the most proton-rich to most neutron-rich ones. The high sensitivity of the neutron-rich heavy fragments to the target neutron content suggests the viability of fission (

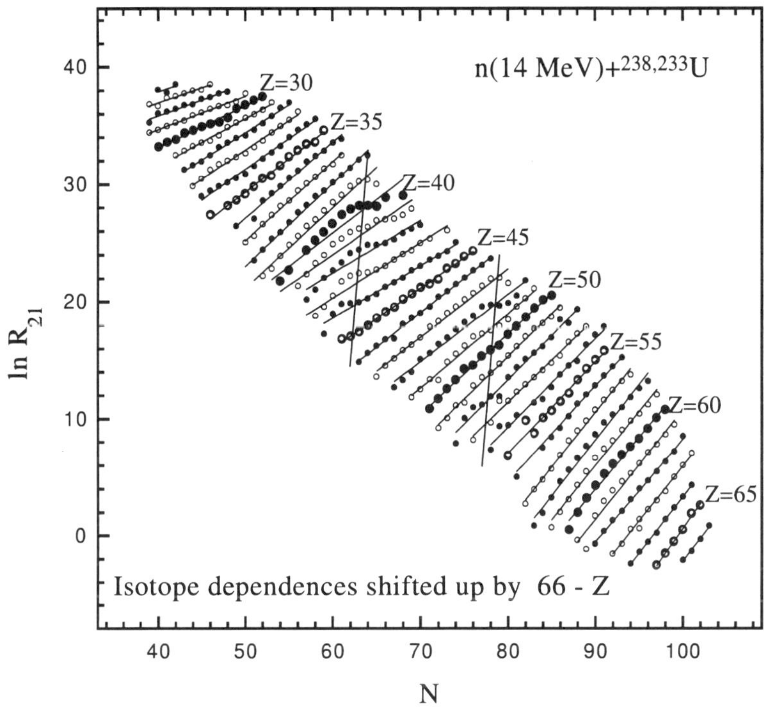

Fig. 8. Ratios of the fragment yields from the fission of $^{238,233}U$ targets induced by 14 MeV neutrons.[16] The data are shown as alternating solid and open circles. The labels apply to the larger symbols. The lines represent exponential fits. For clarity, the R_{21} dependences are shifted from element to element by one unit. Nearly vertical lines mark major isoscaling breakdowns.

possibly following a peripheral collision with another n-rich nucleus) as a source of very neutron-rich heavy nuclei for future rare ion beam facilities. The observed breakdowns of the isoscaling behavior around N=62 and N=80 indicate the effect of two major shell closures on the dynamics of scission, one of them being the deformed shell closure around N=64. The isoscaling analysis of the spontaneous fission of $^{248,244}Cm$ further supports such conclusion. The values of the isoscaling parameter appear to exhibit a structure which can be possibly related to details of scission dynamics at various mass splits. The isoscaling studies present a suitable tool for investigation of the fission dynamics of the heaviest nuclei, which can provide essential information about possible pathways to the synthesis of still heavier nuclei.

7. Summary

An overview of the recent progress on production of neutron-rich nuclei in the nucleus-nucleus collisions around the Fermi energy was presented and the possibilities to produce the very neutron-rich nuclei in the region of mid-heavy to heavy nuclei were examined in both the peripheral and central collisions. The production cross-section trends of neutron-rich nuclei were described in the peripheral collisions when taking into account the isospin asymmetry of the nuclear periphery. For central collisions an isospin-dependent component of the excitation energy of the cold fragments was introduced. Possible scenarios applicable for the new generation of rare nuclear beam facilities such as Eurisol were discussed. Isoscaling was investigated as a possible tool to predict the production rates of exotic species in nuclear reactions induced by both stable and radioactive beams.

This work was supported through grant of Slovak Scientific Grant Agency VEGA-2/5098/25.

References

1. M. Veselsky, *Nucl. Phys. A* **705**, p. 193 (2002).
2. G. A. Souliotis *et al.*, *Nucl. Instr. Meth. B* **204**, p. 166 (2003).
3. G. A. Souliotis *et al.*, *Phys. Rev. Lett.* **91**, p. 022701 (2003).
4. G. A. Souliotis *et al.*, *Phys. Lett. B* **543**, p. 163 (2002).
5. J. P. Bondorf *et al.*, *Phys. Rep.* **257**, p. 133 (1995).
6. L. Tassan-Got and C. Stefan, *Nucl. Phys. A* **524**, p. 121 (1991).
7. M. Veselsky and G. A. Souliotis, *Nucl. Phys. A* **765**, p. 252 (2006).
8. M. Veselsky and G. A. Souliotis, *arXiv.org:* **nucl-th/0607032** (2006).
9. EURISOL Feasibility Study RTD, `http://www.ganil.fr/eurisol/Final_Report.html`.
10. M. Veselsky *et al.*, Preliminary report on the benefit of the extended capabilities of the driver accelerator, `http://www-w2k.gsi.de/eurisol-t11/documents/Extended_driver_report-June-14-2006.pdf`.
11. K. Helariutta *et al.*, *Eur. Phys. J A* **17**, p. 181 (2003).
12. K. Summerer and B. Blank, *Phys. Rev. C* **61**, p. 34607 (2000).
13. A. S. Botvina, O. V. Lozhkin and W. Trautmann, *Phys. Rev. C* **65**, p. 044610 (2002).
14. M. B. Tsang *et al.*, *Phys. Rev. Lett.* **86**, p. 5023 (2001).
15. G. A. Souliotis *et al.*, *Phys. Rev. C* **68**, p. 24605 (2003).
16. M. Veselsky, G. Souliotis and M. Jandel, *Phys. Rev. C* **69**, p. 44607 (2004).
17. M. Veselsky, *arXiv.org:* **nucl-th/0607033** (2006), accepted for publication in Phys. Rev. C.

SIGNALS OF ENLARGED CORE IN ^{23}AL*

Y. G. MA[1†], D. Q. FANG[1], C. W. MA[1,2], K. WANG[1,2], T. Z. YAN[1,2], X. Z. CAI[1], W. Q. SHEN[1], Z. Y. SUN[3], Z. Z. REN[4], J. G. CHEN[1], J. H. CHEN[1,2], G. H. LIU[1,2], E. J. MA[1,2], G. L. MA[1,2], Y. SHI[1,2], Q. M. SU[1,2], W. D. TIAN[1], H. W. WANG[1], C. ZHONG[1], J. X. ZUO[1,2], M. HOSOI[5], T. IZUMIKAWA[6], R. KANUNGO[7], S. NAKAJIMA[5], T. OHNISHI[8], T. OHTSUBO[6], A. OZAWA[9], T. SUBA[8], K. SUGAWARA[5], K. SUZUKI[5], A. TAKISAWA[6], K. TANAKA[8], T. YAMAGUCHI[5], I. TANIHATA[7]

1. Shanghai Institute of Applied Physics, Chinese Academy of Sciences, Shanghai 201800, China

2. Graduate School of the Chinese Academy of Sciences, Beijing 100039, China

3. Institute of Modern Physics, Chinese Academy of Sciences, Lanzhou 730000, China

4. Department of Physics, Nanjing University, Nanjing 210008, China

5. Department of Physics, Saitama University, Saitama 338-8570, Japan

6. Department of Physics, Niigata University, Niigata 950-2181, Japan

7. TRIUMF, 4004 Wesbrook Mal, Vancouver, British Columbia V6T 2A3, Canada

8. Institute of Physical and Chemical Research (RIKEN), Wako-shi, Saitama 351-0198, Japan

9. Institute of Physics, Tsukuba University, Ibaraki 305-8571, Japan

---

* This work is partially supported by the National Natural Science Foundation of China (NNSFC) under Grant No. 10405032, 10535010, 10405033, 10475108, Shanghai Development Foundation for Science and Technology under contract No. 06QA14062, 06JC14082, 05XD14021 and 03QA14066 and the Major State Basic Research Development Program in China under Contract No. G200077404..

† Corresponding author. E-mail address: ygma@sinap.ac.cn

The longitudinal momentum distribution ($P_{//}$) of fragments after one-proton removal from ^{23}Al and reaction cross sections (σ_R) for ^{23,24}Al on carbon target at 74AMeV have been measured using 135AMeV ^{28}Si primary beam on RIPS in RIKEN. $P_{//}$ is measured by a direct time-of-flight (TOF) technique, while σ_R is determined using a transmission method. An enhancement in σ_R is observed for ^{23}Al compared with ^{24}Al. The $P_{//}$ for ^{22}Mg fragments from ^{23}Al breakup has been obtained for the first time. FWHM of the distributions has been determined to be 233±14 MeV/c. The experimental data are discussed by use of the Few-Body Glauber Model (FBGM). Analysis of $P_{//}$ indicates a dominant d-wave configuration for the valence proton in the ground state of ^{23}Al. The possibility of an enlarged ^{22}Mg core for proton-rich nucleus ^{23}Al is demonstrated..

Studies on the structure of nuclei far from the β-stability line have become one of the frontiers in nuclear physics for more than two decades. Since the pioneering measurements of the interaction cross sections (σ_R) and observation of an remarkably large σ_R for ^{11}Li [1], it has been shown that there is exotic structure like neutron halo or skin in light neutron-rich nuclei. Experimental measurements of reaction cross section (σ_R), fragment momentum distribution of one or two nucleons removal reaction ($P_{//}$), quadrupole moment and Coulomb dissociation have been demonstrated to be very effective methods to identify and investigate the structure of halo nuclei. The neutron skin or halo nuclei ^{6,8}He, ^{11}Li, ^{11}Be, ^{19}C etc. [1-3], have been identified by these experimental methods. Due to the centrifugal and Coulomb barriers, the identification of a proton halo is more difficult compared to a neutron halo. The quadrupole moment, $P_{//}$ and σ_R measurements indicate a proton halo in ^{8}B [4,5], whereas no enhancement is observed in the measured σ_I at relativistic energies [1,6]. Evidence of proton halo in the first excited state of ^{17}F has been shown in the capture cross section measurement for ^{16}O(p,γ)^{17}F reactions [7], but there is no anomalous increase in the experimental σ_R for the proton halo candidate ^{17}F [8]. The proton halo in ^{26,27}P and ^{27}S has been predicted theoretically [9]. And the measurements of $P_{//}$ have shown a proton halo character in ^{26,27,28}P [10]. The experimental search for heavy halo nuclei plays a significant role for the investigation of nuclear structure and the improvement of nuclear theory since the properties of those exotic nuclei are expected to be different from stable nuclei.

Proton-rich nucleus ^{23}Al has a very small separation energy (S_p=0.125 MeV) [11] and is a possible candidate of proton halo. An enhanced reaction cross section for ^{23}Al has been observed in a previous experiment on the Radioactive Ion Beam Line in Lanzhou (RIBLL) [12]. To reproduce the σ_R for ^{23}Al, the assumption of a considerable $2s_{1/2}$ component for the valence proton around the ^{22}Mg core within the framework of the Glauber model is necessary [12]. Thus a

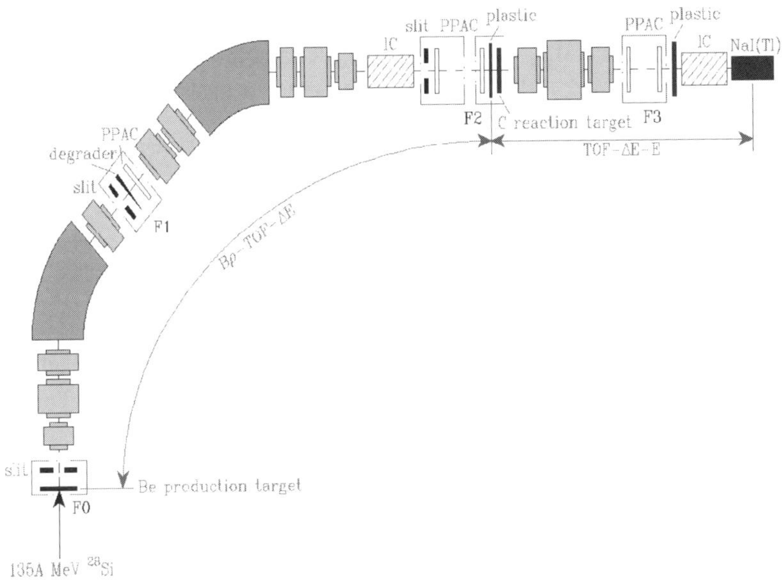

Fig.1: Experimental setup at the fragment separator RIPS.

long tail in the proton density distribution has been extracted for ^{23}Al which indicates an exotic structure. While the spin and parity (J^{π}) for ground state of ^{23}Al has been deduced to be $5/2^{+}$ in a recent measurcment of the magnetic moment [13]. This result favors the *d*-wave configuration for the valence proton in ^{23}Al. But it does not eliminate the possibility of a *s*-wave valence proton if the ^{22}Mg core is in the excited state ($J^{\pi}=2^{+}$). Therefore it will be very important to determine the configuration of the valence proton for ^{23}Al. As we know, the longitudinal momentum distribution of fragments carries structure information of the projectile, especially the configuration of valence nucleon. However, there are no such experimental data for ^{23}Al up to now. In this paper we will report the simultaneous measurement of σ_R and $P_{//}$ for ^{23}Al and also the σ_R for ^{24}Al.

The experiment was performed at the RIKEN Projectile Fragment Separator (RIPS) in the RIKEN Ring Cyclotron Facility. The experimental setup is shown in Fig. 1. Secondary beams were generated by fragmentation reaction of 135 A MeV ^{28}Si primary beam on a ^{9}Be production target in F0 chamber. At the first dispersive focus F1, an Al wedge-shape degrader (central thickness; 583.1 mg/cm^2, angle; 6 mrad) was installed. And a delay-line readout Parallel Plate Avalanche Counter (PPAC) was placed to measure momentum broadening of the beam. Then the secondary beam was focused onto the achromatic focus F2. Two delay-line readout PPACs were installed to determine the beam position and angle. An ion chamber (200ϕ ×780mm) was used to measure the energy loss (ΔE) [14]. And an ultra-fast plastic scintillator (0.5 mm thick) was placed before a carbon reaction target (377 mg/cm^2 thick) to measure the time-of-flight (TOF) from the PPAC at F1. The particle identification before the reaction target was done by means of Bρ-ΔE-TOF method.

After the reaction target, a quadrupole triplet was used to transport and focus the beam onto F3 (~ 6 m from F2). Two delay-line readout PPACs were used to monitor the beam size and emittance angle. Another plastic scintillator (1.5~mm thick) gave a stop signal of the TOF from F2 to F3. A smaller ion chamber (90ϕ ×650mm) was used to measure the energy loss (ΔE). The total energy (E) was measured by a NaI(Tl) detector. The particles were identified by TOF-ΔE-E method. An example of the typical particle identification spectra at F3 for ^{23}Al is shown in Fig.2. In this spectrum, fragments with different charge were already subtracted by the TOF and ΔE method.

In order to estimate and subtract reactions of the projectile in material other than the carbon target, measurements without the reaction target were also performed and the beam energy was reduced by an amount corresponding to the energy loss in the target.

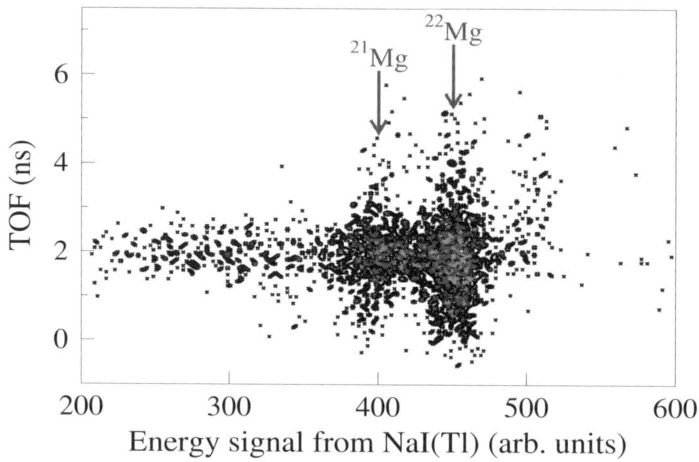

Fig.2: Particle identification at F3 by the bidimensional plot between TOF from F2 to F3 and energy signal from NaI(Tl) (corrected with TOF). Fragments with different charge were already subtracted using the TOF and ΔE correlation spectrum.

Under the assumption of a sudden valence-nucleon removal, the momentum distribution of fragments can be used to describe that of the valence proton which carry significant information about the nuclear structure. The $P_{//}$ of fragments from breakup reactions was determined from the TOF between the two plastic scintillators installed at F2 and F3. The position information measured by the PPAC at F1 was used to derived the incident momentum. The momentum of fragment relative to the incident projectile in the laboratory frame was transformed into that in the projectile rest frame using the Lorentz transformation.

For one-proton removal reactions of ^{23}Al, the signal (reactions in the carbon target) is mixed up with the background (reactions in the detectors and material other than the reaction target) in the $P_{//}$ spectrum of the fragment from target-in measurement as shown in Fig.2. This background was carefully reconstructed based on the ratio of fragments to unreacted projectile identified in the target-out measurement and the broadening effect of the carbon target on $P_{//}$. Then the estimated background was subtracted from the mixed spectrum. The obtained momentum distribution of the ^{22}Mg fragments from ^{23}Al breakup in the carbon target at 74 AMeV is shown in Fig.3. A Gaussian function was used to fit the distributions. The full width at half maximum (FWHM) was determined to be 233±14 MeV/c after unfolding the Gaussian-shaped system resolution. The FWHM is consistent with the Goldhaber model's prediction (FWHM=212 MeV/c with σ_0=90 MeV/c) within the error bar [15]. Since the magnetic fields of the quadruples between F2 and F3 were optimized for the projectile for measurement of σ_R at the same time. The momentum dependence of the transmission from F2 to F3 for fragments was simulated by the code MOCADI [16]. The effect of transmission on the width of $P_{//}$ distribution was found to be negligible which is similar with the conclusion for neutron-rich nuclei [17].

Reaction cross section is determined using the transmission method, by events of projectile before (incident) and after (unreacted) the reaction target for target-in and target-out measurements:

$$\sigma_R = \frac{1}{t} \ln \left(\frac{\gamma_0}{\gamma} \right) \qquad (1)$$

where γ and γ_0 denote ratio of the unreacted outgoing and incident projectiles for target-in and target-out cases, respectively; *t* the thickness of the reaction target, i.e., number of particle per unit area.

The σ_R of ^{23,24}Al at 74 AMeV were obtained to be 1609 ± 79 mb and 1527 ± 112 mb, respectively. The errors include the statistical and systematic uncertainties. The probability of inelastic scattering reaction was estimated to be very small (<1%), e.g., the inelastic cross section is only around 15 mb for ^{23}Al which is much smaller than the error of σ_R.

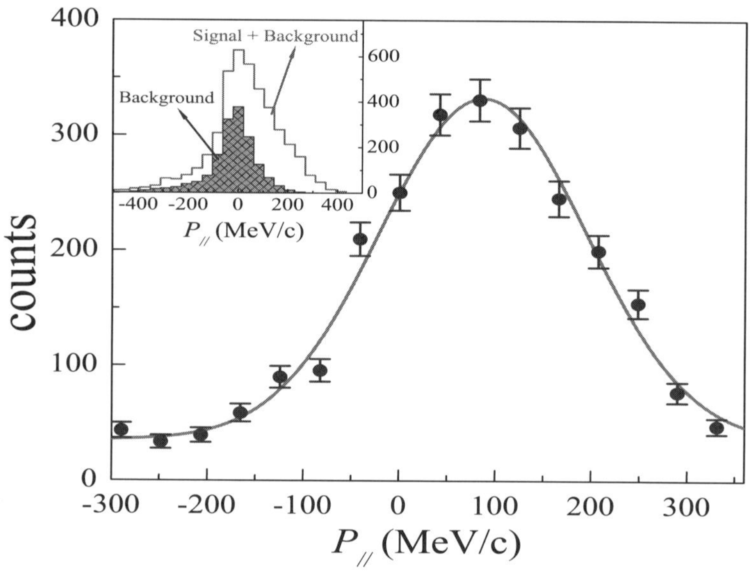

Fig.3: $P_{//}$ distribution of fragment ^{22}Mg after one-proton removal from ^{23}Al. The closed circles with error bars are the present experimental data, the solid curve is a Gaussian fit to the data. Spectra in the inset are the original $P_{//}$ of ^{22}Mg from target-in (solid line) and target-out (shadowed area) measurements, respectively.

Results of both the present and previous data are shown in Fig. 4. Since the energy is different in two experiments, the previous σ_R data at ~30AMeV [12] were scaled to the present energy (74AMeV) using the phenomenological formula [18]. First the radii parameter (r_0) in this formula was adjusted to reproduce the σ_R at ~ 30AMeV, then the same r_0 was used to calculate the σ_R at 74AMeV. As shown in Fig. 4, the σ_R of $^{23,24}Al$ from present and previous experiments are in good agreement. And we observed an enhanced σ_R for ^{23}Al in our data again.

To interpret the measured reaction cross section and momentum distribution data, we performed a Few-Body Glauber Model (FBGM) analysis for $P_{//}$ of $^{23}Al \rightarrow$ ^{22}Mg processes and σ_R of $^{23,24}Al$ [19]. In this model, a core plus one

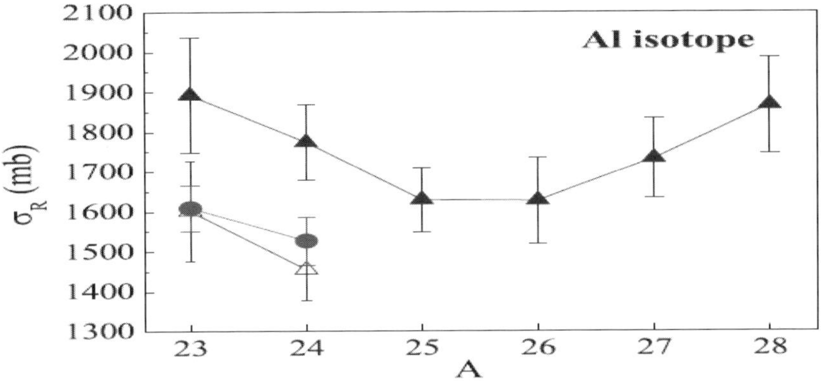

Fig.4: The mass dependence of σ_R for Al isotopes. The solid circles are results of the present experiment (E=74AMeV), the solid triangles are the previous experimental data (E ~ 30AMeV) [12], and the open triangles are the previous data scaled to 74AMeV.

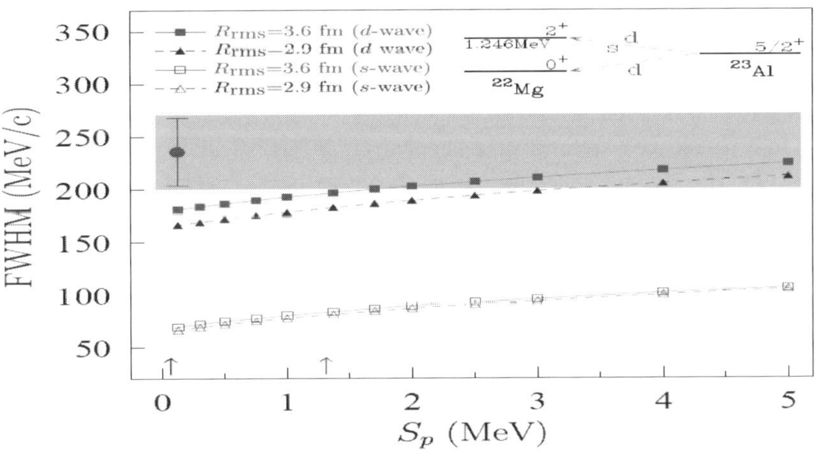

Fig.5: The dependence of FWHM for the $P_{//}$ distribution after one-proton removal of ^{23}Al on the separation energy of the valence proton. The solid circles with error bars are results of the present experiment, the shaded area refers to error of the data. The solid and open squares are the FBGM calculations for the *d* and *s*-wave configuration of the valence proton with the core R_{rms}=3.6 fm. The solid and open triangles are for the core R_{rms}=2.9 fm. The lines are just for guiding the eyes. The two arrows refer to the separation energy of 0.125 MeV and 1.37~MeV (the excitation energy for the first excited state of ^{22}Mg plus the experimental one proton separation energy of ^{23}Al).

proton structure is assumed for the projectile. The total wave function of the nucleus is expressed as

$$\Psi = \sum_{ij} \psi^i_{\mathrm{core}} \phi^j_0, \tag{2}$$

where ψ_{core} and ϕ_0 are the wave function of the core and valence proton; *i,j* denote the different configurations for the core nucleus and the valence protons, respectively.

For the core, HO-type functions were used for the density distributions. The size parameters were adjusted to reproduce the reaction cross sections. The wavefunction of the valence neutron was calculated by solving the eigenvalue problem in a Woods-Saxon potential:

$$\frac{d^2 R(r)}{dr^2} + \frac{2\mu}{\hbar^2}\left[E - U(r) - \frac{l(l+1)\hbar^2}{2\mu r^2}\right] R(r) = 0 \tag{3}$$

where R(r) denotes the radial wave function of valence proton, μ the reduced mass of proton, l the angular quantum number and $\hbar$ the reduced Planck constant. The potential U(r) includes Wood-Saxon, Coulomb and spin-orbit potentials. The separation energy of the last proton is reproduced by adjusting the potential depth. In the calculation the diffuseness and radii parameter were chosen to be 0.67 fm and 1.27 fm, respectively.

In a recent measurement of *g* factor using a β-NMR method, the spin and parity for the ground state of ^{23}Al is shown to be $5/2^+$. This gives a strong restriction on the possible nuclear structure of nucleus. Assuming the ^{22}Mg+p structure, three different configurations are possible for $J^\pi=5/2^+$ of ^{23}Al: 0+ ⊗1d5/2$, 2^+$⊗1d5/2 and 2^+⊗2s1/2 [13]. The *s*-wave configuration is also possible for the core in the excited state.

The momentum distributions for the valence proton in the *s* or *d*-wave configuration are calculated by use of the FBGM. In calculation of the wavefunction for the valence proton, the one proton separation energy S_p=0.125MeV is used for ^{22}Mg in the ground state. While for ^{22}Mg in the excited state ($J^\pi=2^+$, E_x=1.25MeV), $E_x + S_p$ is used for the separation energy of the valence proton. To see the separation energy dependence of $P_{//}$, more values are calculated besides S_p and $E_x + S_p$. And two effective root-mean-square matter radii ($R_{rms} \equiv \langle r^2\rangle^{1/2}$) are adopted for ^{22}Mg to see the core size effect on $P_{//}$. The obtained FWHM of $P_{//}$ are shown in Fig.5. In this figure, we can see that the width for the *s* and *d*-wave are obviously separated. The width for the *s*-wave is much lower than the experimental data, while that of the *d*-wave is close to the experimental FWHM. With the increase of S_p, the width of $P_{//}$ increases slowly.

That means $P_{//}$ will become wider for ^{22}Mg in the excited state. The effect of the core size on $P_{//}$ is negligible for the *s*-wave configuration but not for the *d*-wave configuration. The larger size core will give a wider $P_{//}$ distribution. From comparison of the FBGM calculation with the experimental data in Fig.5, it clearly indicates that the valence proton in ^{23}Al is dominantly in the *d*-wave configuration. And the possibility for the *s*-wave should be very small. This is consistent with the shell model calculations and also the Coulomb dissociation measurement [13,20].

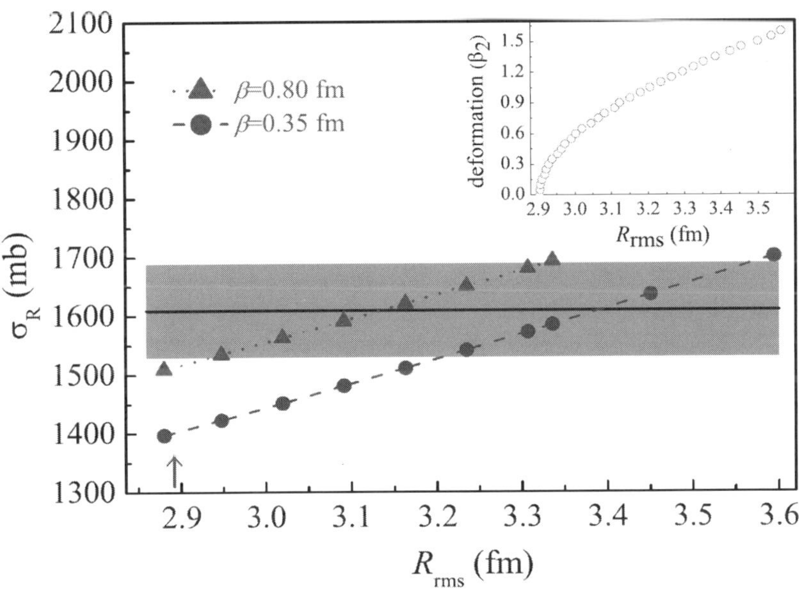

Fig.6: The dependence of σ_R at 74AMeV on the core size R_{rms}. The horizontal line is the experimental σ_R value, the shadowed area is the error of σ_R. The solid circles and triangles denote the FBGM calculations with the range parameter β=0.35 fm and β=0.80 fm, respectively. The size of ^{22}Mg obtained by fitting the σ_I data at around 1AGeV is marked by an arrow. The inset shows the relationship between the quadrupole deformation parameter (β_2) and size of the core if the density distribution is supposed to be homogeneous.

From above discussions of $P_{//}$, the valence proton is determined to be in the *d*-wave. For the calculation of reaction cross section using the FBGM, the core size should be fixed. First we extracted the R_{rms} of ^{22}Mg by reproducing the experimental σ_I data at around 1AGeV [21]. The extracted R_{rms} is 2.89±0.09

(fm). But the calculated σ_R for ^{23}Al is much lower than the obtained σ_R data. Similar puzzle is also encountered for some neutron-rich nuclei. The large σ_I cannot be reproduced by the FBGM even for the valence neutron in the *s*-wave for ^{19}C and ^{23}O. One way is to enlarge the core size to reproduce the experimental reaction cross section [22]. Here we changed the core size by adjusting the width parameters. The dependence of σ_R on R_{rms} of the core is shown in Fig.6. In the FBGM calculations, the parameters α and σ_{NN} in the profile function

(*b* is the impact parameter) are taken from Ref. [19]. The range parameter is

$$\Gamma(b) = \frac{1-i\alpha}{4\pi\beta^2}\sigma_{NN}\exp(-\frac{b}{2\beta^2})$$

calculated by the formula which is determined by fitting the σ_R of ^{12}C + ^{12}C from low to relativistic energies [23]. β is 0 and 0.35 fm for 1AGeV and 74 AMeV, respectively. The σ_R of ^{23}Al is very sensitive to the size of ^{22}Mg. To reproduce the measured σ_R of ^{23}Al, the calculated results indicates an enlarged ^{22}Mg core with R_{rms}=3.37±0.18 (fm). It is 17±7 % larger than the size of ^{22}Mg extracted using the σ_I around 1AGeV [21].

But there is another possibility for the Glauber model. Global underestimation of the reaction cross section (10-20% for stable nuclei and 30-50% for exotic nuclei) was found at intermediate energies in the Glauber model when the densities of the projectile are determined by fitting theσ_I at relativistic energies [24]. To correct the underestimation in the Glauber model, different method has been tried. For an example, transport model has been developed to calculate the total reaction sections [25]. In this method, initial nuclear densities of projectile and target, nuclear equation of state and in-medium nucleon-nucleon cross section can be adjusted to reproduce the σ_R [25]. Also, in the framework of Glauber model, the energy dependent phenomenological correction factor and finite-range effect were introduced [2,23]. But these corrections are performed for almost light stable nuclei and no systematic study has been done for proton-rich nuclei with A>20. The reaction cross section of ^{24}Al is calculated with the size of ^{23}Mg core determined by fitting σ_I at around 1AGeV [21].

But the calculated σ_R for ^{24}Al is only 1430 mb and underestimation still exists (~10%). Since scope of the underestimation in the Glauber model is large even for stable nuclei, underestimation may still exist in the finite range Glauber calculations with β=0.35 fm at 74AMeV. To correct the possible underestimation, we adjusted the range parameter to reproduce the σ_R of ^{24}Al and obtained β=0.8 fm. Using this range parameter, the σ_R of ^{23}Al is calculated and shown in Fig.6. The calculated results indicate the core size of R_{rms}=3.13±0.18 fm (8±7% larger than the size of ^{22}Mg deduced by the σ_I data). Even if a larger range parameter is used in this calculation, a relatively larger core is also obtained for ^{23}Al compared to ^{24}Al. The obtained size of ^{22}Mg is different for the two range parameters, but both calculations suggest an enlarged core inside ^{23}Al.

As shown in the inset of Fig.6, the R_{rms} of the core changes quickly with the increase of the quadrupole deformation parameter β_2. This simple relationship between R_{rms} and β_2 indicates that a deformed core inside ^{23}Al is one of the possible reasons for the enlarged size of ^{22}Mg. And the excited state in ^{22}Mg is calculated within the RMF framework [26]. The R_{rms} of the excited state is obtained to be around 2.4% larger than that of the ground state. Thus the core excitation effect may also contribute to the large size for ^{22}Mg.

In summary, the momentum distribution of fragments after one-proton removal for ^{23}Al and reaction cross sections for ^{23,24}Al were measured. An enhancement was observed for the σ_R of ^{23}Al . The $P_{//}$ distributions were found to be wide and consistent with the Goldhaber model's prediction. The experimental $P_{//}$ and σ_R results were discussed within a Few-Body Glauber Model. We determined the valence proton to be a dominant *d*-wave configuration in the ground state of ^{23}Al. The possibility of an enlarged ^{22}Mg core was revealed in order to explain both the σ_R and $P_{//}$ distributions. The effect of deformation and also core excitation were suggested to be two of the possible contributions for the large size of the core in ^{23}Al. This invokes further investigations both experimentally and theoretically.

Acknowledgments

The authors are very grateful to all of the staff at the RIKEN accelerator for providing stable beams during the experiment. The support and hospitality from the RIKEN-RIBS laboratory are greatly appreciated by the Chinese collaborators.

References

1. I. Tanihata, et al., *Phys. Rev. Lett.* 55, 2676 (1985)
2. M. Fukuda, et al., *Phys. Lett.* B 268, 339 (1991).
3. D. Bazin, et al., *Phys. Rev. Lett.* 74, 3569 (1995);
   T. Nakamura, et al., *Phys. Rev. Lett.* 83, 1112 (1999) ;
   A. Ozawa, et al., *Nucl. Phys.* A 691, 599 (2001).

4. T. Minamisono, et al., *Phys. Rev. Lett.* 69, 2058 (1992);
W. Schwab, et al., *Z. Phys.* A 350, 283 (1995).
5. R.E. Warner, et al., *Phys. Rev.* C 52, R1166 (1995);
F. Negoita, C. Borcea, F. Carstoiu, *Phys. Rev.* C 54, 1787 (1996);
M. Fukuda, et al., *Nucl. Phys.* A 656, 209 (1999).
6. M.M. Obuti, et al., *Nucl. Phys.* A 609, 74 (1996).
7. R. Morlock, et al., *Phys. Rev. Lett.* 79, 3837 (1997).
8. A. Ozawa, et al., *Phys. Lett.* B 334, 18 (1994);
K.E. Rehm, et al., *Phys. Rev. Lett.* 81, 3341 (1998).
9. B.A. Brown, P.G. Hansen, *Phys. Lett.* B 381, 391 (1996);
10. A. Navin, et al., *Phys. Rev. Lett.* 81, 5089 (1998).
11. G. Audi, A.H. Wapstra, *Nucl. Phys.* A 565, 66 (1993).
12. X.Z. Cai, et al., *Phys. Rev.* C 65, 024610 (2002);
H.Y. Zhang, et al., *Nucl. Phys.* A 707 , 303 (2002).
13. A. Ozawa, et al., *Phys. Rev.* C 74, 021301R (2006).
14. K. Kimura, et al., *Nucl. Inst. Meth.* A 538, 608 (2005).
15. A.S. Goldhaber, *Phys. Lett.* B 53, 306 (1974).
16. N. Iwasa, et al., *Nucl. Instrum. Methods* B 126, 284 (1997).
17. D.Q. Fang, et al., *Phys. Rev.* C 69, 034613 (2004);
T. Yamaguchi, et al., *Nucl. Phys.* A 724, 3 (2003).
18. W.Q. Shen, et al., *Nucl. Phys.* A 491, 130 (1989).
19. Y. Ogawa, et al., *Nucl. Phys.* A 543. 722(1992);
Y. Ogawa, et al., *Nucl. Phys.* A 571, 784 (1994);
B. Abu-Ibrahim, et al., *Comput. Phys. Comm.* 151, 369 (2003).
20. T. Gomi, et al., *Nucl. Phys.* A 758, 761c (2005) .
21. T. Suzuki, et al., *Nucl. Phys.* A 630, 661 (1998).
22. R. Kanungo et al., *Nucl. Phys.* A 677 , 171 (2000) ;
R. Kanungo et al., *Phys. Rev. Lett.* 88, 142502 (2002)
23. T. Zheng, et al., *Nucl. Phys.* A709, 103 (2002)
24. A. Ozawa, et al., *Nucl. Phys.* A 608, 63 (1996) .
25. Y. G. Ma et al., *Phys. Lett.* B 302, 386 (1993);
Y. G. Ma et al., *Phys. Rev.* C 48, 850(1993).
26. Z. Z. Ren, et al., *Phys. Rev.* C 57, 2752 (1998) ;
J.G. Chen, et al., *Eur. Phys. J.* A 23, 11 (2005).

NEW INSIGHT INTO THE FISSION PROCESS FROM EXPERIMENTS WITH RELATIVISTIC HEAVY-ION BEAMS

A. KELIC, M. V. RICCIARDI, K.-H. SCHMIDT

Gesellschaft für Schwerionenforschung, GSI, 64291 Darmstadt, Planckstr. 1

The results from a series of experiments performed with a novel inverse-kinematics approach using the experimental installations of GSI, Darmstadt, gave a new global insight into the nuclide production in fission reactions. Combined with previous results, this large body of data has been used to develop a new semi-empirical macroscopic-microscopic fission model.

1. Introduction

Several years ago, a research program has been initiated for studying the nuclear fission process with a new experimental approach by using relativistic heavy-ion beams provided by the SIS18 synchrotron accelerator of GSI, Darmstadt. The essential feature of this approach consists in providing the fissile nucleus to be investigated as a projectile, either as a primordial nuclide directly from the ion source of the accelerator complex or as a secondary beam. Fission is induced by electromagnetic or nuclear interaction in a suitable target. By inverting the kinematics with respect to conventional fission experiments, in which the fission products of the target nucleus are measured, the fission products of the projectile nucleus appear with considerably higher kinetic energies, and thus the detection and identification of all fission products becomes feasible. The present contribution provides an overview on the new experimental possibilities offered by the installations of GSI. An overview on the experimental results on multimodal fission in the light actinides and on nuclide production cross sections in spallation and fragmentation reactions will be given.

These experiments do not directly respond to the requests on nuclide production in fission reactions from applications in nuclear technology and fundamental research, since only a limited number of key reactions could be studied due to the tremendous experimental effort. Therefore, a nuclear-reaction code has been developed on the basis of the body of measured data and which can be used for reliable predictions of nuclide production with different projectile and target nuclei and at different beam energies. As an important part of this code, a new fission model will be described in some detail. It is based on the powerful macroscopic-microscopic approach. In contrast to other purely theoretical descriptions, however, several critical ingredients, which cannot be predicted with the desirable accuracy, are extracted from the available experimental data.

2. Experimental results on multimodal fission in the light actinides

The secondary-beam facility at GSI was used to produce more than 70 different neutron-deficient actinides and pre-actinides by fragmentation of a ^{238}U beam at 1 *A* GeV in a primary beryllium target. The fragmentation residues were separated and unambiguously identified event by event in atomic number *Z* and mass number *A* using the fragment separator FRS. These secondary beams impinged on a secondary lead target mounted at the exit of the FRS, where fission was induced by electromagnetic excitations and by nuclear collisions. The nuclear charges and the kinetic energy of both fission products were determined from the measurement of their energy loss and time of flight, see Figure 1. The energy loss was measured in a vertically subdivided ionisation chamber with a common cathode. The velocities of both fission residues were deduced by the time-of-flight measurement from a plastic scintillator placed in front of the secondary target to a plastic-scintillator wall located 5 meters behind. The time-of-flight was used to correct the velocity dependence of the energy loss measured with the ionisation chamber and to determine the kinetic energies of the fission residues.

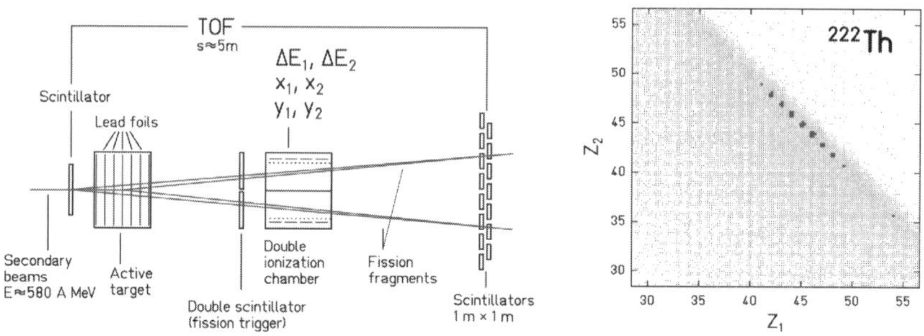

Figure 1: Left part: Schematic drawing of the set up for the fission experiment with secondary beams. Right part: Cluster plot of the nuclear charges of the two fission fragments from ^{222}Th.

A selection on fission events induced by electromagnetic excitation was performed by requiring that the nuclear charges of the two fission fragments sum up to the number of protons in the secondary projectiles. A remaining fraction of nuclear-induced fission events was subtracted on the basis of the nuclear-charge spectrum from fission induced in the scintillator where the cross section for electromagnetic excitation is negligible. Electromagnetic excitations populate states at excitation energies around 11 MeV with a width of a few MeV. In this energy range, shell effects still have a dominating influence on the nuclide distributions produced in fission. A detailed description of the experiment is given in ref. [1].

The elemental yields and the total kinetic energies of fission residues produced by electromagnetic excitation of secondary projectiles from ^{205}At to ^{234}U have been determined. In Figure 2, the elemental distributions of fission residues of 28 secondary beams between ^{221}Ac and ^{234}U are shown. The transition from symmetric fission in the lighter systems to asymmetric fission in the heavier ones is systematically covered for the first time. In the transitional region, around ^{226}Th, triple-humped distributions appear with comparable intensities for symmetric and asymmetric fission.

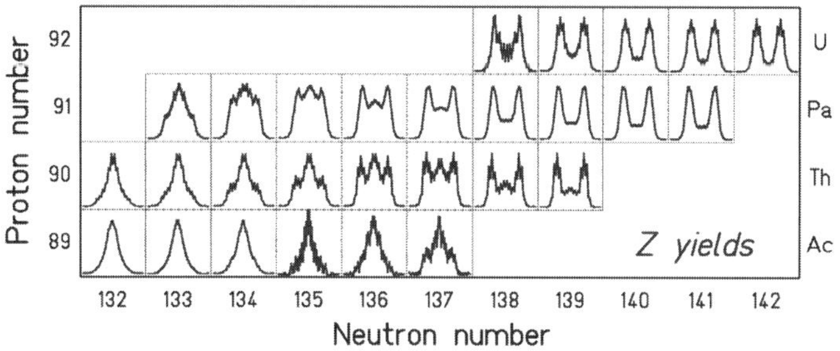

Figure 2: Measured fission-product element distributions in the range Z = 24 to Z = 65 after electromagnetic excitation of 28 secondary beams between ^{221}Ac and ^{234}U on a chart of the nuclides.

The present knowledge on the global characteristics of the fission-fragment mass, respectively element distributions, in low-energy fission, acquired by all the experimental effort is summarized in Figure 3. The secondary-beam experiment has brought a systematic coverage of the transition from predominantly symmetric fission to predominantly asymmetric fission for fissioning systems around mass number 226. Previous investigations on these nuclei were rather hampered due to the lack of suitable target material. The large field of predominantly asymmetric fission in the actinides is also quite well covered by some 50 systems, by neutron- or light-charged-particle-induced fission of available target material or by spontaneous fission of nuclei produced in fusion reactions. The second region of bimodal fission with a transition back to symmetric mass distribution is just observed for the heaviest systems accessible to experiment, e.g. for ^{258}Fm.

3. Experimental results on nuclide production in spallation- and fragmentation-fission reactions with ^{238}U

In another experimental configuration, the fragment separator FRS was used as a tool to separate and identify all reaction products in atomic number *Z* and mass number *A*. However, this configuration is only applicable to beams of primordial nuclides. Due to the limited acceptance of the fragment separator in angle and magnetic rigidity only one fragment in limited magnetic-rigidity range can be detected at a time. However, the full nuclide production can be reconstructed by combining the results obtained with different magnetic fields. As a highly fissile projectile, ^{238}U with energies up to 1 *A* GeV from the GSI accelerator facility was used to induce fission in electromagnetic excitations and nuclear collisions in different targets. In addition to the nuclide identification, see Figure 4, the recoil separator FRS also gives access to the reaction kinematics, and this information allows distinguishing fission events from other reaction channels, see Figure 5.

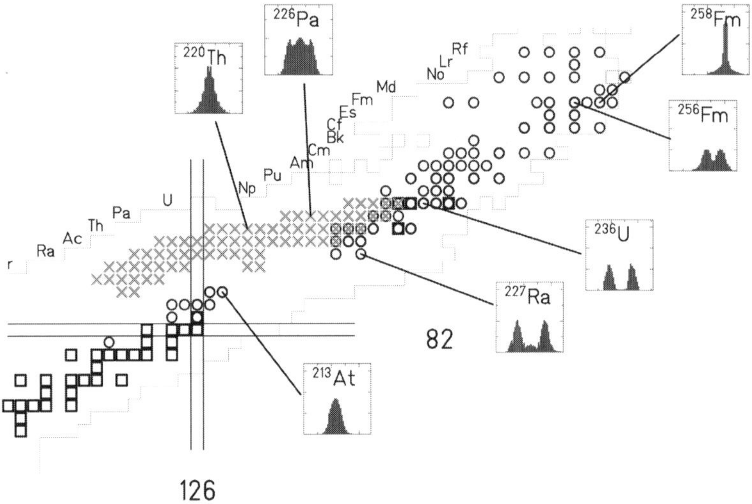

Figure 3: Present knowledge on mass and element distributions of fission products illustrated on the chart of the nuclides. Circles mark the fissioning systems investigated by conventional experiments in normal kinematics. Crosses denote the secondary beams investigated in the secondary-beam experiment performed at GSI in inverse kinematics.

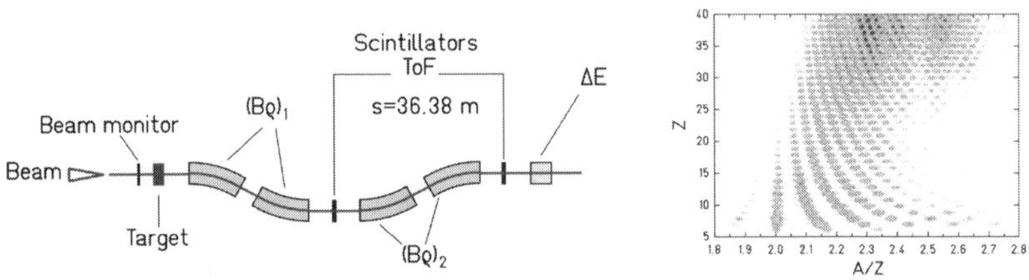

Figure 4: Schematic view on the fragment separator, operated as a magnetic spectrometer to identify the reaction products from the target (left part). Identification pattern of the light residues produced in the reaction ^{238}U (1 *A* GeV) + ^{1}H (right part, from ref. [8]).

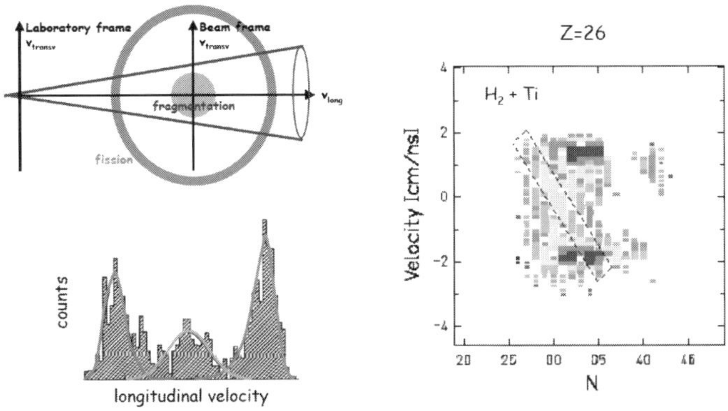

Figure 5: Demonstration of the kinematical cut due to the limited angular acceptance of the fragment separator. Residues emitted under small angles form different patterns: Fission residues appear in two humps from forward and backward fission, while evaporation residues appear in a single Gaussian-like distribution (from ref. [8]).

From the different studies on fission of ^{238}U, e.g. [2, 3, 4, 5], we present the results of two experiments with a 1 *A* GeV beam on a lead target [6] and on an hydrogen target [7, 8, 9, 10], respectively. A systematic overview on the full nuclide production in the reaction ^{238}U + ^{1}H and on the fragmentation-fission residues in the reaction ^{238}U + Pb is shown in Figure 6.

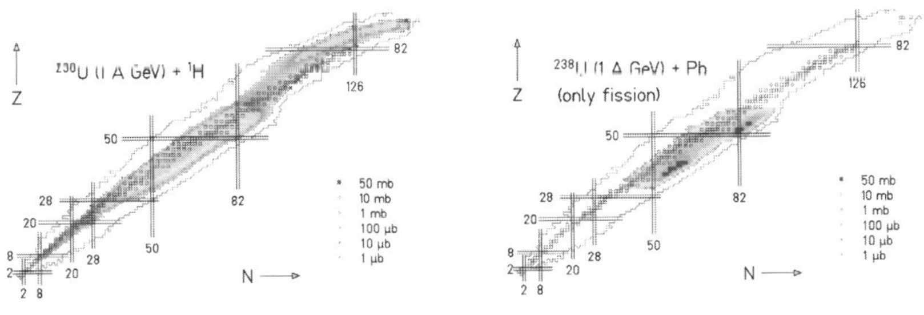

Figure 6: Measured nuclide production cross sections from the reactions ^{238}U (1 *A* GeV) + ^{1}H (left part, from refs. [7, 8, 9, 11]) and ^{238}U (1 *A* GeV) + Pb (right part, from ref. [6]). For ^{238}U + ^{1}H, where spallation-evaporation and spallation-fission residues are rather separated in mass, all residues above Z = 6 are shown. For ^{238}U + Pb, only fragmentation-fission residues are shown, because they strongly overlap with the spallation-evaporation residues. This experiment covered all elements between Z = 32 and 59.

One can observe that the nuclide distributions of the two reactions are rather different. The ^{238}U + ^{1}H system mainly produces symmetric fission. The isotopes are situated close to beta stability. The ^{238}U + Pb system mainly produces asymmetric fission, typical for low-energy fission in the actinide region. The main production is very neutron rich. This difference is explained by the different interactions leading to fission. While the proton, with its energy of 1 GeV in the projectile frame, mostly interacts with the uranium projectile by violent nuclear collisions, leading to a broad excitation-energy distribution which reaches up to several hundred MeV, the lead target nuclei also excite the uranium projectile by electromagnetic excitations with a large cross section at large impact parameters without any nuclear interaction. In this system, nuclear interactions with the lead target nuclei, which have a kinetic energy of 208 GeV in the projectile frame, in most cases lead to multifragmentation or total disassembly of the projectile and thus do not contribute to fission.

4. Semiempirical macroscopic-microscopic fission model

In the following we present our model for the prediction of the nuclide distribution in fission. The model is imbedded in the dynamic de-excitation code ABLA [8], which considers the competition between evaporation of neutrons, light charged particles and intermediate-mass fragments on one side and fission on the other side. For excitation energies above the corresponding threshold also break-up and the simultaneous emission of several fragments is considered [12]. Fission is treated as a dynamical process, taking into account the role of dissipation in establishing quasi-equilibrium in the quasi-bound region by the implementation of a time-dependent fission-decay width [13]. When the system passes the fission barrier and proceeds to fission, it is characterised by mass and atomic number, excitation energy and angular momentum. It is the aim of our model to follow the descent from saddle to scission of the system and to predict the probability that its ends up in one of the many possible splits in Z and A. We would like to mention that preliminary versions of this model have been presented previously [14, 15].

Most model descriptions of the fission process follow one of the following two roads: Either the evolution of the fissioning system is described with a purely theoretical model or the measured distributions of nuclides, kinetic energies and other observables are fitted by suitable functions with empirically determined parameters. The first road, i.e. the purely theoretical approach, is very challenging. Due to the complexity of the problem, any theoretical model has to introduce rather severe simplifications. In addition, one has to face the problem that the theoretical models are able to predict the relevant properties of a nuclear system only with a limited accuracy. This is obvious for the potential-energy surface in deformation space. Even in the nuclear ground state configuration, where the single-particle structure is generally studied very well, e.g. by nuclear spectroscopy, the empirical binding energy can only be reproduced with a standard deviation in the order of half an MeV. Such deviations are crucial for fission, e.g. a shift of 500 keV in the ground-state binding energy modifies the spontaneous-fission half life by about 2 orders of magnitude [16]. Similar or even larger uncertainties are expected in the competition between different fission paths in low-energy fission. Thus, these models are very important for improving our understanding of the fission process, but their ability for quantitative predictions seems to be still rather limited. Following the second road, developing purely phenomenological models, one is able to reproduce measured data very well. However, the predictive power of phenomenological models for extrapolations far from explored regions is rather low due to the lack of the essential physics. We have chosen an intermediate approach based on physical models but using adjustments to experimental data. Of course, also our approach needs several simplifications in order to be tractable. We have tried to choose our simplifying assumptions in best accordance with present theoretical knowledge.

Starting with Fong [17] and followed later by Wilkins et al. [18], the statistical model has been taken as the basis of most fission models. Since the available phase space is a very important driving force of any process in nature, we consider the statistical model as the basis also of our model. In both previous models, it was avoided to consider dynamical effects by applying the statistical model at the scission configuration. This is a rather severe simplification, which is certainly not realistic. Depending on the relaxation times of the different collective degrees of freedom, some memory on previous configurations might be present. Therefore, we will discuss this point with some care.

First we start considering fission at high excitation energy, where shells and pairing correlations are negligible. Observed mass distributions of heavy fissioning nuclei above the Businaro-Gallone point (i.e. $Z^2/A > 22$) from high excitation energies can well be described by a Gaussian distribution. This finding has been related to the available number of states above the potential energy as a function of mass asymmetry, since the potential can be approximated by a parabola near the minimum appearing at mass symmetry [19]. The second derivative $c_A = d^2U/(dA_1)^2$ of the potential as a function of the mass of one of the nascent fragments is related to the standard deviation σ_A of the mass distribution by the following relation:

$$\sigma_A^{\ 2} = \frac{T}{2 \cdot c_A} \qquad (1)$$

T is the nuclear temperature, which is related to the excitation energy of the fissioning system $E = a\,T^2$. The coefficient a is the level-density parameter.

This relation is a very important starting point of our model. Firstly, there exists a large body of experimental data on σ_A values, which provides the empirical data basis for a realistic prediction of fission mass distributions when structural effects are negligible. Secondly, the empirical result that the variance $\sigma_A^{\ 2}$ is proportional to the nuclear temperature supports the validity of the statistical model. However, it is difficult to extract from these data, at which moment on the descent from saddle to scission the decision on the width of the mass distribution is taken. We can imagine two possible extremes. In one case, the phase space at the saddle point determines the mass asymmetry of the system, which is more or less frozen on a fast descent to scission. In the other case, the mass asymmetry degree of freedom adjusts very fast to the potential and thus it is finally determined at scission. Since a variation of the mass asymmetry is connected with a substantial transport of nucleons and, thus,. the inertia should be large, we tend to support the first possibility. Following this idea, we take the systematics established in ref. [19] using the temperature at saddle in equation (1) for deducing the second derivative c_A of the potential from the experimental data. In fact, Itkis and Rusanov deduced a direct relation of c_A with the fissility parameter Z^2/A of the fissioning nucleus. Thus, we have the first quantitative relation we use in our model to calculate the width of the mass distribution in case of sufficiently high excitation energies.

Considering the fission process at lower excitation energies, our approach has to be substantially extended in order to include the appearance of fission channels. Early ideas for this concept are formulated in ref. [20]. Our experiments on multimodal fission in the light actinides [1] have considerably enriched our knowledge on the variation of structural effects in fission as a function of the nature of the fissioning system in terms of mass and atomic number. In particular, the gradual transition from double-humped mass distributions around ^{238}U to single-humped mass distributions around ^{208}Pb has systematically been mapped. This experiment provides valuable information on the fission characteristic of specific fissioning systems at rather well defined excitation energies. We have used this specific information for developing the description of fission channels in our model, which should then be able to calculate the nuclide production in spallation-fission and fragmentation-fission reactions as the superposition of fissioning systems in many de-excitation processes from a large number of excited pre-fragments and a wide range of excitation energies.

Following the hypothesis that the mass asymmetry degree of freedom is essentially frozen at saddle, the probabilities for the population of the different fission channels should be decided at the outer saddle. Therefore, we assume that there is a direct correspondence between the shape of the potential at the outer saddle as a function of mass asymmetry and the population of the fission channels. The appearance of each fission channel is linked to a specific minimum in the mass-asymmetry dependent potential at the outer saddle. At this stage we empirically determine the depths and the widths of potential minima of the different fission channels by the weights and the widths of the corresponding components in the empirical nuclide distributions. For this purpose, we need to calculate the number of states available in the different potential minima. This time, the Fermi-gas level density is not realistic: We have to consider the level density in a configuration with a substantial shell effect. For this purpose, we use the analytical relation proposed by Ignatyuk et al. [21].

The description of Ignatyuk et al. requires the knowledge of the macroscopic potential. According to the previous discussion, we represent it by the parabolic potential deduced by Itkis et al. [20] from the widths of the mass distributions at high excitation energies.

As a starting point for the quantitative determination of the shell effects at the outer barrier, we consider the mass distribution of the fission fragments from $^{238}U(n,f)$. As demonstrated in Figure 7, one obtains a rather consistent description, which reproduces the decrease of the relative population of the asymmetric fission channels with increasing excitation energy just by introducing two additional shells corresponding to the Standard 1 and Standard 2 fission channels, and by considering the washing out of the shell effects. Up to now, the shells at the outer barrier are formulated as a function of mass asymmetry. At this stage, we have a look to the results of two-centre shell-model calculations. They reveal that the shell effects at the outer barrier in ^{238}U are qualitatively similar to the shells in the separate fragments. It seems that the structure of the wave functions is quite similar all the way from the outer saddle to scission. This is not valid any more for more compact shapes, since the energetically favoured shape at the inner saddle is triaxial and mass-symmetric. Thus, we can profit from the investigations of Wilkins et al. [18] on the scission-point configuration, who stated that the most important shells behind the Standard 1 fission channel are $N = 82$ and $Z = 50$, while the Standard 2 fission channel is related to the $N \approx 90$ strongly deformed shell. Following these ideas, we attribute the two fission channels to the shells in the nascent heavy fragment mentioned above, while we neglect the influence of shell effects in the light fragment. From our adjusted parameters it appears that the spherical $N = 82$ and $Z = 50$ shells are considerably weaker than the shell effects we know from the ground-state masses around ^{132}Sn. It might be assumed that the additional matter in the neck disturbs the symmetry of the nascent heavy fragment and reduces the shell gaps compared to the ideal spherical configuration we meet in ^{132}Sn. The dominating appearance of the Standard 2 fission channel in $^{238}U(n,f)$ seems to indicate that the deformed $N \approx 90$ shell, which appears less strong in the separate fragments, see the results of the shell-model calculations in ref. [18], is less affected by the neck.

Once we have introduced the $N = 82$ and $Z = 50$ shells, we should consider the N/Z degree of freedom. In order to include this degree of freedom, we replace the common Standard 1 shell effect, which was formulated as a function of mass asymmetry, by two contributions from the $N = 82$ and the $Z = 50$ shells. In fact, we use the known ground-state shell effects of the nuclei around ^{132}Sn with an adjusted reduction factor. This way, the shell effect at saddle responsible for the Standard 1 fission channel varies as a function of N/Z of the fissioning nucleus. It is maximal for fissioning systems, which have the same N/Z as ^{132}Sn.

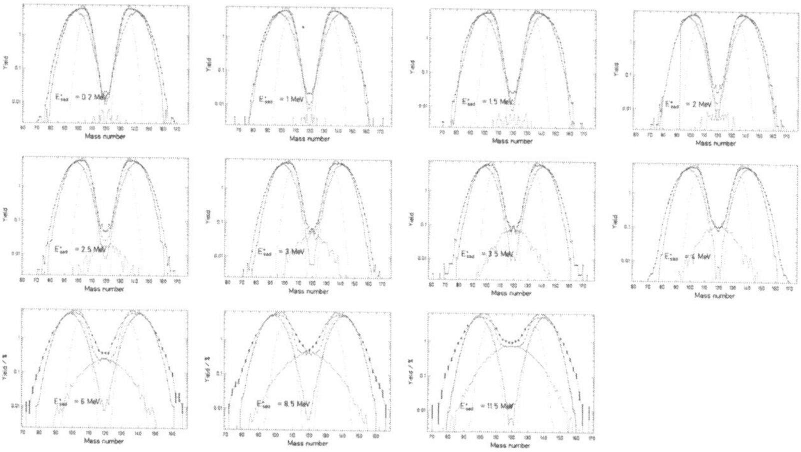

Figure 7: Calculated mass distributions (pink symbols) for neutron-induced fission of ^{238}U in comparison with experimental data (black symbols) [22, 23] for different values of the excitation energy above the fission saddle of the composite system ^{239}U. The calculated individual contributions of the different fission channels are shown in addition: Standard 1 (green), standard 2 (blue), and superlong (orange).

After having adjusted the strengths and the widths of the three shells to the mass distributions of the system ^{238}U(n,f), we are interested to check the predictive power of the model by applying it to other systems. We have chosen five isotopes around ^{226}Th, were we have observed a triple-humped mass distribution. The good agreement with available experimental data in Figure 8 proves that a common description of ^{239}U and a series of thorium isotopes around ^{226}Th is possible by using the same set of parameters.

This result is not too surprising if the assumptions of our model are approximately valid. In particular, the shell effects at the outer saddle should have strong similarities to the shell effects in the separate fragments. Thus, they are only a function of Z and N of the nascent fission fragment, while they should be rather independent from the fissioning system.

Considering this success, we conclude that our model has a remarkable predictive power, once the parameters have carefully been deduced from experimental fission-fragment distributions.

We may formulate the most salient feature of our model as a rather peculiar application of the macroscopic-microscopic approach to nuclear properties. In our consideration of the properties of the fissioning system at the saddle configuration, we may attribute the macroscopic properties to the strongly deformed fissioning system, while the microscopic properties are attributed to the qualitative features of the shell structure in the nascent fragments. This way, the macroscopic and the microscopic properties are strongly separated, and the number of free parameters is independent from the number of systems considered.

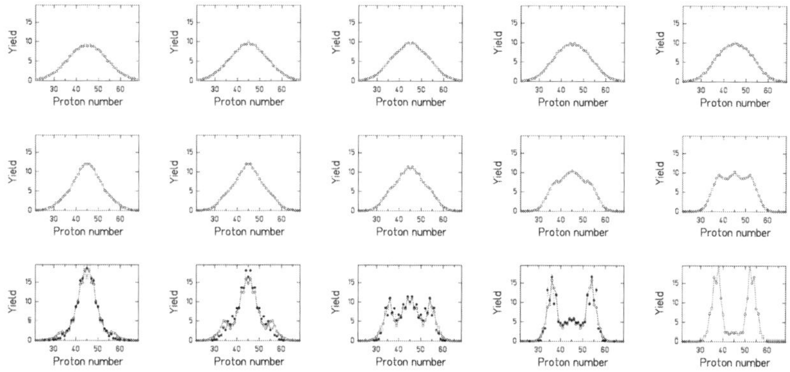

Figure 8: Calculated element distributions for ^{220,223,226,229,232}Th (from left to right) for excitation energies of 10, 20 and 60 MeV(from bottom to top). The model parameters are the same as in Figure 7. Available experimental data from [1] are shown as black symbols for comparison.

For completeness, we should mention that the model not only describes mass distributions but also considers the charge polarization in the nuclide production. Since there is only very little nucleon exchange necessary to exploit the full variation in *N*/*Z* to be expected, we assume that the charge polarization is determined near scission. Quantitatively, the charge polarization is governed by the macroscopic contributions to the energy at scission [24] in most cases. Only the simultaneous influence of the Z = 50 and N = 82 shells lead to a rather important deviation from this trend for the Standard 1 fission channel, which tends to produce nuclides closer to the doubly magic ^{132}Sn.

The full ABLA code, together with a suitable parameterization of the prefragments originating from nuclear-collision stage, was used to calculate the full nuclide distribution in the system ^{238}U (1 *A* GeV) + ^{1}H [8] as shown in Figure 6. Better than comparing the two-dimensional production cross sections on a chart of the nuclides, which almost looks identical to the experimental result, we compare in Figure 9 the mean values and the FWHM of the isotopic distributions. This gives a more qualitative benchmark of the code. Obviously, the complex structural effects are very well reproduced. In particular, one can identify a strong contribution of the Standard 1 fission channel, responsible for the production of neutron-rich isotopes above tin and leading to an increase of the mean *N*/*Z* ratio and an even stronger increase of the width of the isotopic distributions. The gradual decrease of the *N*/*Z* ratio between $Z = 60$ and $Z = 70$ corresponds to the transition from spallation-fission to spallation-evaporation in this range. The data of the system ^{238}U (1 A GeV) + Pb are reproduced with a similar quality. To obtain this good agreement, it is mandatory that the code calculates the fission fragments for a large number of fissioning systems in a large range of excitation energies in a realistic way.

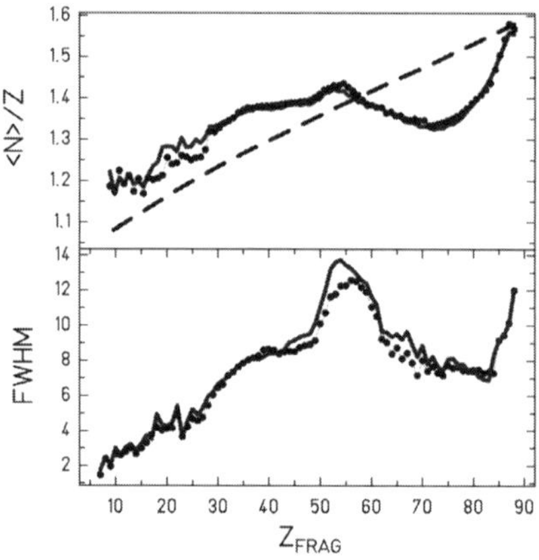

Figure 9: Comparison of the mean values and the FWHM of the isotopic distributions as a function of element number obtained in the reaction ^{238}U (1 *A* GeV) + ^{1}H [8] in experiment (black symbols) and calculation (blue line). Contributions from spallation-evaporation and spallation-fission are included.

5. Conclusion

The systematic measurements performed at GSI with a novel experimental approach have considerably enriched our knowledge on the characteristics of nuclide production in fission. These results were used together with other experimental data to develop a new macroscopic-microscopic fission model with simplified assumptions on the dynamic of the system and an empirical adjustment of the potential-energy landscape of the fissioning system. The good reproduction of the nuclide distributions for a number of different fissioning systems over an extended mass range indicates that the assumptions of the model are rather realistic. Imbedded in a nuclear-reaction code, the model is able to reproduce the nuclide distribution in the spallation of ^{238}U quite well.

References

1. K.-H. Schmidt et al., *Nucl. Phys.* **A 665**, 221 (2000).
2. P. Armbruster et al., *Z. Phys.* **A 355**, 191 (1996).
3. C. Donzaud et al., *Eur. Phys. J.* **A1, 407** (1998).
4 W. Schwab et al., *Eur. Phys. J.* **A2**, 179 (1998).
5. J. Pereira et al., *Phys. Rev.* **C**, in press.
6. T. Enqvist et al., *Nucl. Phys.* **A 658**, 47 (1999).
7. M. Bernas et al., *Nucl. Phys.* **A 725**, 213 (2003).
8. M. V. Ricciardi et al., *Phys. Rev.* **C 73**, 014607 (2006).
9. M. Bernas et al., *Nucl. Phys.* **A 765**, 197 (2006).
10. P. Armbruster et al., *Phys. Rev. Lett.* **93**, 212701 (2004).
11. J. Taieb et al., *Nucl. Phys.* **A 724**, 413 (2003).
12. K.-H. Schmidt et al., *Nucl. Phys.* **A 710**, 157 (2002).

13. B. Jurado et al., *Nucl. Phys.* **A 747**, 14 (2005).
14. J. Benlliure et al., *Nucl. Phys.* **A 628**, 458 (1998).
15. K. Kruglov et al., *Eur. Phys. J.* **A 14**, 365 (2002).
16. Z. Patyk et al., *Nucl. Phys.* **A 491**, 267 (1988).
17. P. Fong, *Phys. Rev.* **C 17**, 1731 (1978).
18. B. D. Wilkins et al., *Phys. Rev.* **C 14**, 1862 (1976).
19. M. G. Itkis et al., *Phys. Part. Nucl* **29**, 160 (1998).
20. M. G. Itkis et al., *Sov. J. Nucl. Phys.* **43**, 719 (1986).
21. A. V. Ignatyuk et al., *Sov. J. Nucl. Phys.* **21**, 255 (1975).
22. F. Vives et al., *Nucl. Phys.* **A 662**, 63 (2000).
23. C. M. Zöller, PhD thesis, Technical University Darmstadt, 1995
24. P. Armbruster, *Nucl. Phys.* **A 140**, 385 (1970).

New results for the intensity of bimodal fission in binary and ternary spontaneous fission of ^{252}Cf

C. Goodin*, D. Fong, J.K. Hwang, A.V. Ramayya, J.H. Hamilton, K. Li, Y.X. Luo
Department of Physics and Astronomy, Vanderbilt University, Nashville, TN 37235
** E-mail: christopher.t.goodin@vanderbilt.edu*

J.O. Rasmussen, S.C. Wu
Lawrence Berkeley National Laboratory, Berkeley, CA 94720

M.A. Stoyer
Lawrence Livermore National Laboratory, Livermore , CA 94550

T.N. Ginter
National Superconducting Cyclotron Laboratory, Michigan State University, East Lansing, MI 48824

S.J. Zhu
Department of Physics, Tsinghua University, Beijing 100084, Peoples Republic of China

R. Donangelo
Universidade Federal do Rio de Janeiro, CP 68528, RG Brazil

G.M. Ter-Akopian, A.V. Daniel, G.S. Popeko, A.M. Rodin,A.S. Fomichev
Flerov Laboratory of Nuclear Reactions, JINR, Dubna, Russia

J.D. Cole
Idaho National Lab, Idaho Falls, ID 83402

Triple coincidence data from the fission of ^{252}Cf were used to deduce the intensity of the proposed "hot" mode in barium channels. $\gamma-\gamma-\gamma$ and $\alpha-\gamma-\gamma$ fission data were analyzed to find the neutron multiplicity distribution for several binary and ternary charge splits. The binary channels Xe-Ru and Ba-Mo were analyzed, as well as the Ba-α-Zr, Mo-α-Xe, and Te-α-Ru ternary channels. An improved method of analysis was used in order to avoid many of the complexities associated with fission spectra. With this method, we were able to place an upper limit of 1.25% for the relative intensity of the second mode in the Ba-Mo case. For the Ba-α-Zr case, we found a relative intensity of 17% for the second mode.

1. Introduction

Observations of prompt γ-rays produced in the spontaneous fission of ^{252}Cf have shown evidence for a "hot" fission mode in the Ba-Mo channel. The evidence for this mode is observed as a higher relative intensity for the 7-10 neutron channels [1-3]. A later analysis [5,6] did not confirm the second mode, but did find an "irregularity" around the eight-neutron channel. In recent years, more complete data on the levels and relative intensities of transitions in barium and molybdenum isotopes have become available. There is also recent evidence that the second mode might be seen in Ba-α-Zr ternary fission [7]. Because fission spectra are often complex and the events of interest are rare compared with other channels, this type of analysis is difficult and prone to errors caused by random coincidences and background. Therefore an improved method which avoids many of these complexities was developed in order to determine the relative intensity of the second mode in both the binary and ternary cases involving barium.

2. Experimental Details

The data for this analysis comes from two experiments with the Gammasphere detector array located at Lawrence Berkeley National Laboratory. The binary data were taken in Nov. 2000. A 62μCi source was placed between two iron foils in order to stop fission fragments. This arrangement was then put in a 7.62 cm polyethylene ball and placed in Gammasphere. A total of 5.7×10^{11} $\gamma-\gamma-\gamma$ events were recorded. A coincidence cube was then constructed using the Radware software package [8]. The ternary data were taken in Dec. 2001. A 35μCi ^{252}Cf source was deposited as a 5 mm spot on a 1.8μ Ti foil covered on both sides by gold foils. Eight ΔE-E detectors were placed around the source to detect light charged particles (LCPs). In this experiment, 9.0×10^5 LCP-γ- γ events were recorded [10] and a γ-γ matrix was constructed. The ternary data were analyzed by using the γ-γ matrix peak fitting software written by Andrei Daniel. Earlier versions of this software were used by Ref. [1-3].

3. Data Analysis

The number of prompt neutrons emitted in a fission event can be determined by finding the mass number of the fragments produced in the event. For example, if the fission fragments of ^{252}Cf are determined by some method to be ^{144}Ba and ^{103}Mo, then five neutrons must have been emitted. The relative intensity of a particular neutron channel can be found from triple coincidence data by double gating on a pair of transitions in the heavy fragment, then measuring the intensity of its partners. This must be done for each isotope of the heavy partner. The yield as a function of neutrons emitted can then be determined by summing the contributions of all possible pairs.

In practice, the yield is found by fitting peaks in double gated spectra to find the area. For example, if a double gate is taken on ^{142}Ba, then the resulting spectra will show the 171.5 keV peak from ^{106}Mo, as well as the unresolved peaks of $^{104,108}Mo$ at 192.0 and 192.7 keV, respectively. The number of counts in the ^{106}Mo peak gives the relative intensity of the ^{142}Ba-^{106}Mo part of the four neutron channel. This number must be corrected for detector efficiencies and internal conversion coefficients. In principle this must be done for all possible sets of transitions to the ground state in each barium and molybdenum isotope in order to count all events. However, for this analysis only the strongest transitions in each barium and molybdenum isotope were measured. Since the relative intensities of other transitions are now known in the isotopes of interest, these values were used instead of attempting to measure each peak. Tables I and II show the yield matrices for the Ba-Mo and Ba-α-Zr splits calculated in this way. Note that the values given in the table are scaled and do not match the values in the following figures, which are normalized so that the total area of the Gaussian fit is equal to one. Summing along diagonals of this matrix gives the relative intensity for each channel. For example, the intensity of the 9-neutron channel in the Ba-Mo split would be the sum along the diagonal of the ^{141}Ba-^{102}Mo, ^{140}Ba-^{103}Mo, ^{139}Ba-^{104}Mo, and ^{138}Ba-^{105}Mo channels.

When analyzing the Ba-Mo and Xe-α-Mo splits, special care was taken in the measurement of the relative intensities of the ^{104}Mo and ^{108}Mo isotopes. With energies of 192.0 and 192.7 keV, respectively, the 2^+ to 0^+ transitions are not resolved in the double gated spectra. Instead, the intensities of the 6^+ to 4^+ 519 keV transition in ^{104}Mo and the 4^+ to 4^+ 414 keV transition in ^{108}Mo were measured. Since the relative intensities and detector efficiencies for these transitions are known, the individual yields of $^{104,108}Mo$ could be determined in this way. However, this introduces an-

Table 1. Relative Yield Matrix for the Ba(rows)-Mo(columns) charge split

	102	103	104	105	106	107	108	109
138				0.4(3)	0.4(1)		0.78(36)	
139			0.41(8)	0.4(1)	2.3(2)	2.7(6)	4.8(8)	2.5(1)
140		0.4(1)	0.2(1)	2.9(9)	11.8(6)	11(1)	22(5)	2.5(2)
141		1.2(9)	1.8(8)	13.0(7)	37(1)	26(3)	39(3)	3.2(3)
142	0.007(3)	6(1)	27.0(7)	45.9(9)	92(2)	41(2)	51(33)	5.5(2)
143	4(1)	18.6(8)	59.4(9)	93(3)	111(2)	32(3)	59(9)	2.8(8)
144	10.0(3)	51(1)	102(4)	99(2)	71.7(2)	10(1)		
145	21(1)	53(2)	79(2)	46(2)	21.9(9)			
146	16.7(7)	30.6(6)	39(3)	11.0(2)	4.5(2)			
147	4.3(2)	14.5(4)	5.1(4)	2.1(2)				
148	2.1(2)	1.8(2)	0.5(2)					

Table 2. Relative Yield Matrix for the Ba(rows)-α-Zr(columns) split

	98	99	100	101	102	103	104
140			2.2(7)		2(1)		4(1)
141					7(1)	4(2)	3(1)
142	3(1)	5(2)	1.9(5)	5(2)	7.2(8)	2.5(9)	2.0(6)
143	2(1)	5(3)	3(1)	78(3)	7(1)		
144	2.4(7)	3(1)	13.6(8)	4(1)	10.7(7)		
145	1.3(8)	7(3)	2(1)	12(3)	5(1)	6(2)	
146	4.2(8)		4.4(7)	4(1)			
147	3(2)	12(4)	4(1)				

other source of error because the relative transition intensities depend on the population of the upper levels, which can depend on the neutron channel. The population of a given level can vary by as much as 30% between neutron channels in some cases.

A related problem is that only three and higher fold coincidences are included in the analysis. Because the populations of upper states are low for some fragment pairs (like the 4^+ in ^{138}Ba), this can lead to a reduction in the relative intensity of these pairs when comparing only three-fold coincidences. The difficulties of measuring the intensity of the ^{103}Mo-^{138}Ba are discussed in [5].

The ternary analysis was similar to the binary analysis in that only the strongest coincidence was used to deduce the intensity of each channel. However, the ternary analysis was not done using the Radware software package. Instead, a two dimensional peak fitting code was used. The code fits coincidence peaks in the $\gamma-\gamma$ matrix and subtracts a smooth background. Figure 1 shows an example of the 2 dimensional, α-gated coincidence spectrum, where the horizontal and vertical axes are energies, and the color scale rep-

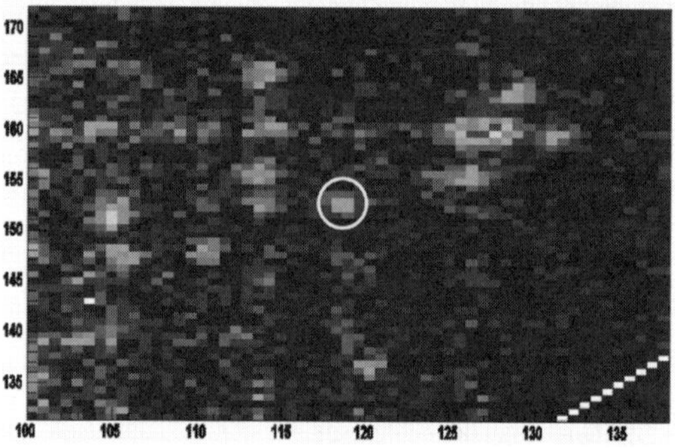

Fig. 1. α-gated coincidence matrix used in ternary analysis.

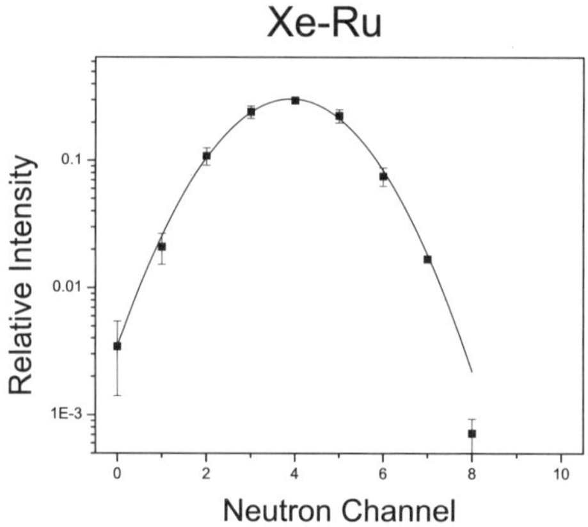

Fig. 2. Neutron distribution for the Xe-Ru charge split fit by a single Gaussian

resents counts/channel. Coincidence peaks, as well as ridges corresponding to coincidences between strong transitions and Compton background, can be seen. The circled peak is the coincidence between the 117.7 KeV transition in ^{143}Ba and the 151.8 KeV transition in ^{102}Zr, which is one part of the 3N channel for the Ba-α-Zr split.

After fitting the strongest coincident peaks in the $\gamma - \gamma$ matrix, all the other intensities can be calculated in a manner similar to the binary analysis from the intensity of the strongest coincidence. Because ternary fission is rare (on the order of one per few hundred binary events), statistics for this analysis were much lower, producing a greater statistical error. Since

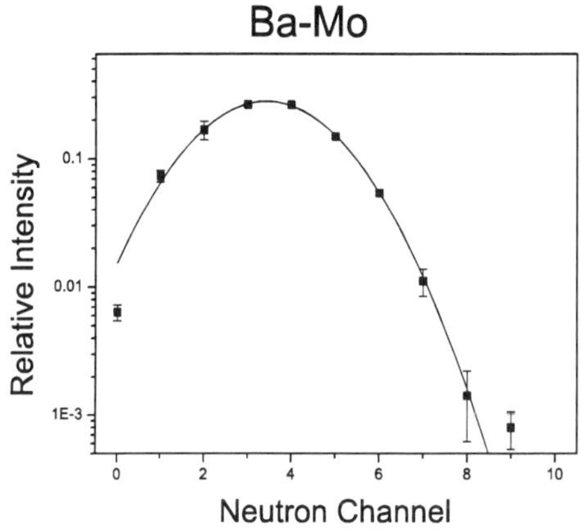

Fig. 3. Neutron distribution for the Ba-Mo split

α emission constitutes about 90% of ternary fission events, it was sufficient to consider only $\alpha - \gamma - \gamma$ events, rather than all LCP-$\gamma - \gamma$ events.

Although the authors of Ref. [1-3] derived the yields for fragment pairs by summing up the observed values of all peaks corresponding to γ transitions to the ground states of odd A fragments, the wealth of spectroscopic information now available, as well as the high statistics collected in the Gammasphere experiment conducted in 2000 make it worthwhile to revise the calculated yields of fission fragment pairs. Although taking only the 2^+ to 0^+ transitions is a good approximation for most even-even nuclei, there are some cases (^{141}Ba, for example) in which the less intense transitions make up a non-negligible percentage for the overall intensity of the channel, particularly in the 7 to 10-neutron channels.

The new method also has the advantage that most of the error from random contamination in transitions can be avoided. For example, because the 240.7 keV, 2^+ to 0^+ transition in ^{110}Ru is so intense in the fission of ^{252}Cf, it will add counts to the 241.1 keV peak in ^{103}Mo, even in double gated spectra. This effect becomes especially problematic in channels with very low statistics such as the ^{103}Mo-^{139}Ba 10-neutron channel. Such random contaminations can be avoided almost completely in this analysis.

To extract a relative intensity from coincidence spectra, it is necessary to normalize the measured intensity of the channel to the absolute yield of one of the isotopes of the pair. For the binary analysis, the values given by [9] for the heavy fragment were used. However, no absolute yields have been calculated for ternary fission. It was therefore necessary to use an ad-hoc

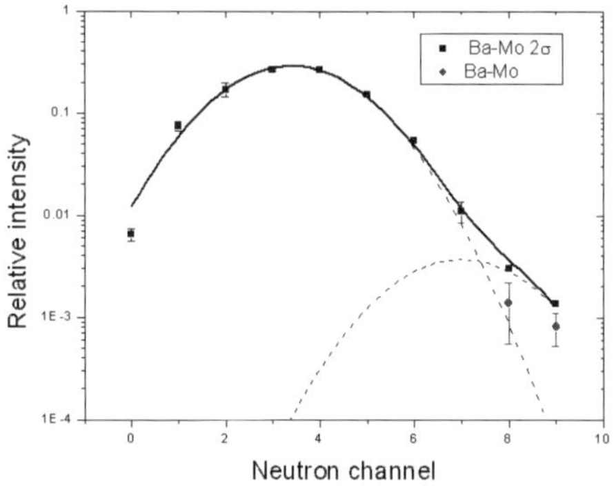

Fig. 4. Upper limit for the intensity of the 2nd mode in the Ba-Mo split.

normalization factor for the ternary cases. The ad-hoc factor was calculated from the binary distributions for each neutron channel by averaging the relative increase in each channel due to normalization. The calculated ternary normalization factors where negative 11% for even neutron channels and plus 20% for odd neutron channels. Because each neutron channel has contributions from several pairs of isotopes, the net effect of normalization should be roughly the same in the binary and ternary cases even though the individual yields of each isotope may be different. The primary drawback to this method is that this type of normalization might mask the appearance of a second mode in the ternary cases. This can be checked by comparing the normalized and unnormalized spectra to see if there is enhanced intensity in the 7-10 neutron channels.

4. Results

The Xe-Ru binary channel was the first to be analyzed. The resulting neutron distribution is fit well by a single Gaussian; there is no second mode, as shown in Fig. 2. However, in the Ba-Mo case, there is still a higher than expected yield for 9-neutron channel, as shown in Fig. 3. This could be evidence for a hot mode, but it is difficult to interpret this result in the usual way since this distribution is not fit well by a double Gaussian. An upper limit for the intensity of the second mode in this case was determined by taking the 2σ upper limits for the 8 and 9 neutron channels and fitting this distribution to a double Gaussian with the requirement that the first mode be centered around 3 neutrons and the second mode be centered at >6 neutrons. The result is shown in Fig 4. If there is indeed a second mode

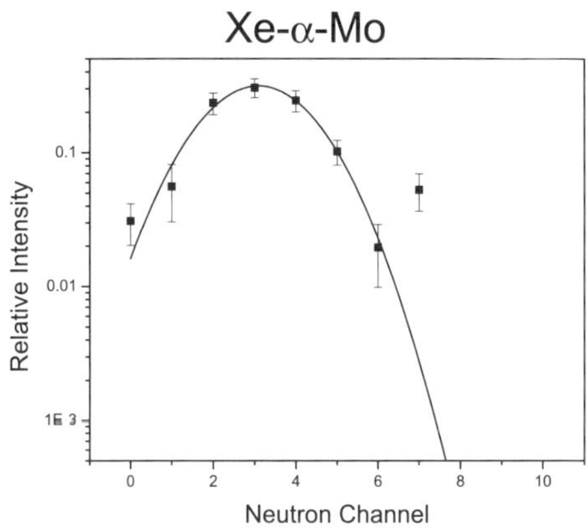

Fig. 5. Neutron distribution for the Xe-α-Mo ternary split with Gaussian fit

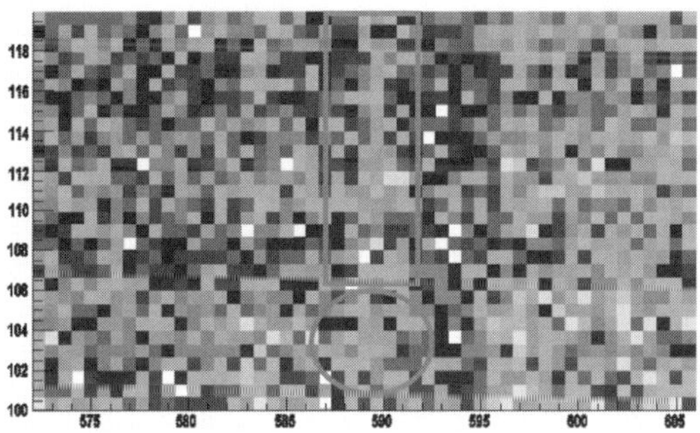

Fig. 6. Example of coincidences between Compton events and strong 589 keV transition in ^{141}Ba.

in the Ba-Mo split, this analysis suggests that the intensity is below 1.25% of the primary mode.

For the ternary analysis, the Te-α-Ru, Ba-α-Zr, and Mo-α-Xe splits were analyzed. Figure 5 shows the result for the Mo-α-Xe case. The enhanced 7 neutron emission is probably due to random coincidences of the 588.6 keV transition in ^{141}Ba with the Compton background. Figure 6 shows the α gated $\gamma - \gamma$ spectrum. The ridge associated with random coincidences is outlined by the rectangular box, while the peak of interest is outlined by the circle. It should be clear from the figure that, because ^{141}Ba is highly

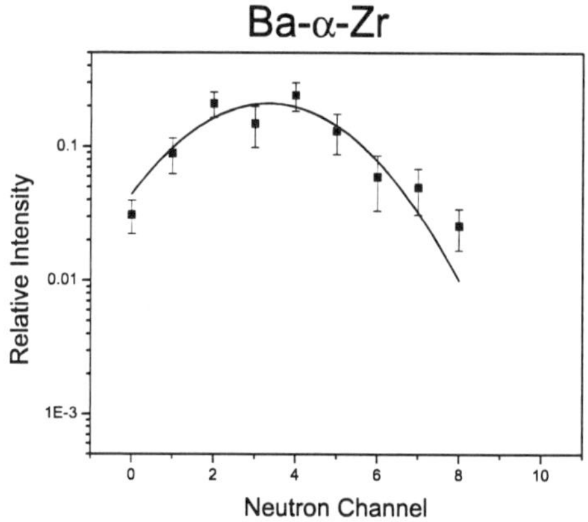

Fig. 7. Neutron distribution for the Ba-α-Zr split

populated, random coincidences are intense for this transition. Therefore, the enhanced 7 neutron emission shown by Fig. 5 is not considered a real effect.

The results for the Ba-α-Zr case are shown in Fig. 7, where the data are fit by a single Gaussian. One interesting feature is the greater width of the Gaussian fit. The FWHM of the fit is about 3.8, whereas in the ternary other cases the width is about 2.7. This is also true in the binary case, where the width of the Ba-Mo distribution is about 2.8 and the other binary cases have a width of 2.7. The increased width of the fit may be indicative of a lower fission barrier in the potential for the barium modes or of some other unique feature of the potential.

It is also possible to fit a double Gaussian to the Ba-α-Zr distribution. If the widths of the peaks are restricted to be the same as the other cases (FWHM=2.7), the distribution can be fit by two equal width Gaussians. To perform the fit shown in Fig. 8, it was also necessary to hold the peak position of the primary peak constant at 3.34 and restrict the position of the second peak to greater than 5 neutrons emitted. Therefore, the only variable parameters in the fit were the relative areas of the peaks and the x-position of the secondary peak. The resulting distribution fits the data almost as well as the single Gaussian distribution. The relative intensity of the secondary peak is about 17% of the primary peak. This is in agreement with our previous result (Ref. 7), but is not definitive because of the restrictive fitting parameters. However, it is interesting to note that, with reasonable assumptions about the widths and locations of the peaks, the distribution can be fit by a double Gaussian. Otherwise, there is not explanation for the much larger FWHM of the Gaussian fit for this case.

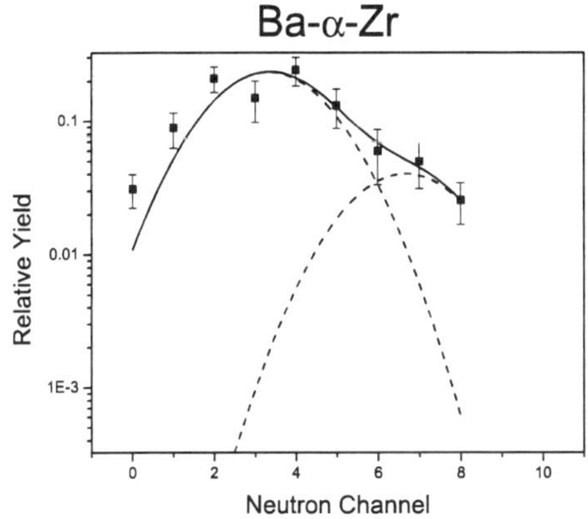

Fig. 8. Neutron distribution for the Ba-α-Zr ternary split fit to a double Gaussian distribution

5. Conclusion

In conclusion, the relative intensity as a function of neutrons emitted was determined for two binary channels (Ba-Mo and Xe-Ru) and three ternary channels (Ba-α-Zr, Xe-α-Mo, Te-α-Ru). By using a simplified method designed to reduce errors related to random coincidences and low peak to background ratios, no definitive hot mode was observed in the binary Ba-Mo split. However, an increased FWHM for the prompt neutron distribution was observed for the two barium channels, as well as an enhanced 9 neutron emission for the binary case. An upper limit for the relative intensity of the second mode was set at 1.5% in the Ba-Mo split. There is also evidence for an intense second mode in the Ba-α-Zr case with a relative intensity of 17%, although the distribution is fit well by a single Gaussian if the FWHM is allowed to be about 40% larger than in the other binary and ternary cases. A "hot" second mode in Ba-α-Zr could arise from a hyper-deformed shape of ^{144}Ba at scission, as suggested by [2] in interpreting the

Ba-Mo hot fission mode. Future work will include using data from a recent [11] experiment to deduce the total kinetic energy of the fission fragments and and an alternative method to resolve the ^{104,108}Mo peaks.

Acknowledgments

The authors would like to acknowledge the essential help of I. Ahmad, J. Greene and R.V.F. Janssens in preparing and lending the ^{252}Cf source we used in the year 2000 runs. The work at Vanderbilt University, Lawrence Berkeley National Laboratory, Lawrence Livermore National Laboratory, and Idaho National Laboratory are supported by U.S. Department of Energy under Grant No. DE-FG05-88ER40407 and Contract Nos. W-7405-ENG48, DE-AC03-76SF00098, and DE-AC07-76ID01570. The Joint Institute for Heavy Ion Research is supported by U. of Tennessee, Vanderbilt University and U.S. DOE through contract No. DE-FG05-87ER40311 with U. of Tennessee. The authors are indebted for the use of ^{252}Cf to the office of Basic Energy Sciences, U.S. Department of Energy, through the trans-plutonium element production facilities at the Oak Ridge National Laboratory.

References

1. G.M. Ter-Akopian *et al.*, Phys. Rev. Lett. **73**, 1477 (1994).
2. G.M. Ter-Akopian *et al.*, Phys. Rev. Lett. **77**, 32 (1996).
3. G.M. Ter-Akopian *et al.*, Phys. Rev. **C55**, 1146 (1997).
4. D.C. Biswas *et al.*, Eur. Phy. J. **A7** 189 (2000).
5. S.C. Wu *et al.*, Phys. Rev. **C62**, 041601 (2000).
6. S.C. Wu *et al.*, Nucl. Instrum. Meth. Phys. Res. **A480**,776 (2002).
7. D. Fong *et al.*, Fifty-Fifth Intern. Conference on Nuclear Spectroscopy and Nuclear Structure, St. Petersburg, Russia, 2005. Organizers: Y.T. Oganessian, K.A. Gridnev, L.V. Krasnov, A.K. Vlasnikov. To be published in *Bulletin of Russian Academy of Science, Series of Physics* (2006).
8. D.C. Radford, Nucl. Intstrum. Meth. Phys. Res. **A361**, 297 (1995).
9. A. Wahl, At. Data and Nucl. Tables **39**, 1 (1988).
10. D. Fong *et al.*, Proceedings of the Third International Conference on Fission and Properties of Neutron-Rich Nuclei, Sanibel Island, Florida, November 3-9, 2002., eds.J.H.Hamilton, A.V.Ramayya and H.K.Carter, pp.454-459,World Scientific Singapore (2003).
11. A.V. Daniel *et al.*, Physics of Atomic Nuclei **69** 8, 1405-1408 (2006).

RARE FISSION MODES: STUDY OF MULTI-CLUSTER DECAYS OF ACTINIDE NUCLEI

D.V. KAMANIN for FOBOS collaborations

Joint Institute for Nuclear Research, 141980 Dubna, Russia; Moscow Engineering Physics Institute, 115 409 Moscow, Russia; Department of Physics of University of Jyväskylä, FIN-40014 Jyväskylä; Hahn-Meitner-Institut-GmbH, Glienicker Strasse 100, D-14109 Berlin, German; Khlopin-Radium-Institute, 194021 St. Petersburg, Russia; Institute for Nuclear Research RAN, 117312 Moscow, Russia

We present a brief review of the results obtained by our collaboration in the frame of the program aimed at searching for new type of multibody decay of actinides, which was arbitrarily called as "collinear cluster tripartition" (CCT). First indications of new decay mode obtained for ^{248}Cm (sf) and ^{252}Cf (sf) let one to suppose that at least ternary almost collinear decay of the initial nucleus into the fragments of comparable masses appear to occur with the probability of about 10^{-5} per binary fission. The process is strongly influenced by shell effects in the decay partners. The results under discussion were obtained by the "missing mass" method i.e. only two of the decay products were detected in coincidence while the conservation laws indicate a presence of at least third partner.

1. Experiment at the modified FOBOS setup

First indications onto unusual multibody decays of ^{248}Cm (sf) and ^{252}Cf (sf) we have obtained in the experiments performed at the 4π-spectrometer FOBOS [1-3]. In order to improve reliability of identification of the CCT events the ordinary FOBOS setup has been modified and covered by the belt of neutron detectors. The experimental layout of the modified FOBOS spectrometer is shown in Figure 1.

Due to the low cross-section of the process and some additional requirements addressed to the spatial arrangement of the detectors involved the two-arm configuration containing five big and one small standard FOBOS modules in each arm was used. . Every module consists of position-sensitive avalanche counter (PSAC) and Bragg ionization chamber (BIC). Such scheme of the double-armed TOF-E (time-of-flight vs. energy) spectrometer covers ~29% of the hemisphere in each arm and thus the energies and the velocity vectors of the coincident fragments could be detected. In order to provide "start" signal for all the modules only wide-aperture start-detector capable to span a cone of ~100° at the vertex could be used. Even more essential requirement for the proper detection of the multibody events consists in combination of "start" detector with the radioactive source. Such three-electrode wide-aperture avalanche counter was especially designed for providing a "start" signal.

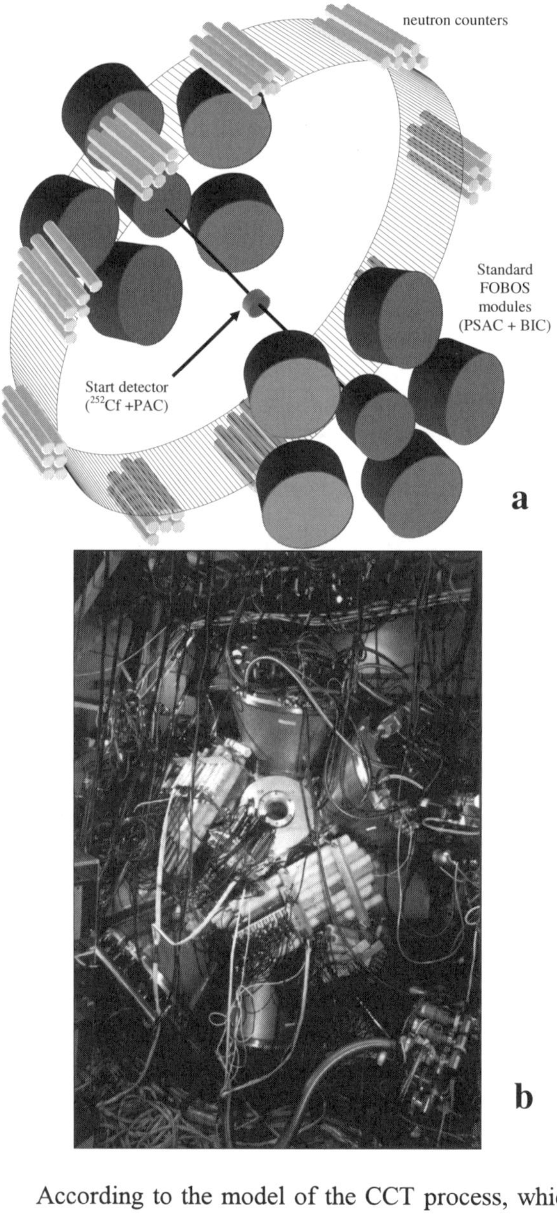

a

b

Figure 1. Schematic view of the modified FOBOS setup (a). FOBOS spectrometer surrounded by the belt of neutron counters (b).

According to the model of the CCT process, which could be referred from the initial experiments, the middle fragment of the three-body pre-scission chain borrows almost the whole deformation energy of the system. Being presumably in rest it would be an isotropic source of post-scission neutrons of a high multiplicity (~10) in the lab system. On the contrary, the neutrons emitted from the moving fission fragments are focused along the fission axis. In order to exploit this phenomenon for revealing the CCT events the "neutron belt" was assembled in a plane being perpendicular to the symmetry axis of the spectrometer, which serves as the mean fission axis at the same time. The centre of this belt coincides with the location of the FF source. The neutron detector consists of 140 separate hexagonal modules [4] comprising a ^{3}He-filled proportional counters which cover altogether ~35% of the complete solid angle of 4π. The number of tripped ^{3}He neutron counters was added to the data stream as an additional parameter for each registered fission event.

According the mathematical model of the neutron registration channel worked out [5, 6] the registration efficiency for those neutrons emitted from an isotropic source was found to be very closed to its geometrical limit, while the registration efficiency for neutrons emitted from the fission fragments registered by the FOBOS modules amounted to ~4% because they are focused along the fission axis which is perpendicular to the plane of the neutron counter belt. The registration probability for more than one neutron from ordinary spontaneous fission in this geometry amounts to 1%, however, the same probability for the CCT events runs up to 85%. The registration probabilities for more than two neutrons are 0.3% and 62%, respectively. Thus the neutron belt proves to be an effective instrument for revealing fission events accompanied by the isotropically emitted neutrons.

The mass-mass plot of the coincident fragments with the high multiplicity of neutrons (at least 3 of them should be detected) is shown in Figure 2a. It is easy to recognize the rectangular-shaped structure below the locus of conventional binary fission. This structure becomes more conspicuous (Figure 2b) if the velocity cut shown in Figure 3a is applied to the distribution.

The rectangle in Figure 2b, which is bounded by the clusters from at least three sides. Corresponding magic numbers are marked in this figure at the bottom of the element symbols. More complicated structures (marked by the arrows a, b, c in Figure 2c are observed in the mass-mass plot if the events with two fired neutron counters are also taken into play. Omitting for a moment physical treating of the structures observed, we attract ones attention to the specific peculiarity of some lines constituted the structures "b" and "c". The sum of the masses along them remains constant; see the dashed line in the lower left corner of Figure 2c for comparison.

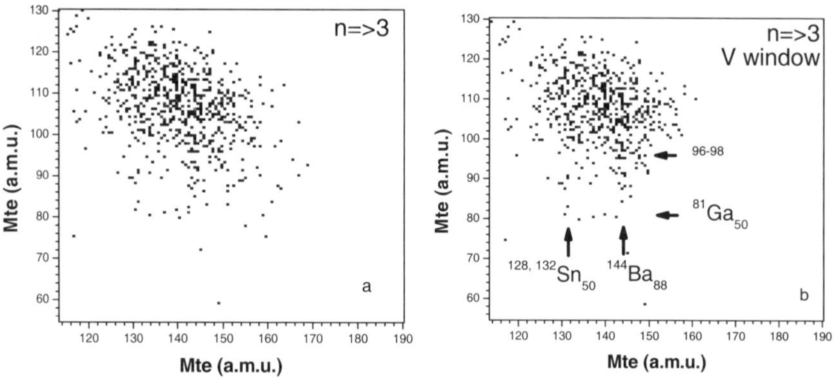

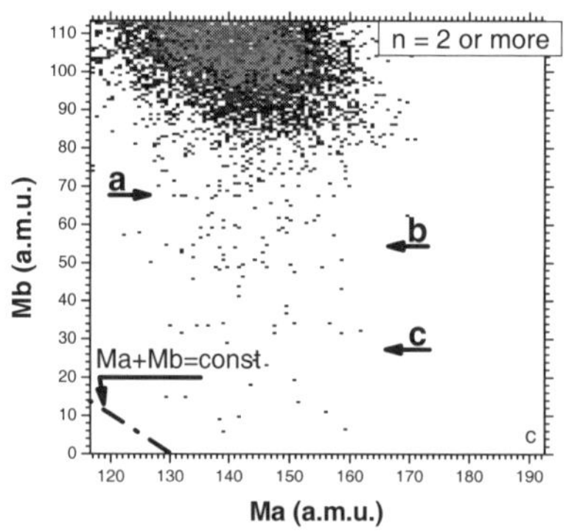

Figure 2. a) The mass-mass plot of the complementary fragments with at least 3 neutrons detected; b) The same plot after filtering the fragment velocities in the rectangular box shown in Figure 3; c) The same as (a), but the lowest number of tripped neutron counters let down to 2.

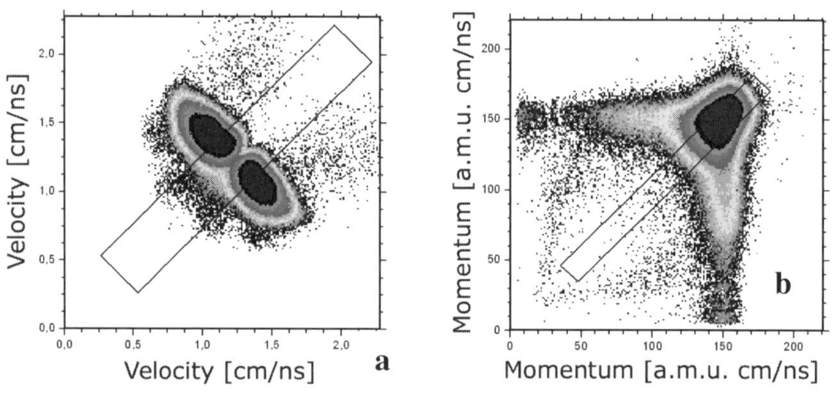

Figure 3. Velocity matrix of the complementary fragments (a) and momentum-momentum plot (b). The events falling into the rectangular boxes in Figures 3a, b were used to compose the final mass-mass plot in Figure 4.

Figure 4a represents a similar structure to that shown in Figures 2a, b except that it is not gated by neutrons and both the velocity and the momentum windows are used here to reveal the mass-symmetric partitions. The corresponding momentum distribution of the fragments and the selection applied are shown in Figure 3b. The plot in Figure 4b obtained on conditions of the momentum selection solely is not so clear. However like in the previous case the rectangular structure observed is bounded by the magic fragments, namely ^{68}Ni (the spherical proton shell Z=28 and the neutron sub shell N=40) and, probably, ^{84}Se (the spherical neutron shell N=50). Each structure revealed maps an evolution of the decaying system onto the mass space.

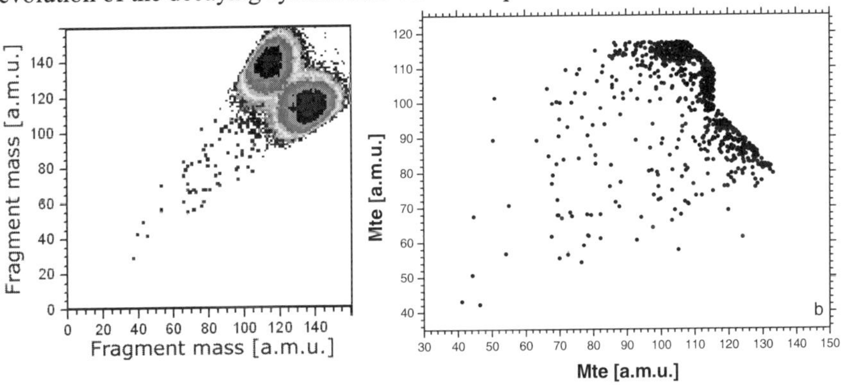

Figure 4. Mass-mass matrix of the complementary fragments selected by requirement of their approximately equal velocities and momenta (a), the same matrix if only momentum selection is assumed (b).

It should be stressed that the observed neutron multiplicity (the number of tripped neutron counters) for the events from the rectangle in Figure 4 is low. This fact contradicts the expectations put forward earlier that a middle fragment in the chain should be the source of neutrons of high multiplicity. The discrepancy reported may be an indication of more complicated decay scenario to be restored.

The following conclusions can be drawn from the results presented above:

- the multi-fragment (at least ternary) fission is experimentally confirmed;
- clustering of the decaying system, i.e. pre-formation of the magic constituents inside its body is decisive for the effect observed;
- collinear pre-scission configuration predicted by theory is proved to be a preferable one for true tripartition.

2. Comparative study of the effect at different spectrometers

The next series of results to be reported were obtained in three different experiments [7] devoted to searching for collinear tripartition of the ^{252}Cf nucleus.

In the first experiment (Ex1, Figure 5 a), performed at the FOBOS spectrometer installed in Flerov Laboratory (JINR, Dubna), about 13*10^6 coincident binary fission events were recorded. The TOF of the fragments was measured over a flight path of 50cm between the "start" detector (3) based on the micro-channel plates (MCP) placed next to the ^{252}Cf-source (1) and the "stop" position-sensitive avalanche counters (PSAC, 4). The energies of the coincident fragments, which passed through the PSACs were measured in the Bragg ionization chambers (BIC, 5) with entrance windows supported by a grid (6) with 70% transparency. The geometrical structure of the grid is hexagonal, the side view is shown in the insert "a" of Figure 5.

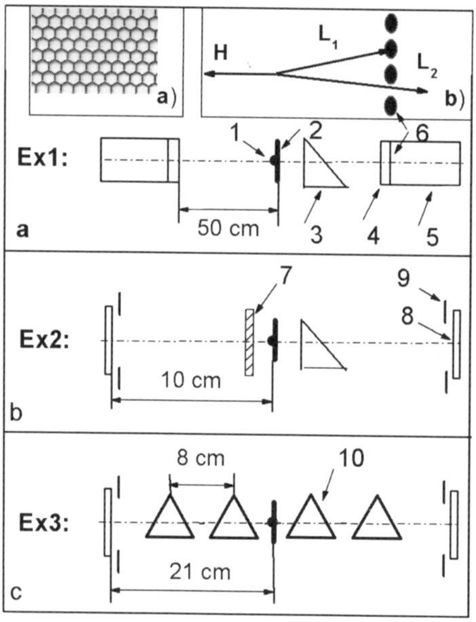

Figure 5. The scheme of coincident measurements of two fragments of the collinear tri-partition partners for the three experiments. First experiment Ex1 (a) was performed at the FOBOS setup. Here 1 – Cf source, 2- source backing, 3- microchannel plate (MCP) based timing "start" detector, 4- position sensitive avalanche counter as "stop" detector, 5- ionization chamber with the supporting grid 6 on the entrance window. The side view of the grid is shown in the insert "a". Second and third experiments Ex2, Ex3 (b and c) were performed at the spectrometers based on MCP detectors 2, 10 and PIN diodes 8 bounded by frame 9. Insert a) the scheme of detecting of the tripartition partners is shown in the insert "b". After passage of the dispersion foil two light fragments (L_1 and L_2) obtain a small angle divergence due to multiple scattering. One of the fragments (L_1) can be lost hitting the separating block, while the fragment L_2 reaches the energy detector.

This mechanical structure of the detectors is essential for the registration of the effect described below. In Figure 5 (insert b) the primary heavy fragment (H) is emitted to the left from the free side of the ^{252}Cf-source; the two light fragments (L1 and L2) are emitted into the same direction. As explained below scattering processes will separate the two light fragments in a small angular separation, and only one of them is likely to be registered. If both fragments enter only the total energy is measured.

A similar source of ^{252}Cf was used in further experiments performed in the Accelerator laboratory of the University of Jyväskylä, Finland (JYFL). In the second experiment (Ex2, Figure 5 b) we used a different TOF-E-spectrometer based on one MCP "start" detector and two PIN diodes (8), the latter provided both time and energy signals. An active area of each PIN diode was bounded by the frames (9). The flight-paths here were 10 cm for each detector arm. An Al-foil (7), 5 μm thick has been placed just near active ^{252}Cf layer. In this experiment $2*10^6$ binary events were registered.

In the third experiment (Ex3, Figure 5c) two pairs of the MCP-based timing detectors (10) provided signals for measuring TOF′s with flight paths of 8 cm each. The fragment energy was measured by PIN diodes. The total transparency of each arm amounted to 70% due to the grids of the electrostatic mirrors (four per detector, instead of two as in the Ex1 and Ex2) of the timing detectors. In this third experiment $2*10^6$ of binary events were collected.

In Figure 6a we show in a logarithmic scale the two-dimensional (2D) distribution of the two registered masses of the coincident fragments in the experiment (Ex1) at the FOBOS set up. The "tails" in the mass distributions marked (3)-(6) in Figure 6a, extending from the loci (1) and (2) used to mark the conventional binary fission, are mainly due to the scattering of the fragments on both the foils and on the grid-edges of the "stop" avalanche counters and the ionization chambers. The only small, but important, asymmetry between the two arms to be emphasized consists in a very thin source backing for the "rear side" and the start detector foil located in the arm "b" only (Figure 6a). An astonishing difference in the shapes of the "tails" (3) and (4) attracts attention. There is a distinct bump, marked (7), on the latter "tail" (4), oriented approximately parallel to the line defining a constant sum of masses, Ma+Mb=const, i.e. tilted by 45^0 with respect to the abscissa axis. The explanation of this bump is the essence of our analysis. The bump is located in a region corresponding to a large "missing" mass. The statistical significance of the events in the structure (7) can be deduced from Figure 6b. There the spectra of total masses, M_{total}=Ma+Mb, for the "tails" (4) and (3), spectrum "a" and spectrum "b", respectively, are compared. The difference spectrum of "b" and the tail (3) is marked "c", the integral of these events is $4.7*10^{-3}$ relative to the conventional fission events contained in the locus (2), shown in Figure 6a.

In order to explain the differences in the “tails” (4) and (3) mentioned above following scenario is proposed, the geometry is shown in Figure 5 (insert b). In ternary fission the three fragments are emitted collinearly and two of the fragments are emitted in one direction but become separated with an angle less than 1^0 after passing a dispersing media, due to scattering. These materials are the backing of the source (located only on the side of tail (4) or the Al foil placed deliberately in the path). If both fragments pass on and enter into the (BIC), we register a signal corresponding to the sum of the energies of the two fragments. The event is registered as binary fission with almost usual parameters. In the other scenario only a proper energy (mass) of one of the light fragments is measured, because the second one is stopped (lost) in the supporting grid of the ionization chamber, or for the other cases in the frame of the PIN diode playing the role of the separating element.

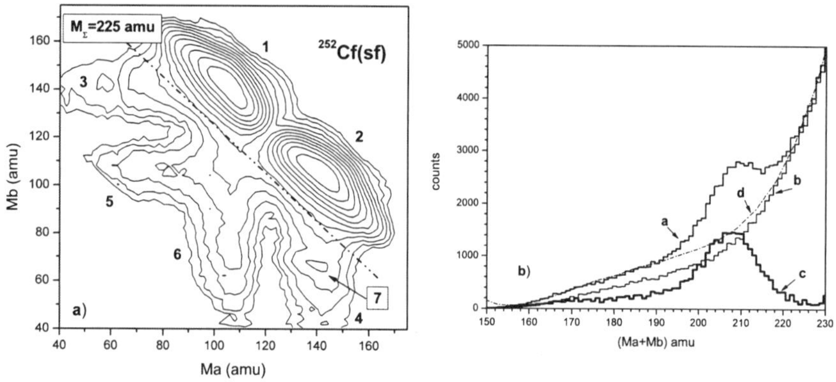

Figure 6. Experimental evidence of the collinear tripartiton of ^{252}Cf obtained at the FOBOS setup. (a) Contour map (in logarithmic scale) of the mass-mass distribution of the collinear fragments detected in coincidence in the opposite arms (marked by letters “a” and “b”) of the spectrometer. The loci of conventional binary fission events 1, 2 are prolonged by the “tails” marked as 3-6 due to the scattered fragments. Bump 7 located below the line of the sum Ma+Mb=225 amu is analyzed in Figure 6b. There the spectra of total masses for the “tails” (4) and (3), spectrum “a” and spectrum “b”, respectively, are compared. The difference spectrum is marked “c”. Is a polynomial fit Curve “d” is a polynomial fit using the points outside of the gross peak on spectrum “a”.

The scenario proposed above is confirmed by the results obtained in the experiments Ex2 and Ex3 (Figures 5b, c). Figure 7 depicts the spectra obtained in the same manner as it was done for the Ex1. As can be referred from the figure the bumps similar to this labeled by number 7 in Figure 6 demonstrate a multipeak structure which manifests itself due to better mass resolution in Ex2 and Ex3. The origin of these peaks can be understood from the fine structure of the bump 7 in two-dimensional presentation (Figure 8).

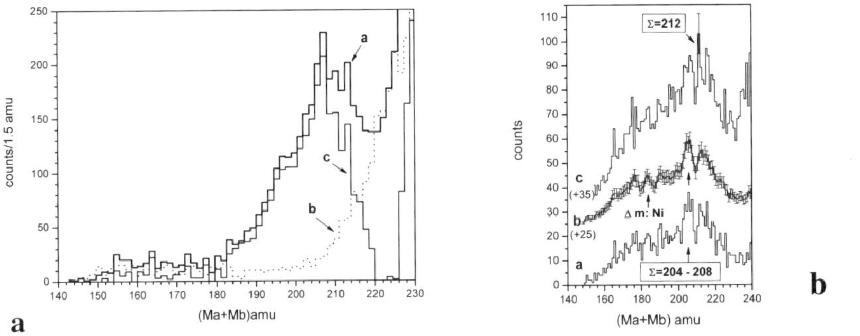

Figure 7. Spectrum of sum of masses (Mb+Ma) from experiment Ex2 (**a**), for two registered fragments for the gate similar to the "tail"4 from Figure 5a, spectrum "b" corresponds to the "clean" opposite arm free from dispersion foil, "c" is the difference spectrum. Spectrum of sum of masses of two detected fragments obtained in our third experiment Ex3 (**b**): (a) from the arm facing the source backing, (b) the same spectrum smoothed by means of averaging of counts in three adjacent channels (shifted up by 25 counts). (c) The sum of spectrum "a" and a complementary spectrum obtained in the second arm of the spectrometer (shifted up by 35 counts). The sums marked in the panels correspond to different pairs of magic nuclei (see text). The peak in spectrum "b" marked by arrow is due to the doublet of "missing" ^{70,68}Ni fragments.

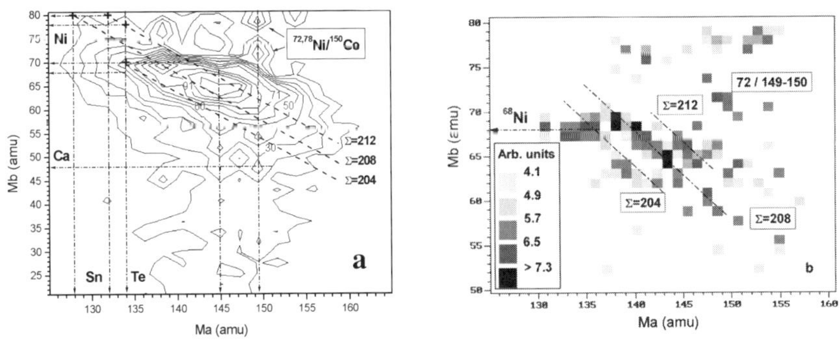

Figure 8. a) The figure depicts as a 2D-contour map (Mb versus Ma) the difference between the "tails" 4 and 3, of the events measured with the FOBOS-detector system shown in Figure 5a; note the expanded scale for the lighter mass fragments. Dashed lines tilted by 45^0 with respect to the Ma axis correspond to the fixed total mass of detected fragments (see the text for more details). Part b) the same as in a, however, passed through a filter which emphasizes the two dimensional structures.

3. Discussion

From the observed mass spectra we will have to consider a ternary fission process with one heavier and two lighter fragments. The missing masses in the sum spectra of the experiment (Ex1) suggest subsystems with particular masses. The same mass values are observed as distinct peaks in Figure 7; these are also seen as ridges in Figure 8b. We note that from these data the shell closures in proton and neutron number are decisive for the formation of the emitted subsystems. As can be deduced from Figure 8a the ridges (marked by the dashed lines) go through crossing points corresponding to different combinations of two fragments with "magic" nucleon numbers (marked by the dash-and-dot arrows). These marked points could be related to mass values with magic subsystems well known from binary fission [8] as follows: 204→ ^{70}Ni+^{134}Te or ^{72}Ni+^{132}Sn ("missing" ^{48}Ca), 208→ ^{80}Ge+^{128}Sn ("missing" $^{44}S_{28}$) and for M_{total}=212 → ^{80}Ge+^{132}Sn or ^{78}Ni+^{134}Te or ^{68}Ni+^{144}Ba.

It should be noted that the central peak in the Figure 7a (marked as Σ = 204-208) is likely a triplet which includes the peak centered at 206 a.m.u. It could be related to magic subsystem: 206→ ^{72}Ni+^{134}Te ("missing" $^{46}Ar_{28}$). Thus, three subsystems from these proposed above consist of three magic clusters each. The ridges discussed are crossed as well by the horizontal ridge (seen via bunching of contour lines in Figure 8a), this effect can be linked with the isotopes of ^{68,70}Ni which are also magic [9]. This observation would imply that the detected light fragment from the two L_1, L_2 fragments (see Figure 5b) is always a Ni-isotope. Due to the symmetry of the detector setup, namely that the two L_1, L_2 fragments are always detected in coincidence with the same heavy fragments, one must also observe events with a "missing" Ni-fragment. This is indeed observed, the peak corresponding to "missing" masses of 70 and 68 a.m.u. is well seen in Figure 7 (curve b). Thus the different peaks in the "missing"- mass spectrum consistently correspond all to the ternary decay scenario proposed.

References

1. Ortlepp H.-G. et al., *Nucl. Instr. and Meth.* A403 (1998) 65
2. Yu.V. Pyatkov et al., *Preprint JINR* P15-98-263, Dubna, 1998 (in Russian)
3. Yu.V. Pyatkov et al., *Proc. Int. Conf. "50Years of Shells"*, 21-24 April 1999, Dubna, p.301
4. Sokol E.A. et al., *Nucl. Instr. and Meth.* A 400 (1997) 96
5. Kamanin D.V. et al., *Physics of Atomic Nuclei* 66 (2003) 1655

6. Pyatkov Yu.V. et al, *Physics of Atomic Nuclei* 66 (2003) 1631
7. Yu.V. Pyatkov et al., *Preprint JINR* E15-2005-99, Dubna, 2005
8. Wilkins B.D. et. al. *Phys. Rev.* C 14 (1976) 1832
9. Rochman D. et. al. *Nucl. Phys.* A735 (2004) 3

Energy Distribution of Ternary α-Particles in ^{252}Cf(sf)

M. Mutterer [1,2*], Yu.N. Kopatch [3], S.R. Yamaletdinov [2], V.G. Lyapin [2,4†], J. von Kalben [1], S.V. Khlebnikov [2,5], M. Sillanpää [2], G.P. Tyurin [2,5], W.H. Trzaska [2,4]

[1]Institute of Nuclear Physics, Univ. of Technology, Darmstadt, Germany;
[2]Department of Physics, University of Jyväskylä, Jyväskylä, Finland;
[3]Joint Institute for Nuclear Research, Dubna, Russia;
[4]Helsinki Institute of Physics, Helsinki, Finland;
[5]V.G. Khlopin Radium Institute, St. Petersburg, Russia

We have dedicated a new experiment to remeasuring the ternary α-particle energy spectrum in ^{252}Cf spontaneous fission using an array of unshielded silicon detectors and unambiguously discriminating α-particles from neighboring isotopes by time-of-flight techniques using fission fragments as the start. Compared to many previous experiments, which entailed detection thresholds of 6 to 9 MeV α-particle energy due to the ΔE-E method applied and use of protection foils on detectors, the energy distribution of ternary α-particles could, for the first time, be measured down to 1 MeV. Furthermore, the energy spectrum of ^{6}He could be analysed, albeit with weak statistics, and the yield ratio ^{6}He/^{4}He was deduced. For ^{4}He, an excess in the yield as compared to a Gaussian shape is observed at energies below 9 MeV whose magnitude is in close agreement with data presented by Tishchenko et al. in 2002. On the other hand, the data formerly published by Loveland in 1974 exceed the low-energy yield of ternary α particles by up to a factor of two. We present our results on the ternary α-particle spectrum, and also the data on ^{6}He, and discuss prospects for future measurements.

Keywords: Ternary fission; ^{252}Cf(sf); E distributions of ternary ^{4}He and ^{6}He; TOF - E method.

PACS: 24.75.+i, 25.85. -w

*corresponding author, e-mail: mutterer@email.de
†deceased

1. Introduction

Since the discovery of ternary fission in the forties of last century there have been numerous experiments devoted to the energy distribution of the ternary particles, especially ternary (so-called "long-range") α particles of 16 MeV mean energy which exhaust ≈ 90 % of the total ternary fission yield (see, e.g. reviews [1,2]). There is common agreement that the shape of the α spectrum closely resembles a Gaussian but shows some yield in excess of at low energy [3]. The yield at low-energy (sometimes called "short-range" α particles) is particularly significant. Low-energy α particles presumably derive from low α-particle initial energies or above-average stretched scission configurations of the main fragments. They thus may provide important insight into both, the emission mechanism of ternary particles and the scission stage of the fission process. However, particle-unstable ternary particles (e.g., ^{5}He and ^{8}Be [4]) may also give rise to low-energy α's in sequential processes, and thus mask the true ternary particle emission. Without doubt, precise experimental data are very much needed for addressing the problem further.

From a literature study it became clear that most experiments performed up to now were not optimized for the study of lowest particle energies, even for the dominant α's . In fact, ΔE-E detector telescopes (or ΔE-E detectors coupled to electromagnetic spectrometers) were usually applied for particles discrimination, and the spectral distributions measured have low-energy thresholds not below 6 MeV α energy, at the best (e.g., Wagemans et al. [3]). In most cases thresholds were even higher, in the order of 8 to 9 MeV, inasmuch as absorber foils were used for shielding the detectors from the 300 to 500 times more frequent fission fragments and, in the case of ^{252}Cf, from the roughly 10 000 times more frequent 6 MeV α particles from radioactivity. Such rather high threshold values compared to the 16 MeV mean energy do not only cut away the interesting low-energy part of the spectrum but leave also substantial ambiguity in the energy assignment of the above threshold events due to uncertainty in absorber thicknesses and related energy losses. The only measurement with unshielded energy detectors reported up to now has been the recent work by Tishchenko et al. [5], where the 4π Berlin Silicon Ball (BSiB) with 162 detectors was facing a weak ^{252}Cf sample (30 fissions/s), and particle separation was performed by simultaneous time-of-flight registration of ternary particles and fission fragments. Here, the detection threshold was pushed down to 2.0 to 2.5 MeV, but separation of α particles, tritons and ^{6}He was only weak due to the short flight paths of 10 cm inside the ball. There have been also attempts to measure low-energy ternary α's by non-electronic methods e.g., a mass spectroscopic measurement after using Pb catcher foils for the reaction $^{235}U(n_{th}.f)$ [6], and solid state nuclear track detectors (SSNTD) for $^{252}Cf(sf)$ [7]. While in [6] a massive "short-range" component below 7.7 MeV was stated, no intensity in excess of a Gaussian shape was found in [7] in this region.

We have dedicated a new counter experiment to remeasuring the ternary α-particle spectrum from ^{252}Cf fission by improved time-of-flight (TOF) techniques with unshielded silicon detectors placed at a distance of about 20 cm to the source and registering the fission fragments with a channel plate detector (CP) at a close distance of about 2.5 cm.

2. Layout of Experiment

A photo of the experimental setup is shown in Fig. 1. The ^{252}Cf sample, produced at the Radium Institute at St. Petersburg, had an activity of about 500 fissions/ sec. It was prepared by self-transfer method of ^{252}Cf from a mother source, purified by ion exchange from any stable and long-lived radioactive impurities. The transfer was limited to a 0.1 cm^2 area on the support backing. The backing of the source consisted of the 22.2 ± 3.4 $\mu g/cm^2$ aluminum oxide (Al_2O_3) film with 10 $\mu g/cm^2$ layer of gold evaporated. Typical energy loss of fission fragments emerging perpendicularly through the backing material was ≤ 1.5 MeV. The fragment energy loss in the deposit layer , i.e. when emitted from the front surface of the source, was estimated to be less than 60 keV. The californium deposit (Cf_2O_3) is estimated to be 3.5 $\mu g/cm^2$ thick.

For measuring ternary particles an array of 10 silicon p-i-n diodes (380 μm thickness, and 30 x 30 mm^2 area) was placed at a distance of $\approx$ 20 cm from the source, together with a set of silicon p-i-n diodes of 500 μm thickness and 20 x 20 mm^2 area. With dedicated preamplifiers and low-noise timing filter amplifiers in the timing channels the energy threshold could safely be reduced to $\approx$ 0.5 MeV. This is the lowest cut-off value ever achieved in experiment. As the start signals, fission fragments emitted from the source were registered in a 30 mm diameter CP detector placed at right angle to the particle detectors, at 2.5 cm distance to the source. Discrimination of fission fragment start signals from the 30 times more frequent 6 MeV alphas was achieved by coincident registration of the complementary fragments in 10 silicon p-i-n diodes of 20 x 20 mm^2 area (roof detectors) mounted in a semicircular configuration at a distance of 6 cm to the sample, opposite to the CP. It has to be noted that having the particle detectors at right angles to the direction of emission of the fission fragments does bias the experiment to detect mainly ternary particles emitted at the instant of scission, i.e. the so-called equatorial particles. The detector geometry covers an angular range between about 70 ° to 110 ° for the mutual angle $\Theta_{\alpha L}$ between the ternary particle and the light-mass group of fission fragments. The experiment was set up in the large scattering chamber LSC at the Jyväskylä Accelerator Laboratory, Finland, and data were collected over a period of about 6 weeks. Due to the large total area of the silicon detectors the dose per unit area was rather low, so that radiation damage was not essential. Radiation damage in the roof detectors was more serious but had no direct influence on the energy and time registration for the ternary particles. The edges of all detectors were shielded by proper diaphragms for minimising scattering of fission fragments with an energy degradation to promote background pulses in the energy region of the ternary particles. Energy calibration of the silicon detectors was performed with alpha lines from a spectroscopic thin ^{226}Ra source and a BNC PB5 precision pulser.

Fig. 1. Experimental set-up for measuring the ternary α-particle spectrum from ^{252}Cf fission by the time-of-flight-E method. The assembly of ^{252}CF source, channel plate start detector and fragment energy detectors (roof detectors) is seen at the left-hand side. Ternary particle detectors are placed at 20 cm distance. Data from the sideward located detectors triplets have not yet been included in the present analysis.

3. Data Analysis

Currently data analysis is still not entirely terminated, so all results to be presented are still considered as preliminary ones. Analysis has been limited to events measured with the array of ten 380 μm thick ion-implanted silicon detectors, 30 x 30 mm^2 in size, which face the open side of the sample (array seen at the right-hand side in Fig. 1). Also, only fission events with fragments registered in both, the CP and the central pair of roof detectors, have been analyzed. The measured difference in the fragment flight times to the roof detector and CP, respectively, versus the fission fragment energy E_{ff} registered in the roof detector was used to correct the measured TOF spectra of ternary particles for the difference in flight time between heavy and light fragment masses from source to CP. Figure 2 shows a scatter plot of TOF vs. E of ternary particles summed over the 10 silicon p-i-n diodes selected.

The intense bunch in the centre of Fig. 2 corresponds to ternary alphas, and the weaker bunches of the neighboring isotopes ^{3}H and ^{6}He, below and above the α particle distribution, are nicely separated from it. The three bunches in the upper left corner are identified as ^{27}Al, ^{16}O and ^{12}C scattered off from the source backing or the roof detector surface by fission fragments. Between these groups and the ternary ^{6}He a few events from heavier ternary particles, mainly ^{8}He and ^{10}Be, are visible. It has to be noted that the TOF-E pattern in Fig. 2 is particularly clean from fragment background, which is expected to be true also for the low-energy region of our major interest, below 9 MeV. The vertical line at 6.1 MeV represents random coincidences with the 10^4 times more frequent α particles from ^{252}Cf radioactive decay which are of low enough probability to be safely subtracted in the time window of the ternary α particle distribution.

It is interesting to see also a small 6.1 MeV peak about 2 ns above the pattern for the ternary α's which is attributed to start signals from x-rays or conversion electrons in the CP when the ^{252}Cf radioactive decay proceeds through the excited state of ^{248}Cm. This peak falls accidentally into the TOF-E pattern of ternary ^{6}He ruling out an analysis of the ^{6}He spectrum in a small energy gap around 6 MeV. At the high-energy side the ternary α particles spectrum is cut off at 27.5 MeV due to the limited detector thickness of 380 μm. Since the cut-off takes place at a yield level of ≈ 3 % relative to the maximum yield at 16 MeV there is only minor influence of it on the high-energy half of the spectrum. For ternary ^{3}H, the TOF-E pattern reaches its highest energy value due to the detector thickness already at 11.5 MeV (see Fig. 2) bending back at higher ^{3}H energies and

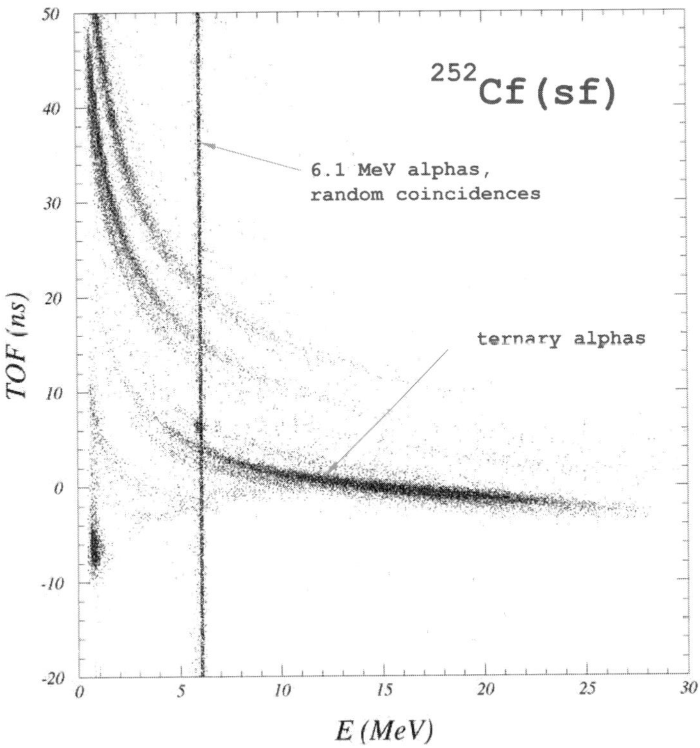

Fig. 2. Scatter plot TOF vs. E of ternary particles in ^{252}Cf(sf), as measured with 10 silicon p-i-n diodes of 380 μm thickness and 30 x 30 mm^2 in size, located at 20 cm distance from the source. Time is in ns with respect to the flight time of 16 MeV α particles; energy E is in MeV.

interfering with the respective pattern for the protons.

4. Results and Discussion

Figure 3 is our preliminary energy distribution of ternary α particles, which covers the wide energy range from 1 to 27.5 MeV. Above about 9 MeV our data are in fair agreement with most previously measured spectra [3–5,7,8]; over the full energy range a comparison with the results by Tishchenko et al. [5] and Loveland [8] is depicted in the right-hand plot in Fig. 3. In the high-energy region from 22 to 27 MeV our data are slightly below the values measures in [3,5]. This may be due to variances in the calibration

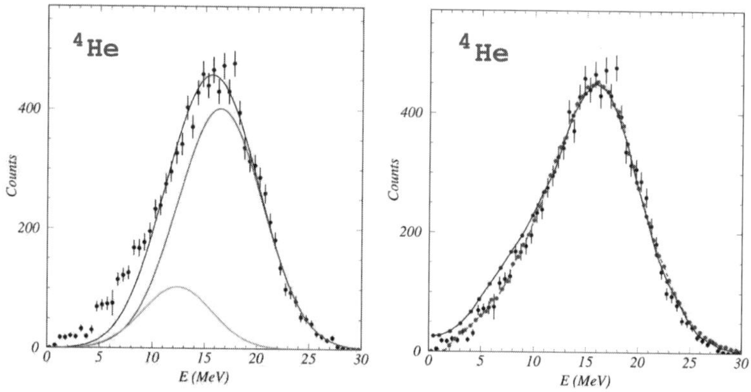

Fig. 3. Tentative energy distribution of ternary α particles from ^{252}Cf fission. Total number of counts recorded is 10,654. **Left-hand side:** Two-Gaussian curves fitted to the data above 9 MeV, taking residual α particles from ^{5}He decay into account, according to the results of Kopatch et al. [4](see text). **Right-hand side:** Comparison with data by Tishchenko at al [5](dashed line) and Loveland [8] (full line).

procedures but can likely be caused also by our constraint in the α particle emission angle $\Theta_{\alpha L}$. In view of the measured data E_α vs. angle $\Theta_{\alpha L}$(see Fig. 2 in [9]) it may well be that a minor part of the yield at higher energy is suppressed by the angular dependent coincidence efficiency of the present detector geometry. The issue is currently studied with the aid of a simulation calculation, using data on $\Theta_{\alpha L}$ vs. E_α measured previously by Heeg et al. [10]. On the other hand, if ternary particle spectra were measured without fragment coincidences, as was, e.g., the case in [3], the high-energy end of the α spectrum is slightly enhanced compared to the spectrum of equatorial α's by the interference with the rare but higher energetic polar ones. The resulting slightly weaker slope is then mirrored to the low-energy side when the analysis is made in terms of Gaussian fits.

Shown in the left plot of Fig. 3 are also two Gaussian curves which result from a fit to the present data above 9 MeV, and consider besides true ternary α particles (the dominant Gaussian) also the about 17% contribution of residual α particles from the decay of ternary ^{5}He, as was recently measured by Kopatch et al. [4]. It is obvious from the data presented in Fig. 3, that the ternary α spectrum shows more low-energy α particles than would be predicted by the Gaussian shapes. Apparently, the spectral shape measured previously at energies $E > 9$ MeV [3,4] can not be extrapolated meaningful to low energies. Of particular interest is that there are ternary α particles emitted from ^{252}Cf fission with energies as low as 1 MeV, with the energy distributions indicating a flat shelf or shoulder at energies below 5 MeV. We would like to note that, in the present experiment, there is not any material between the ^{252}Cf source and the surface of the detectors that could slow down the ternary particles, even at low energy. All ion-implanted silicon detectors in use have aluminium front windows of nominally 140 nm thickness. The corresponding effective dead-layer was determined with angular dependent α spectroscopy to be 369(11) nm of silicon equivalent thickness [11], which results in an energy correction of 110 keV for 1 MeV α particles, and 30 keV at 10 MeV.

Within experimental errors, the present spectrum is in good agreement with the low-energy data obtained recently by Tishchenko et al. [5] between 2.5 and 9 MeV. However, the ternary α particles spectrum reported in the latter work contains the about 4 % admixture of ^{6}He with a maximum yield around 12.3 MeV (see below) which may slightly change the spectral slope at the low-energy side. On the other hand, the data reported as early as 1974 by Loveland [8] exceed the real low-energy yield of ternary α particles by up to a factor of two. We have not been able to clarify the discrepancy of both recent works with Loveland's early result which has for a long time been the only ^{252}Cf data on low-energy ternary α's available in literature. It also remains unclear to us how the method of ΔE-E applied in [8] could provide precise data in the energy region below the threshold energy of the ΔE silicon detector used. The present experiment, and that of Ref. [5], show a smaller low-energy tailing as suggested from the data in [8], being now comparable in magnitude with the tailing known from ^{235}U(n_{th},f) [3,12].

We have finally made an attempt to extract also the energy spectrum of ternary ^{6}He from our data shown in Fig. 2, leaving out the energy region around 6 MeV. Our preliminary energy distribution of the ternary ^{6}He is plotted on the right-hand side in Fig. 4, the total number of events collected over the 6 weeks period of the measurement being 489. This is a rate of about 10 events per day. To our knowledge it is the first time that ternary ^{6}He particles from ^{252}Cf(sf) were measured over their full energy range. The spectrum turns out to be asymmetric as well. However, the crucial point for the reliability of these low-rate data is the question about a possible interference with some background in the analysis window. Summing up the spectrum shown in Fig. 4, with interpolating the missing values around 6 MeV, and relating it to the sum of α-particles shown in the left-hand side of Fig. 4, has yielded for the ratio ^{6}He/^{4}He a value of 0.0450(20). Fitting the spectrum, for energies above 9 MeV, with a single Gaussian

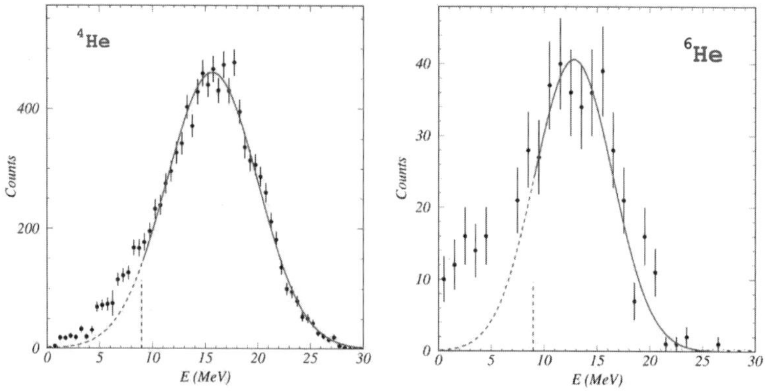

Fig. 4. **Left-hand side:** Tentative energy distribution of ternary α particles from ^{252}Cf fission. Total number of counts recorded is 10,654. **Right-hand side:** Tentative energy distribution of ternary ^{6}He particles. Total number of counts is 489. Solid lines are Gaussian curves fitted to the data above 9 MeV.

curves and taking the area under the Gaussian as the estimate for the ^{6}He yield gives for the ^{6}He/^{4}He ratio a value of 0.0365(18), which is in line with most values deduced earlier from experiments with similar threshold energy (e.g., Refs. [13–15]), indicating that a background contribution to the present spectrum, which is difficult to determine precisely, is not essential. Furthermore, the analysis of the spectrum from the rare ^{6}He particles gives us confidence that an essential interference of fragment background with the about 25 times more intense ternary α spectrum, shown in Figs. 3 and 4, can safely be neglected.

Encouraged from the progress achieved with the present TOF-E measurement on the ternary α spectrum we are planning a new experiment with an about 10 times stronger ^{252}Cf source on a thin Ni backing and an improved fragment trigger with using the sample as the conversion foil of a micro channel plate (MCP) start detector. This will improve time resolution, the statistical accuracy in the low-energy regime, and allow studying dependence of the energy distribution on particle emission angle. We also want to scrutinize whether combination of TOF with pulse-shape discrimination techniques in suitable reverse-mounted silicon surface barrier (SB) detectors [16] can be applied for the discrimination of ternary particles according to both, their mass and nuclear charge. This would permit registration of ternary particle spectra up to carbon isotopes down to low energies. Application of the present technique for neutron induced fission reactions where interference with alphas from radioactive decay is generally less, e.g. in $^{235}U(n_{th},f)$, is also envisaged.

Acknowledgments

The present work was supported by the Academy of Finland, Center of International Mobility, and by the INTAS Grant No. 03-51-6417. One of us (M.M.) wants to thank the Academy of Finland for a research grant. Fruitful discussions with F. Gönnenwein are gratefully acknowledged.

References

1. C. Wagemans, in The Nuclear Fission Process, edt. C. Wagemans (CRC Press, Boca Raton, Fl. USA, 1991), Chap. 12.
2. M. Mutterer, and J. Theobald, in Nuclear Decay Modes, edt. D.N. Poenaru (IOP, Bristol, UK, 1996), Chap. 12.
3. C. Wagemans, J. Heyse, P. Jansen, O. Serot, and P. Geltenbort, Nucl. Phys. A 742 (2004) 291.
4. Yu.N. Kopatch, M. Mutterer, D. Schwalm, P. Thirolf, and F. Gönnenwein, Phys. Rev. C65 (2002) 044614.
5. V.G. Tishchenko, U. Jahnke, C.-M. Herbach and D. Hilscher, Report HMI-B 588, Nov. 2002.
6. G. Kugler, and W.B. Clarke, Phys. Rev. C 5 (1972) 551.
7. H. Afarideh, K. Randle, and S.A. Durrani, Ann. Nucl. Energy 15 (1988) 201; also: Intern. Journ. of Radiation Applications and Instrumentation D 15 (1988) 323.
8. W. Loveland, Phys. Rev. C 9 (1974) 395.
9. F. Gönnenwein, M. Mutterer, and Yu. Kopatch, Europhysics News 36/1 (2005) 11.
10. P. Heeg, J. Pannicke, M. Mutterer, P. Schall, J.P. Theobald, K. Weingärtner, K.H. Hoffmann, K. Scheele, P. Zöller, G. Barreau, B. Leroux, and F. Gönnenwein, Nucl. Instr. Meth. in Phys. Research A278 (1989) 452.
11. A. Spieler, Diploma Thesis, TU Darmstadt, 1992, unpublished.
12. F. Caitucoli, B. Leroux, G. Barreau, N. Carjan, T. Benfoughal, T. Doan, F. El Hage, A. Sicre, M. Asghar, P. Perrin, and G. Siegert, Z. Physik 298 (1980) 219.
13. Z. Dlouhý, J. Švanda, R. Bayer, and I. Wilhelm, Proc. Int. Conf. on Fifty Years Research in Nuclear Fission (Berlin, 1989), Report HMI-B 464, p. 43.
14. V. Grachev, Y. Gusev, and D. Seliverstov, Sov. J. Nucl. Phys. 47 (1988) 622.
15. G.M. Raisbeck and T.D. Thomas, Phys. Rev. 172 (1968) 1272.
16. M. Sillanpää, S. Khlebnikov, Y. Kopatch, M. Mutterer. W. H. Trzaska, G. Tyurin, and V. Lyapin. Proc. Annual Meeting of the Finnish Physical Society, 2006, Tampere, Finland, Poster 07/33.

PRELIMINARY RESULTS OF EXPERIMENT AIMED AT SEARCHING FOR COLLINEAR CLUSTER TRIPARTITION OF $^{242}Pu^{*}$

Yu.V. PYATKOV[†] for HENDES and FOBOS collaborations

Moscow Engineering Physics Institute, 115 409 Moscow, Russia; Joint Institute for Nuclear Research, 141980 Dubna, Russia

First results of experiment aimed at searching for collinear cluster tripartition channel in the reaction $^{238}U+^{4}He$ (40 MeV) are presented. Such unusual decay mode was observed earlier in ^{252}Cf (sf). A two-arm TOF-E (time-of-flight vs. energy) spectrometer with micro- channel plate detectors and mosaics of PIN diodes was used. Among ternary events detected there are some presumably due to the decay of Pu shape-isomers built on the pairs of magic clusters. Fission of these states results in forming of long lived di-nuclear molecule like systems which can disintegrate via inelastic scattering on the materials on the flight path..

1. Experiment

In series of experiments using different time-of-flights spectrometers we observed unusual decay mode of ^{252}Cf (sf) which was treated as "collinear cluster tripartition" [1-4]. So far experimental manifestations of this decay channel were obtained in the frame of the "missing mass" method. It means that only two almost collinear fragments were detected in coincidence and they were much smaller in total mass than initial nucleus. It is reasonable to suppose that "missing" mass corresponds to the mass of undetected fragment (or fragments) flying apart almost along a common fission axis. Shell effects in the resultant fragments seem to be decisive for the process of interest.

Evidently, a direct detection of all decay partners is pretty desirable for reliable identification of unusual reaction channels. In order to solve this problem a setup of high granularity should be used. Such kind of spectrometer

[†] Work partially supported by Russian Foundation for Basic Research, grant 05-02-17493, CRDF, grant MO-011-0.

installed at the JYFL, (Jyväskylä, Finland) was chosen for studying the reaction $^{238}U+^{4}He$ (40 MeV). The scheme of the experimental setup is shown in Figure 1.

The spectrometer includes two arrays 19 PIN-diodes each, two MCP (micro-channel plate)-based start detectors and specially designed target holder. Each PIN diode provides both energy and timing “stop” signals. The size of the individual PIN-diode is 3 x 3 cm^2, the depth of the depleted layer is about 200 μm. MCP aperture is 30 mm (diameter of the entrance window) and the thickness of the converter carbon foil does not exceed 30 $\mu g/cm^2$. Target holder with two targets (100 $\mu g/cm^2$ layer of ^{238}U evaporated on 50 $\mu g/cm^2$ thick Al_2O_3 backing each) was installed on the axle in the center of reaction chamber and can be rotated. In the working position the angle between the target and beam direction is 30 °. The beam (FRC= 14.820 MHz what gives ~ 67 ns interval between the beam bursts) was focused on the target into a spot of 5 mm in diameter by means of two collimators.

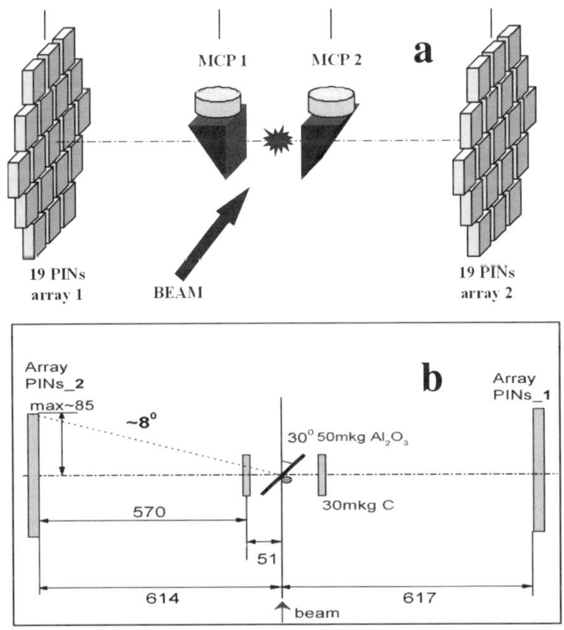

Figure 1. Overall scheme of the experimental setup (a) and additional information concerning its parameters (b).

One of the specific problems appearing in using of PIN-diodes for the heavy ion spectrometry is the necessity to take into account so-called pulse-height defect (PHD). PHD is commonly defined as a difference between the energies of a heavy ion and of an alpha particle yielding the same pulse height. PHD parameterization proposed in [5] was used in calculation of the Mte masses i.e. masses obtained in the frame of TOF-E (time-of-flight vs. energy) method. Corresponding calibration parameters were obtained by fitting to the known fission fragments (FF) spectra of the velocities and masses for ^{252}Cf (sf) (see [6] for details). Experimentally evaluated mass resolution achieved does not exceed 2.5 a.m.u. (fwhm).

Next setup peculiarity to be pointed out is connected with a position of the "start" detectors namely in 51 mm from the target (Figure 1b). In the multibody decay the most faster from the fragments hitting "start" detector gives "start" signal to be common for the corresponding arm. Consequently a velocity of this fragment only will be measured correctly while others fragments detected in the same arm get shifted velocity values. True (emission) velocities can be calculated according formulas in Table 1 where Vexp, Vemis – are, respectively, the experimental and emission velocities.

Table 1. Corrections to the fragments velocities.

Variant №	Kinematical scheme of the decay mode	Formula for calculation of emission velocity
1	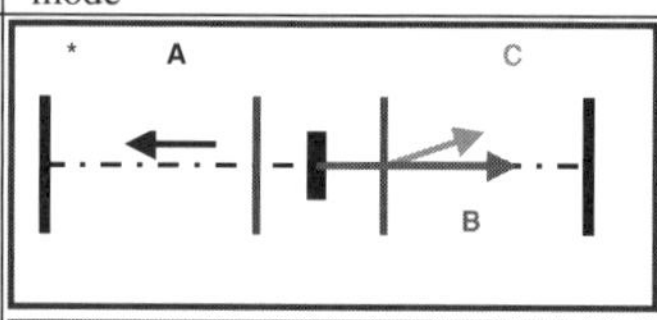	2 fragments formed in the MCP detector fly in the same direction: <u>no corrections,</u> i.e. Vemis= Vexp
2	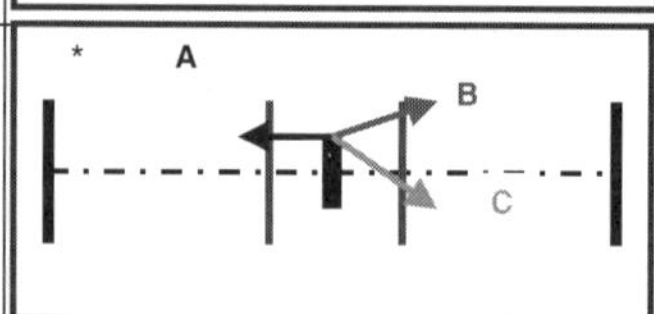	2 fragments (fragment C to be faster)were formed in the target: ^{B}Vemis= 61.6/(56.6/ ^{B}Vexp + 5.1/ Vc)
3	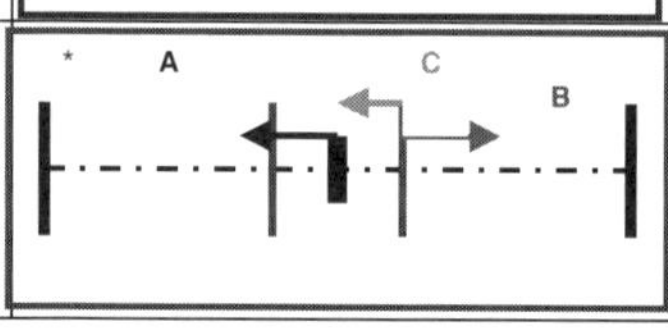	2 fragments were formed in the MCP detector but fly in opposite directions: ^{C}Vemis= 66.8 / [56.6/ ^{C}Vexp - 5.1(1/ Vb-1/ Va)]

Altogether, about 40 millions binary fission events were collected. Besides binary, ternary and quaternary coincidences were also detected in the experiment. Data processing is still in progress but first results bearing on events with multiplicity 3 are presented below.

2. Experimental results

Preliminary analysis [6] showed that the main part of the triple events detected are due to the random coincidences of the FF originated from conventional binary fission with both scattered α-particles from the beam and ions of oxygen and aluminum knocked from the target backing by the beam. Such events form pronounced loci in Figure 2a (marked by the arrows). Both loci (1 and 2) are of the same nature but are linked with adjacent bunches. For the sake of convenience the fragments in each ternary event were resorted in order of decreasing of fragment mass namely Ma to be the heaviest one and so on. A long locus in the upper part of Figure 2a should be excluded from the further analysis. It follows from Figure 2b were gating Ms<250 a.m.u. was used.

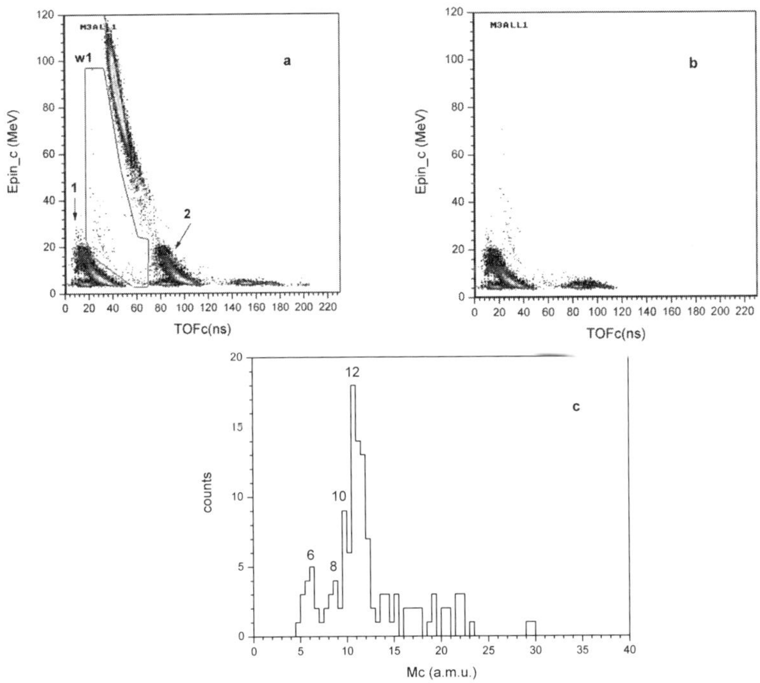

Figure 2. TOF versus energy detected in PIN for the lightest fragment in each ternary event (a). The same under condition that a total mass of all three fragments detected does not exceed 250 a.m.u. (b). Spectrum Y(Mc) for the events included in window w1 (c). Each channel on the mass axis corresponds to 0.5 a.m.u.. The numbers above the peaks correspond to their centers.

Thus only the events in the window w1 will be analyzed below. Their projection onto Mc axis is shown in Figure 2c.

For searching for unusual events Figure 3a was used. Here a difference in time when "start" detectors were tripped by the fragments flying apart is plotted on vertical axis. This time is known to be approximately proportional to masse ratio of the FF's from binary fission.

Grouping of the points attracts attention. Corresponding groups are marked in the figure by contours w2-w4. In its turn for instance set of events w2 looks like as some distinct families of points (marked as w5-w8 in Figure 3b) on Ma-Mb plane. Pair of magic fragments which total mass is equal to the mass of compound nucleus (242 a.m.u.) "starts" each family. We have analyzed them event by event. Corresponding results are presented in tables below.

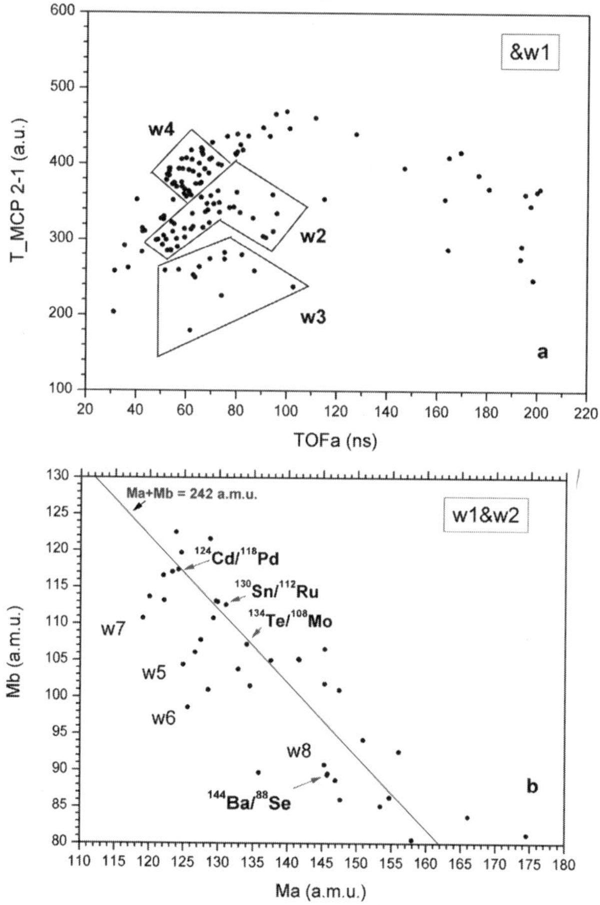

Figure 3. Time-of-flight versus time between "start" signals (a) for the events gated by the window w1 (Figure 2a). Mass-mass distribution for the events selected using condition w1&w2 (b). See text for detail.

Table 2. Information which was taken into account at derivation of the decay scheme for each triple event analyzed (example for an event from the family w5 in Figure 3b).

Point number	parameter		A	B	C	Dcal	Decay scheme
4	PIN number		205	111	208		^{128}Sn + ^{108}Mo + ^{6}He ↓ missing ----- - ^{114}Ru ↓ ^{12}C from "start" detector
	M (amu)		127.4	107.8	10.8	^{6}He	
	V(cm/ns)		0.867	1.22	2		
	Efr (MeV)		49.9	83.5	22.6		
	Parm1(a.u.)	131.6					
	Parm2 (a.u.)	132.3					
	Ms (amu)	246.1					
	Ma+b (amu)	235.2					

The following designations are used in the table above. Each column A, B, C, D (in order of decreasing of fragment mass) involves parameters of specific fragment. PIN number – lets one to know which PIN diode was tripped and in which arm. M, V, Efr – are, respectively, FF mass, velocity and true energy deposited in PIN diode (PHD was taken into account). Parm 1,2 – are total momentum of the fragments detected in a corresponding arm. Ms-is a total mass of all fragments detected in this event. Ma+b – total mass of two heaviest fragments. Dcal – is a mass of third fragment calculated using mass conservation law under condition that reference mass of the decaying system to be 242 a.m.u.(a number of prescission neutrons emitted in a specific event is unknown).

For the event under discussion the following arguments were taken into account in order to restore a most reasonable decay scheme could provide the parameters observed in the experiment. For three fragments detected momentum conservation law is met very well. At the same time the total mass Ms is too large even in compare with the maximal possible mass of decaying system (242 a.m.u.). One can surmise that the fragment C could be a Carbon ion knocked from the MCP detector foil. It is known from the Table 2, that fragments A and C fly in the same direction (PIN numbers of both start from the amount "2"). Using Rutherford formula for elastic scattering one obtains that the energy of C-ion scattered at zero angle reference to the velocity vector of fragment A having energy Ea'= Efr_a + Efr_c should be 22.72 MeV, what agrees well with the experimental value 22.6 from Table 2. Experimental values of masses of the fragments A and B are very close to the masses of known magic nuclei ^{128}Sn and ^{108}Mo respectively. It is believed that just these nuclides to be the decay partners. Conservation of nuclear charge and mass requires that the third fragment should exist such as ^{6}He by composition. It can be as well ^{4}He and two neutrons, 2 ^{3}H and so on.

Table 3. Summary on Sn/Ru mode (gate w5 in Figure 3b).

$_{50}\mathbf{Sn}_{82}$ (β2~0, "G' ", "G") / $_{44}\mathbf{Ru}_{Nucd=69.1}$ (β2~0.55,"C' ")			
Point number	Decay scheme	Point number	Decay scheme
1	$^{129}\mathbf{Sn} + {}^{113}\mathbf{Ru}$ - binary fission ↓ $^{9}\mathbf{Be}$ from "start" detector	**5**	$^{126}\mathbf{In} + {}^{3}H + n + {}^{106}\mathbf{Nb} + {}^{6}\mathbf{Li}$ missing --------------- -------------- ^{130}Sn ^{112}Ru
2	$^{130}\mathbf{Sn} + {}^{112}\mathbf{Ru}$ - binary fission ↓ ^{12}C from "start" detector	**6**	$^{129}\mathbf{Sn} + {}^{103}\mathbf{Zr}^{\ 2)} + {}^{6}\mathbf{He} + {}^{4}He$ missing ---------------------- ^{113}Ru
3	$^{129}\mathbf{Sn} + {}^{113}\mathbf{Ru}$ - binary fission ↓ $^{19}\mathbf{F}$ from "start" detector	**7**	$^{124}\mathbf{Cd} + {}^{6}He + {}^{8}\mathbf{Be} + {}^{104}\mathbf{Zr}$ missing -------------- -------------- ^{130}Sn ^{112}Ru
4	$^{128}\mathbf{Sn} + {}^{108}\mathbf{Mo} + {}^{6}He$ ↓ missing ------------------------ ↓ ^{114}Ru ^{12}C from "start" detector		

The upper line of the table provides the information concerning deformations and nucleon compositions of the shells involved while capital letters in the brackets correspond to the shell minima loci in [7]. Unchanged charge density (Z_{ucd}) hypothesis was used for calculation for instance number of neutrons in the nucleus with magic number of protons ($_{44}\mathbf{Ru}_{Nucd=69.1}$) and vice versa. The fragment which mass was corrected according the formulas 2 or 3 from Table 1 will be marked by a corresponding number (for instance, $^{103}\mathbf{Zr}^{\ 2)}$). Bold symbols in the decay schemes correspond to really detected fragments.

Table 4. Summary on Cd/Pd mode (gate w6 in Figure 3b).

$_{48}$**Cd** $_{Nucd=75.4}$ (β2~0.8, "K' ") / $_{46}$**Pd** $_{Nucd=72}$ (β2~0.8, "K' ")			
Point number	Decay scheme	Point number	Decay scheme
8	^{124}**Cd** + ^{118}**Pd** - binary fission ↓ ^{12}C from "start" detector	**12**	^{122}**Cd** +2n + ^{118}**Pd** - binary fission ↓ missing ^{12}C from "start" detector
9	^{124}**Cd** + ^{118}**Pd** - binary fission ↓ ^{9}**Be** from "start" detector	**13**	^{122}**Cd** + ^{6}**He** + ^{113}**Ru** + n ---------------- missing ^{119}**Pd**
10	^{122}**Ag** + ^{122}**Ag** ↓ ^{19}**F** from "start" detector	**14**	^{120}**Ag** + ^{2}H + ^{6}He + ^{114}**Ru** ----------- ------------ ↓ -------------- ^{122}Cd ^{120}Pd ^{8}**Li**
11	^{108}**Mo** $^{2)}$ + ^{14}C + 2n + ^{118}**Pd** missing ------------------------ ^{124}Cd	**15**	^{119}**Pd** + ^{111}**Tc** + ^{12}**B** $^{3)}$ -------------- ^{123}Cd

Table 5. Summary on Te/Mo mode (gate w7 in Figure 3b).

$_{Zucd=52}$**Te** $_{82}$ (β2~0, "G") / $_{Zucd=42}$ **Mo** $_{66}$ (β2~0.58, "C")			
Point number	Decay scheme	Point number	Decay scheme
16	^{134}**Te** + ^{108}**Mo** - binary fission ↓ ^{12}**C** from "start" detector	**19**	^{126}**In** + ^{8}Li + ^{10}Be + ^{98}**Sr** missing ----------- -------------- ^{134}Te ^{108}Mo ↓ ^{12}**C** from "start" detector
17	^{132}**Sn** + 2p + 2n + ^{98}**Sr** $^{2)}$ + ^{8}**Be** missing (^{4}He) ------------ ------------------ ^{134}Te ^{108}Mo	**20**	^{128}**Sn** + ^{6}**He** + ^{101}**Y** + ^{7}Li missing ------------- ------------ ^{134}Te ^{108}Mo
18	^{134}**Te** + ^{102}**Zr** + ^{6}He ↓ missing --------------- ↓ ^{108}Mo ^{12}**C** from "start" detector		

Table 6. Summary on Ba/Sr mode (gate w8 in Figure 3b).

$_{Zucd=56}$ **Ba** $_{88}$ (β2~0.65, "H") / $_{38}$**Sr** $_{Nucd=60}$ (β2~0.38, "B' ")			
Point number	Decay scheme	Point number	Decay scheme
21	^{144}**Ba** + ^{88}**Se** + ^{10}Be ↓ missing ---------------	**23**	^{144}**Ba** + ^{9}**Be** $^{3)}$ + ^{89}**Se** ------------ ^{98}Sr

	↓ **^{98}Sr** ^{12}C from "start" detector		
22	^{144}Ba + ^{3}H + **^{7}Li** $^{3)}$ + **^{88}Se** -------------------- ------------ ^{98}Sr **^{147}La**	**24**	**^{144}Ba** + **^{88}Se** + ^{10}Be ↓ missing -------------- ↓ ^{98}Sr **^{16}O** from "start" detector

Table 7. Some comments to the tables 4, 6.

all 3 fragments were detected in the events below		
Event numbe r	Decay scheme	Comments
13	Molecule after scission ←----------- **^{122}Cd** + **^{6}He** + **^{113}Ru** → ----------------- ^{119}Pd	There are no shifts in the FF's velocities, thus (see scheme 1 in table 1) **^{6}He** was born at the MCP detector due to the decay of the molecule **^{122}Cd** + **^{6}He** in inelastic scattering on the carbon foil.
14	**^{120}Ag** + ^{2}H + ^{6}He + **^{114}Ru** ← ---------- → ------------ ↓ -------------- ^{122}Cd ^{120}Pd **^{8}Li** →	There are no shifts in the FF's velocities, thus **^{8}Li** was born at the MCP detector. Namely, in the scission point decaying system consisted of two clusterized nuclei of Pd. After scission **^{120}Ag** nucleus and the molecule **^{8}Li+^{114}Ru** fly apart. The latter decays due to inelastic interaction with the nucleus of the converting foil. The decay products continue to fly in the same direction.
15	**^{119}Pd** + **^{111}Tc** + **^{12}B** $^{3)}$ -------------- ^{123}Cd	Scheme 3 from table 1 was used in reconstruction of the decay scheme, thus **^{12}B** was born at the MCP detector in a two stage process: 1. ← **^{119}Pd** - ^{123}Cd → immediately after scission; ↓ 2. ← **^{12}B**, **^{111}Tc**→ after inelastic scattering of ^{123}Cd on the carbon foil of MCP detector.
22	^{144}Ba + ^{3}H + **^{7}Li** $^{3)}$ + **^{88}Se** -------------------- ------------ ^{98}Sr **^{147}La**	Also two stage process presumably took place: 1. ^{144}Ba + ^{98}Sr to be clusterized as ^{3}H + ^{7}Li + ^{88}Se in the scission point; ^{144}Ba + ^{3}H form molecule to be equal by composition to **^{147}La,** which was really detected. ^{7}Li + ^{88}Se form also molecule till the 2-nd stage: 2. inelastic scattering of the latter on the carbon foil: ←**^{7}Li** , **^{88}Se**→ (partners fly apart)
23	**^{144}Ba** + **^{9}Be** $^{3)}$ + **^{89}Se** ----------- ^{98}Sr	1. ← **^{144}Ba** - ^{98}Sr → after scission; 2. ← **^{9}Be** , **^{89}Se**→ after inelastic scattering of ^{98}Sr on the carbon foil of the MCP detector

3. Discussion

We observe some distinct "families" of events based on two magic clusters each: Sn/Ru, Cd/Pd, Te/Mo, Ba/Sr. Experimental values of masses of the detected clusters prove to be unshifted or in other words agree with calibration obtained at ^{252}Cf (sf) sours i.e. without beam. At the same time the FF's masses for the conventional binary fission (evidently prompt) are a little bit shifted [6] likely due to influence of the beam on the MCP based detectors. Such difference can be understood if the ternary events analyzed above originate from the decays of isomeric states of Pu isotopes [8], the decays to be delayed by their nature thus appearing predominantly between the bunches.

Analyzing the decay schemes presented in the tables 3-6 one can refer a following general rule. Each initial cluster clusterises during an elongation of the fissioning system (secondary clusterization) forming lighter magic cluster and at least one light particle. The mechanism seems to be close to this standing behind a well known Ikeda rule [9]. A rupture of so prepared multi-component system can occur along any one of several boundaries of the nuclei involved (see, for instance, events 13, 22 from the Table 7). Light "remains" of secondary clusterization likely may unit into heavier nucleus (see event 14 from the Table 7).

After scission disintegration of such at least di-nuclear "molecule" can appear to occur via inelastic scattering on the target backing or carbon foil of the "start" detector. It should be stressed that we deal with long lived bounded states bearing in mind a typical time-of-flight (~ 5 ns) between the target and "start" detector where a decay of a molecule is happened (see Table 7).

At the moment it must not be ruled out that the ternary decays observed are of the same nature as known "polar emission" of α-particles and protons [10].

4. Conclusions

Summing up first results presented we came to following conclusions.

1. Presumably we deal with multi-body decays of some shape-isomers of Pu isotopes. They are built on pair of magic nuclei (clusters) each.
2. In the scission point each state above clusterises (secondary clusterization) forming also two lighter magic nuclei and at least two light clusters such as ^{6}He, ^{10}Be and so on.
3. Scission results in formation of one or two long lived "nuclear molecules" (or isomers) based on magic nucleus and light clusters.
4. Disintegration of such molecule can appear to occur via inelastic scattering on the target backing or carbon foil of the "start" detector.

Acknowledgments

We are grateful to Professor V.I. Zagrebaev for fruitful discussions.

References

1. Yu. V. Pyatkov et al., *Proc. Int. Conf. "50Years of Shells"*, 301 (1999).
2. W. Trzaska et al., Proc. Seminar on Fission Pont d'Oye V, Belgium, 102 (2003).
3. Yu. V. Pyatkov et al., *Preprint JINR* **E-15-2004-65.**
4. Yu. V. Pyatkov et al., *Preprint JINR* **E15-2005-99.**
5. S. Mulgin et al., *NIM* **A388**, 254 (1997).
6. Yu. Pyatkov et al., Proc. 14-th International Seminar on Interaction of Neutrons with Nuclei (ISINN-14), (2006) (in press).
7. B.D. Wilkins et al., *Phys. Rev.* **C14**, 1832 (1974).
8. S.M. Polikanov and G. Sletten, *Nucl. Phys.* **A151**, 656 (1970).
9. K. Ikeda, Proc. 5th Int. Conf. on Clustering Aspects in Nuclear and Subnuclera Systems, Kyoto, 277 (Contributed papers, 1988).
10. E. Piasecki, L. Nowicki, IAEA-SM-241/F11, p.193

COMPARATIVE STUDY OF THE TERNARY PARTICLE EMISSION IN ^{243}Cm(n_{th},f) AND ^{244}Cm(SF)

S. VERMOTE[†] AND C. WAGEMANS

Dept. of Subatomic and Radiation Physics, University of Gent, B-9000 Gent, Belgium

O. SEROT

CEA Cadarache, F-13108 Saint-Paul-lez-Durance, France

J. HEYSE

EC-JRC-IRMM, B-2440 Geel, Belgium

T. SOLDNER AND P. GELTENBORT

ILL, F-38042 Grenoble, France

In this paper we report on the energy distribution and the emission probability of ^{3}H, ^{4}He and ^{6}He particles emitted in the spontaneous ternary fission of ^{244}Cm (E_{exc} = 0 MeV) and in the neutron induced ternary fission of ^{243}Cm (E_{exc} = 6.8 MeV). Both measurements were performed using suited ΔE-E telescope detectors, at the IRMM (Geel, Belgium) for the spontaneous fission and at the ILL (Grenoble, France) for the neutron induced fission measurements. By combining these results with similar data obtained for the fissioning systems ^{246}Cm and ^{248}Cm, a different impact of the excitation energy on the emission probability of ternary alphas and tritons could be demonstrated, which is ascribed to the influence of the α-cluster preformation probability on the ternary alpha emission.

1. Introduction

Roughly 2 to 4 times every thousand fission events, the two heavy fission fragments are accompanied by a light charged particle. This process is called ternary fission and more specifically, ternary particles ranging from protons to Ar are emitted. However, the emission yields for α particles, tritons and ^{6}He particles are the most important.

The ternary fission process is an important source of helium and tritium gas in nuclear reactors and used fuel elements. Therefore accurate ternary fission yields for ^{4}He and tritons are requested by nuclear industry. Furthermore ternary fission data are of interest for nuclear physics in order to improve our

[†] E-mail address: sofie.vermote@ugent.be

understanding of ternary particle emission and to provide information about the fission process itself.

An interesting characteristic of the process is related to the influence of the excitation energy of the fissioning nucleus on the ternary particle emission probability. This effect can be measured by comparing the ternary particle emission probability for the same compound nucleus at zero excitation energy (in the case of spontaneous fission) and at an excitation energy corresponding to the neutron binding energy (in the case of low energy neutron induced fission).

In previous experiments done by our research group, energy distributions and fission yields for ternary α particles (often referred to as Long Range Alphas or LRA) and tritons (t) have been studied for the compound nuclei ^{246}Cm and ^{248}Cm [1, 2]. In this paper we describe new measurements performed on the neutron induced fission of ^{243}Cm (E_{exc} = 6.8 MeV) and on the spontaneous fission of ^{244}Cm (E_{exc} = 0 MeV). A comparison of the results for the fissioning systems ^{244}Cm, ^{246}Cm and ^{248}Cm will give a better insight in the influence of the excitation energy on the LRA and triton characteristics.

2. Experimental setup

The spontaneous fission of ^{244}Cm has been studied at the Institute for Reference Materials and Measurements (IRMM) in Geel, Belgium. The ^{243}Cm neutron induced measurement was carried out at the PF1b cold neutron guide installed at the high flux reactor of the Institute Laue-Langevin (ILL) in Grenoble, France. The neutron flux at the sample position was about $2x10^9$ neutrons/s.cm^2.

2.1. *Sample characteristics*

The Cm samples were prepared at the Russian Federal Nuclear Center in Arzamas. In both cases a spot of curium oxide with a diameter of 15 mm was deposited on a 30 μm thick Al-foil. The 18.5 μg ^{244}Cm sample used for the spontaneous fission measurement had an enrichment of 99.48%, while a 2 μg ^{243}Cm sample with an enrichment of 98.35% was used for the neutron induced fission measurement.

^{244}Cm has a half life value for spontaneous fission of $(1.32 \pm 0.02)x10^7$ y, while the half life for alpha decay is only 18.10 y. Due to this ratio which is unfavourable to measuring spontaneous fission, it was necessary to use a very active source of 53.7 MBq in order to obtain a reasonable fission counting rate.

For the neutron induced fission measurement of ^{243}Cm a quite active source of 3.7 MBq was used. Because of the radioactive alpha decay of ^{243}Cm (half life for alpha decay of 29.10 y), ^{239}Pu is produced in the sample. Although the Pu was chemically removed from the sample material 2 years prior to our experiment, a correction still had to be made for the ^{239}Pu produced since then, because of its large thermal fission cross section (748 b).

2.2. *Detection system*

For the ^{243}Cm neutron induced fission measurement the sample was placed in the center of a vacuum chamber at an angle of 45 degrees with the incoming neutron beam. A polyimide foil was used to cover the sample in order to prevent the contamination of the chamber by recoil nuclei.

The measurements were performed in two separate steps. In a first one, ternary particles were detected allowing the determination of both energy distributions and counting rates. Therefore two silicon surface barrier ΔE-E telescope detectors were placed on both sides of the sample, and perpendicular to the beam as shown in Fig. 1.

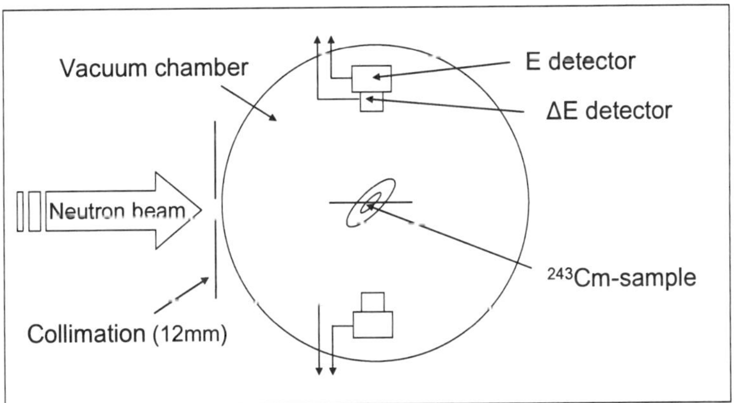

Figure 1: Experimental setup for the ^{243}Cm neutron induced fission measurement.

In addition the ΔE detector facing the sample was covered with a thin aluminium foil of 30 μm in order to stop the decay alphas and the fission fragments. On the other side, these were all stopped by the 30 μm thick aluminium sample backing.

The signals coming from the surface barrier detectors were sent through a pre-amplifier and an amplifier. These signals were digitized in an ADC or Analogue to Digital Converter, and coincident ΔE and E signals were stored on a PC.

The spontaneous fission of ^{244}Cm was measured without neutron beam, so the sample could be placed right in front of the two pairs of detectors as can be seen in Fig. 2.

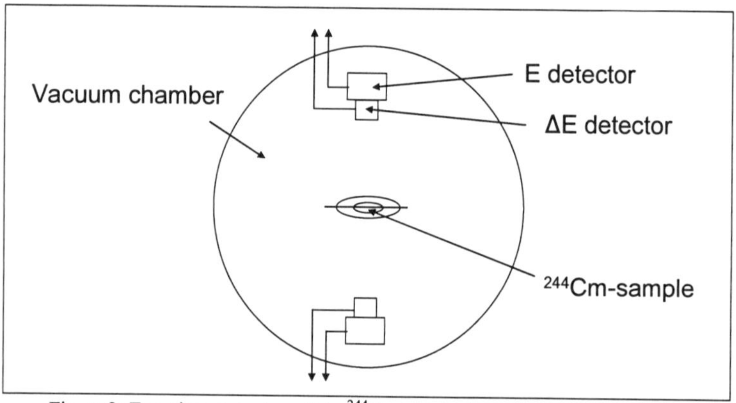

Figure 2: Experimental setup of the ^{244}Cm spontaneous fission measurement.

For both experiments the detector characteristics were chosen in order to have the best setup for simultaneously detecting α particles and ^{6}He particles, or α particles and tritons (Table 1).

Table 1: Thickness of the surface barrier detectors used.

^{243}Cm	Sample side [μm]	Backing side [μm]
ΔE	29.8	49.8
E	500	1500
^{244}Cm		
ΔE	28.9	31.7
E	1500	1500

The ΔE-signal is proportional to the energy deposited by the ternary particle traversing the silicon surface barrier detector, and the E-signal is proportional to the remaining particle energy. Typical examples of (coincident) ΔE and E spectra for ^{243}Cm(n_{th},f) are shown in Fig. 3.

In a second step, binary fission fragments were detected in order to determine the Binary Fission Yield (B). At this stage, the ΔE detector facing the sample was removed, together with the aluminium foil, and replaced by a dummy ring with exactly the same dimensions. In this manner binary fission fragments could be measured with the E-detector under the same detection geometry as the ternary particles.

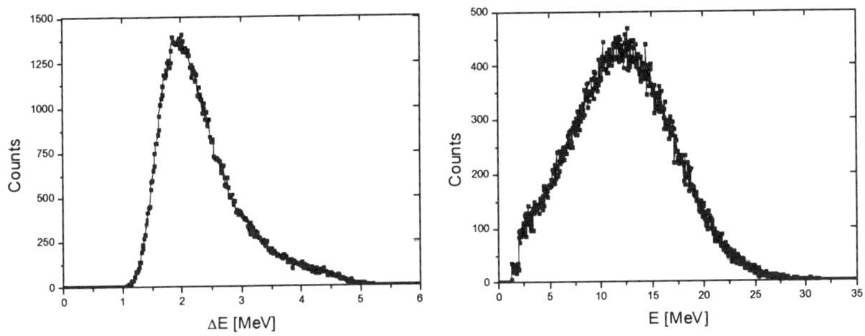

Figure 3: Example of a typical ΔE (left) and E (right) spectrum for $^{243}Cm(n_{th},f)$.

3. Analysis and results

3.1. *Particle identification*

The procedure used to identify the different ternary particles and separate them from the background is the one proposed by Goulding et al. [3]. This method is based on the difference in energy loss of different particles in the same material leading to the relation: $T/a = (E+\Delta E)^{1.73} - E^{1.73}$, with T the thickness of the ΔE detector and a a particle and material specific constant. An example of a T/a spectrum is shown in Fig. 4, together with a 3-dimensional view of E-ΔE.

The selection of the ternary particles was realized by putting a window on the region of interest of the T/a spectrum. This has been done for LRA, tritons and ^{6}He particles. In Fig. 5 a T/a spectrum containing only LRA particles is shown, together with the corresponding E-ΔE spectrum. After the selection, ΔE and E spectra for a given ternary particle were obtained and the total energy distribution could be deduced. The thresholds in energy for each ternary particle are due to the thickness of the ΔE-detector, the electronic noise and the presence of the Al-foil. The average energy and the full width at half maximum of the energy distribution were obtained from a Gaussian fit performed on the experimental data.

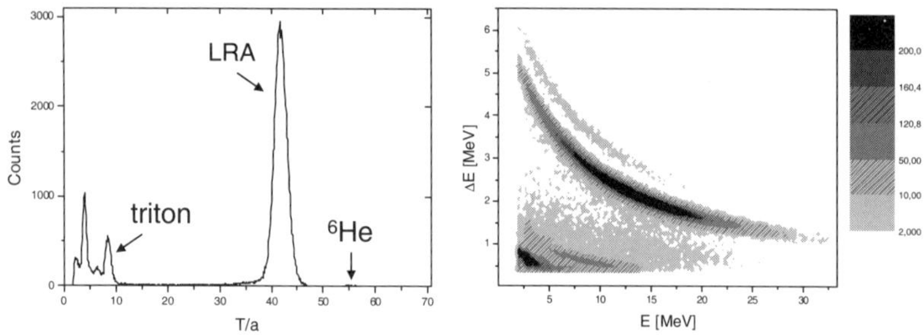

Figure 4: T/a spectrum for all ternary particles (left) and corresponding E-ΔE spectrum (right).

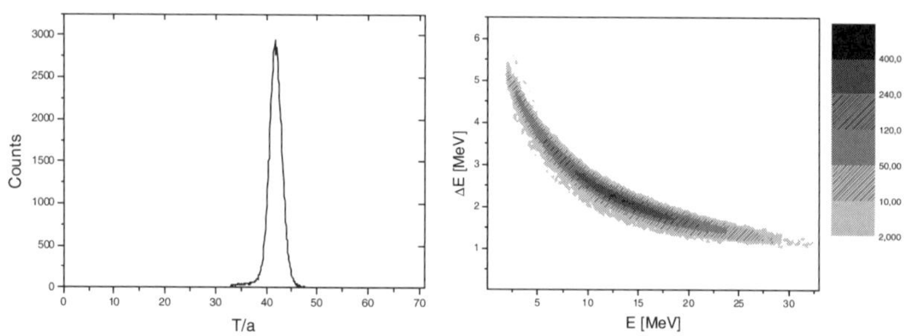

Figure 5: T/a spectrum for LRA particles alone (left) and corresponding E-ΔE spectrum (right).

3.2. *Results for the $^{243}Cm(n_{th},f)$ measurement*

The spectrum obtained from the binary fission measurement is plotted in Fig. 6. The two bumps of the fission fragments have to be separated from the alpha pile-up peak due to the radioactive decay of ^{243}Cm. Then the remaining spectrum is extrapolated. The number of binary fission events was obtained after integration of the extrapolated spectrum, yielding 225.15 ± 0.79 binary fissions per second.

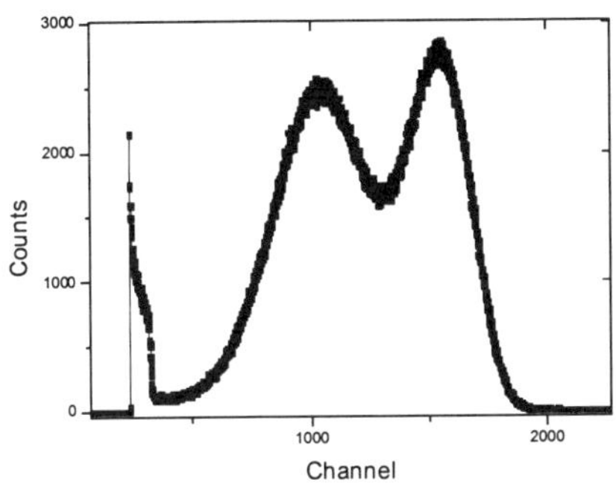

Figure 6: Binary fission spectrum for the $^{243}Cm(n_{th},f)$ measurement.

Fig. 7 shows the spectra for the LRA, ^{6}He and triton measurements. The characteristics of these energy distributions are given in table 2. Emission probabilities relative to ^{4}He are shown in table 3, together with the absolute emission probabilities.

For the LRA particles, a Gaussian fit was performed for particles with an energy above 12.5 MeV. In the case of the ^{6}He particles the fit started at 10 MeV. Due to a difficult separation from the background, the tritons were the most difficult ternary particles to examine. The energy threshold is quite high, so the fit was done starting from 6.5 MeV, imposing a fixed mean energy and full width at half maximum, guided by the expectations based on the systematics.

Table 2: Values for the average energy (E) and full width at half maximum (FWHM) for the different ternary particles measured for the compound nucleus ^{244}Cm.

^{243}Cm	E [MeV]	FWHM [MeV]
LRA	16.13 ± 0.12	10.33 ± 0.44
^{6}He	11.28 ± 0.53	9.79 ± 0.71
tritons	7.8 ± 0.2 (fixed)	7.9 ± 0.2 (fixed)
^{244}Cm		
LRA	16.04 ± 0.13	10.24 ± 0.52
^{6}He	11.22 ± 0.74	9.90 ± 1.14
tritons	7.6 ± 0.2 (fixed)	7.36 ± 0.8

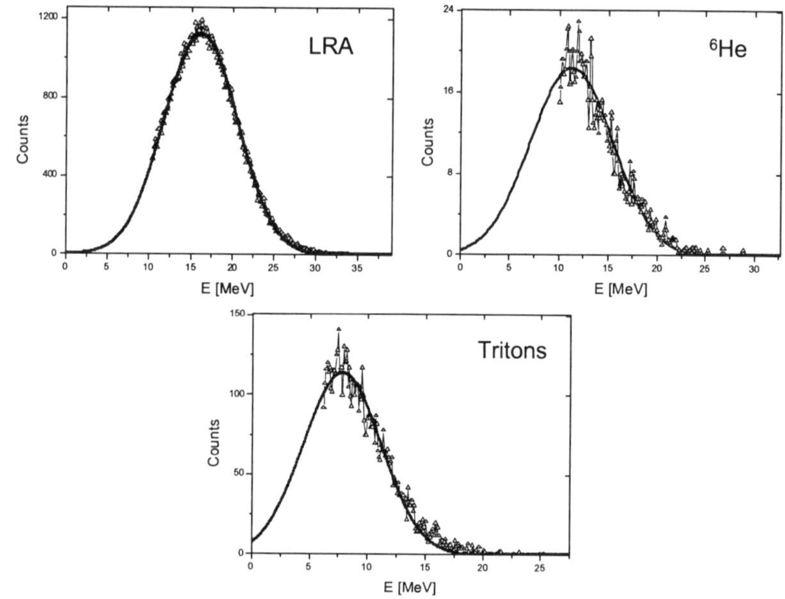

Figure 7: Energy distributions for LRA, ^{6}He and tritons for ^{243}Cm(n_{th},f).

Table 3: Values for the relative and absolute emission probabilities (per fission) for the different ternary particles measured for the compound nucleus ^{244}Cm.

^{243}Cm	rel. em. prob. [%]	abs. em. prob.
LRA	100	$(2.45 \pm 0.12)x10^{-3}$
^{6}He	1.65 ± 0.32	$(4.04 \pm 0.81)x10^{-5}$
tritons	8.43 ± 1.10	$(2.07 \pm 0.29)x10^{-4}$
^{244}Cm		
LRA	100	$(3.14 \pm 0.09)x10^{-3}$
^{6}He	2.05 ± 0.52	$(6.44 \pm 1.64)x10^{-5}$
tritons	6.75 ± 1.43	$(2.12 \pm 0.45)x10^{-4}$

3.3. *Results for the ^{244}Cm(SF) measurement*

The spectra for LRA, ^{6}He particles and tritons are plotted in Fig. 8. For the LRA particles, a Gaussian fit was performed starting from 12.5 MeV. For ^{6}He particles statistics were not so good, but still a fit was done starting from 10.5 MeV. For the tritons, the fit was performed from 6.5 MeV, after fixing the average energy. The values for the energy distributions for these three ternary particles are given in table 2.

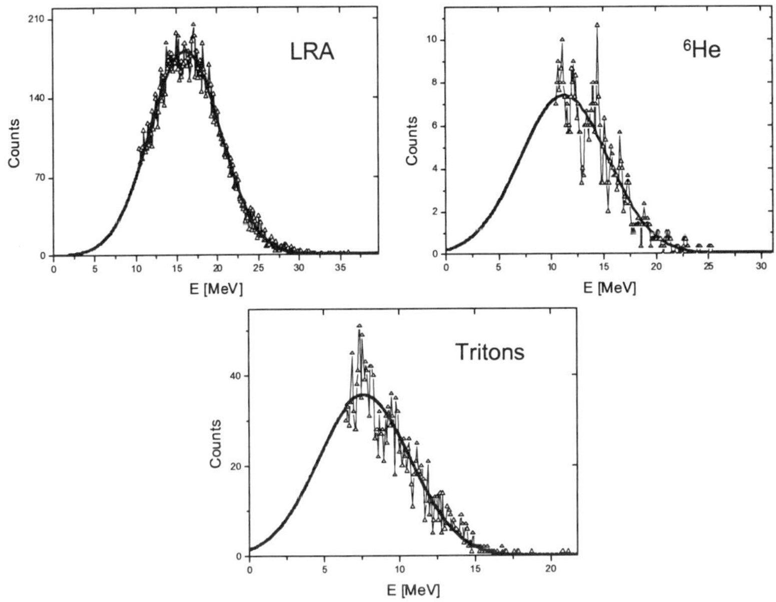

Figure 8: Energy distributions for LRA, ^{6}He and tritons for ^{244}Cm(SF)

In this new measurement on ^{244}Cm only emission probabilities relative to ^{4}He could be obtained, since no binary fission measurement was done due to geometry problems. In order to deduce absolute emission probabilities, this measurement was combined with a value for $LRA/B = (3.14 \pm 0.09) \times 10^{-3}$ determined in a previous experiment [4]. This result was obtained using a ΔE-E telescope consisting of an ionization chamber as ΔE-detector and a surface barrier detector of 2000 mm^2 as E-detector. In this way the absolute emission probabilities for ^{6}He and tritons could be calculated and are shown in table 3.

4. Discussion

Over the last decade our research group was involved in a systematic study of different Cm-isotopes. The compound nuclei ^{246}Cm and ^{248}Cm were examined in the past [1, 2]. Combining these results with our new values for the compound nucleus ^{244}Cm, interesting comparisons can be made.

4.1. *Energy distributions*

The energy distributions for LRA particles and tritons were measured for all six Cm-isotopes. An overview of the values for the average energy and the full width at half maximum is shown in table 4 for the LRA particles and in table 5 for the tritons. Due to poor statistics and too high energy thresholds, the ^{6}He particles were not always studied in the past. Energy distributions from previous experiments are known only for ^{245}Cm, with fixed mean energy [5].

From a comparison of the data reported in tables 4 and 5 we can conclude that the average energy is roughly 8 MeV for the tritons and 16 MeV for the LRA particles. The corresponding FWHM is roughly 7.5 MeV for the tritons and 10 MeV for the LRA particles.

Table 4: Values for the average energy (E) and full width at half maximum (FWHM) for LRA particles measured for all Cm-isotopes studied.

	E [MeV]	FWHM [MeV]
^{243}Cm	16.13 ± 0.12	10.33 ± 0.44
^{245}Cm	16.35 ± 0.15	10.10 ± 0.20
^{247}Cm	16.01 ± 0.13	10.37 ± 0.24
^{244}Cm	16.04 ± 0.13	10.24 ± 0.52
^{246}Cm	16.41 ± 0.20	9.73 ± 0.28
^{248}Cm	15.97 ± 0.12	10.03 ± 0.14

Table 5: Values for the average energy (E) and full width at half maximum (FWHM) for tritons measured for all Cm-isotopes studied.

	E [MeV]	FWHM [MeV]
^{243}Cm	7.8 ± 0.2 (fixed)	7.9 ± 0.2 (fixed)
^{245}Cm	8.40 ± 0.25	7.76 ± 0.38
^{247}Cm	8.55 ± 0.27	7.52 ± 0.33
^{244}Cm	7.6 ± 0.2 (fixed)	7.36 ± 0.8
^{246}Cm	8.05 ± 0.34	7.77 ± 0.47
^{248}Cm	8.86 ± 0.18	7.47 ± 0.29

4.2. *Emission probabilities*

To examine the influence of the excitation energy on the emission probabilities for the different Cm-isotopes two comparisons are shown in Fig. 9. In both plots a distinction is made between the data for spontaneous fission and the neutron induced fission data for the different compound nuclei. In the left figure the LRA emission probability is plotted as a function of the fissility parameter Z^2/A. It can be clearly seen that the increase of the excitation energy always leads to a decrease of the LRA emission probability. This behaviour can be explained by introducing the so called alpha cluster preformation probability factor S_α . The LRA emission process seems to be strongly influenced by this α-cluster preformation probability. When the fissioning nucleus is formed after capture of a neutron, the S_α factor is likely to decrease, in this way explaining the observed decrease of LRA/B with excitation energy.

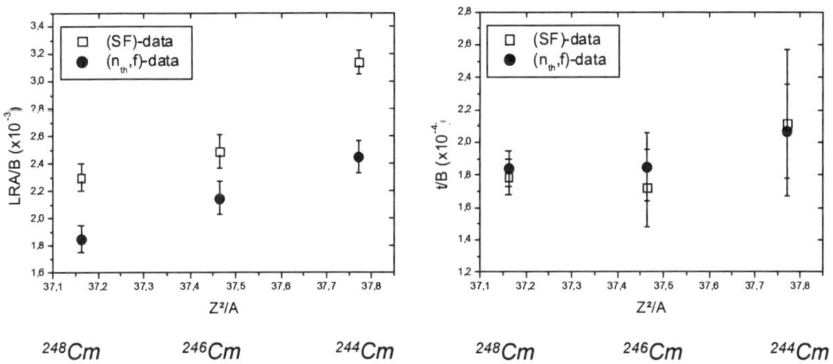

Figure 9: The emission probabilities for LRA (left) and tritons (right) as a function of Z^2/A.

The right part of Fig. 9 shows the triton emission probability as a function of Z^2/A. Here no decrease of the triton emission probability with increasing energy is observed. Since in the triton emission process, no cluster preformation is involved, this is in line with our above explanation for the decrease of LRA/B.

For ^{6}He particles nothing can be concluded due to the high uncertainties on the values shown in table 3.

5. Conclusions and outlook

In the present paper the main characteristics (energy distribution and emission probability) of LRA, tritons and ^{6}He particles emitted in the neutron induced fission of ^{243}Cm and the spontaneous fission of ^{244}Cm are presented. It has been shown that the average energy and the full width at half maximum for both LRA and tritons are consistent with the values already observed for the other Cm fissioning systems. A comparison between the spontaneous fission and the neutron induced fission for all Cm-isotopes permitted to determine the influence of the excitation energy of the fissioning nucleus on the ternary emission probabilities.

As a next step, further measurements to determine both energy distributions and emission probabilities of ^{6}He particles could be very useful in order to get a better insight in the impact of cluster preformation on the emission probability.

References

1. O. Serot, C. Wagemans, J. Wagemans and P. Geltenbort, "Influence of the excitation energy on the ternary triton emission probability of the ^{248}Cm fissioning nucleus", in *Proc. 3rd Int. Conf. on Fission and Properties of Neutron-Rich Nuclei – Sanibel Island, USA,* edited by G.H. Hamilton, A.V. Ramayya and H.K. Carter, World Scientific, 2003, p. 543.
2. O. Serot, C. Wagemans, J. Heyse, J. Wagemans and P. Geltenbort, "New results on the ternary fission of Cm and Cf isotopes", in *Proc. Seminar on Fission – Pont d'Oye V,* edited by C. Wagemans, J. Wagemans and P. D'hondt, World Scientific, 2004, p. 151.
3. F.S. Goulding, D.A. Landis, J. Cerny and R.H. Pehl, *Nucl. Instr. Meth.* **31**, 1 (1964).
4. S. Vermote, C. Wagemans, J. Heyse, O. Serot and J. Van Gils, "Systematic study of the ternary fission of Cm-isotopes: new results on ^{243}Cm(n_{th},f) and ^{244}Cm(SF)", *Fysica 2006, NNV and BPS symposium*, Leiden University, 2006, p. 78.
5. O. Serot et al., "Energy distributions and yields of ^{3}H, ^{4}He and ^{6}He-particles emitted in the ^{245}Cm(n_{th},f) reaction", in *Proc.5th Int. Conf. on Dynamical Aspects of Nuclear Fission,* Casta–Papiernicka, Slovak Republic, World Scientific, 2001, p. 319.

MANIFESTATION OF AVERAGE γ-RAY MULTIPLICITY IN THE FISSION MODES OF ^{252}Cf(SF) AND THE PROTON - INDUCED FISSION OF ^{233}Pa, ^{239}Np AND ^{243}Am

M. BERESOVA[1,5], J. KLIMAN[1,5], L. KRUPA[1,5], A.A. BOGATCHEV[1], O. DORVAUX[2], I.M. ITKIS[1], M.G. ITKIS[1], S. KHLEBNIKOV[5], G.N. KNIAJEVA[1], N.A. KONDRATIEV[1], E.M. KOZULIN[1], V. LYAPIN[3,4], I.V. POKROVSKY[1], W. RUBCHENIA[3,4], L. STUTTGE[2], W. TRZASKA[3], D. VAKHTIN[4]

[1]Flerov Laboratory of Nuclear Reactions, JINR, 141980 Dubna, Russia
[2]Institut de Recherches Subatomiques, CNRS-IN2P3, Strasbourg, France
[3]Deparment of Physics, University of Jyväskylä, FIN-40351, Jyväskylä, Finland
[4]V.G. Khlopin Radium Institute, St.-Petersburg 194021, Russia
[5]Institute of Physics SASc, Dubravska cesta 9, 84228 Bratislava, Slovak Republic

Average preequilibrium $<M_n^{preeq}>$, average statistical prescission $<M_n^{stpre}>$ and postscission $<M_n^{post}>$ neutron multiplicities as well as average γ-ray multiplicity $<M_\gamma>$, average energy $<E_\gamma>$ emitted by γ-rays and average energy per one gamma quantum $<\varepsilon_\gamma>$ as a function of mass and total kinetic energy <TKE> of fission fragments were measured in the proton induced reactions $p+^{232}Th\rightarrow^{233}Pa$, $p+^{238}U\rightarrow^{239}Np$ and $p+^{242}Pu\rightarrow^{243}Am$ (at proton energy E_p=13, 20, 40 and 55 MeV) and spontaneous fission of ^{252}Cf. The fragment mass and energy distributions (MEDs) have been analyzed in terms of the multimodal fission. The decomposition of the experimental MEDs onto the MEDs of the distinct modes has been fulfilled in the framework of a method that is free from any parameterization of the distinct fission mode mass distribution shapes [1]. The main characteristics of symmetric and asymmetric modes have been studied in their dependence on the compound nucleus composition and proton energy. The manifestation of multimodal fission in average γ-ray multiplicity $<M_\gamma>$ of fission fragments was studied in this work.

Keywords: Mass and energy distributions, Multimodal fission

1 Introduction

Recently wide range of mass and energy distribution properties have been interpreted in the framework of the multimodal concept. This concept is based on the assumption that experimental MEDs are a superposition of MEDs of individual fission modes. These modes are caused by the valley structure of the deformation potential energy surface. At present it is supposed that there are four distinct fission modes for the heavy nuclei – symmetric (S) mode and three asymmetric modes – Standard 1 (Sl), Standard 2 (S2) and Standard 3 (S3). S mode fragments are strongly elongated with masses around $A_{CN}/2$. S1 mode is characterized by high kinetic energies of fission fragments. Heavy fragment is spherical with M_H~134, Z_H~50 and N_H~82. Kinetic energies of S2 mode fragments are lower than those of S1 mode.

Heavy fragment with M_H~140 is slightly deformed, influenced by the deformed neutron shell closure N=88. S3 mode comprises deformed heavy fragment and spherical light fragment with N_L~50. It has been found that apart from mass and energy distributions, fission modality influences also the postscission neutron and γ-ray emission. The way that multimodal fission expresses itself in the average γ-ray multiplicity $\langle M_\gamma \rangle$ in the fission of compound nuclei ^{243}Am, ^{239}Np and ^{233}Pa will be presented in this work.

2 Experiment

The experiment was carried out at the Accelerator Laboratory, University of Jyväskylä [2]. The measurements were performed with 13, 20, 40 and 55 MeV proton beams. As a targets 100 mg/cm^2 - layers of fissile isotopes ^{238}U, ^{242}Pu and ^{232}Th evaporated on 60 mg/cm^2 thick Al_2O_3 backing were used. A typical beam spot diameter on the target was 5 mm, the average beam intensity was 10 pnA. Experimental setup included reaction-product spectrometer CORSET, an eight-detector time-of-flight neutron spectrometer DEMON, a High Efficiency Detection System (HENDES) facility and six 7.62 × 7.62 cm NaI(Tl) γ-ray detectors. The velocities and coordinates of the fission fragments were measured with a two-armed time-of-flight spectrometer CORSET [2]. Each arm of the spectrometer consisted of microchannel plate detector with electrostatic mirrors providing start signal, situated 3.5 cm from the target and two stop position-sensitive microchannel plate detectors placed at the distance 18.1 cm from the target. Calibration was fulfilled with the use of ^{226}Ra α-particle source, the fission fragments of ^{252}Cf and elastic scattering peaks directly during the experiment. Extracting of mass and energy distributions of fission fragments and data processing of γ-rays and neutrons was carried out in the way as described in Ref. [2].

3 Method of the Analysis

The decomposition of the experimental MEDs onto the MEDs of the distinct modes was performed using a method proposed in Ref. [1]. According to this method the yields $Y_{i,M}$ of fission modes are found from the condition of the functional minimum

$$\chi^2(M) = \sum_E \left[\varepsilon(E,M)(\sum_i^n \eta_i(M) Y_{i,M}(E) - Y_{exp,M}(E)) \right]^2,$$

where $Y_{i,M}(E)$ is the normalized energy distribution of the *i*-th mode, $\eta_i(M)$ is the relevant weight factor for the *i*-th mode, ε(E,M) is the value that is in inverse proportion to the total error of the $Y_{exp,M}(E)$. In case when derivatives are in linear dependence on the relevant parameters then the finding of the optimal values of $\eta_i(M)$ is reduced to solving the system of n equations:

$$\sum_E \left[\varepsilon(E,M) Y_{i,M}(E) (\sum_j^n \eta_j(M) Y_{j,M}(E) - Y_{\exp,M}(E)) \right] = 0,$$

where i and j correspond to the fission modes S1, S2, S3 and S.

4 Results and Discussion

The results of decomposition of fragment mass distributions performed for compound nuclei ^{233}Pa, ^{239}Np and ^{243}Am are shown in Fig. 1 and Fig. 2, respectively.

Basic characteristics of fission modes - mass yield, average mass of heavy fragment and mass yield dispersion were studied as a function of the incident proton energy for three compound nuclei. These dependencies are presented in Fig. 3. Open symbols represent data taken from Ref. [1]. One can see from Fig. 3 that the relative contribution of S mode increases with increasing proton energy. At the same time S2 mode contribution decreases with proton energy raise. Average masses of heavy fragment do not seem to vary significantly with proton energy increase and dispersion dependencies do not show any considerable changes. The same trends can be found for all of the studied nuclei. When comparing our results with those of Ref. [1] a very good agreement between the two sets of data is evident.

Fragment mass versus TKE matrix for the fission of ^{252}Cf, ^{243}Am at proton energy 13 MeV and ^{239}Np at 20 and 55 Mev is illustrated in Fig. 4. The dotted contours designate experimental mass distributions and solid lines border the regions with at least 51% and 76% contribution of given mode to the total mass yield.

Deeper insight into the multimodal fission is gained by investigating the γ-ray multiplicity of the fragments. Experimental data on $\langle M_\gamma \rangle$ from the regions where the contribution of given mode exceeds 75% of the total yield were processed in coincidence with the fragment data to examine the features of average γ-ray multiplicity $\langle M_\gamma \rangle$ for individual fission modes. In the Figs. 5 and 6 matrices of $\langle M_\gamma \rangle$ per event are given for ^{252}Cf(SF), ^{239}Np, ^{233}Pa and ^{243}Am at proton energies 13, 20, 40 and 55 MeV. It appears that going higher in excitation energy leads to γ-ray emission growth mainly in the regions of S and S3 modes. On the contrary the increase in γ-ray multiplicity for other modes is not so significant. The obtained data are listed in Tab. 1.

The γ-ray multiplicity $\langle M_\gamma \rangle$ of fission modes manifests the nuclear shell structure. Our results are in compliance with the generally accepted ideas about multimodal fission: S mode characteristics are influenced by strongly elongated

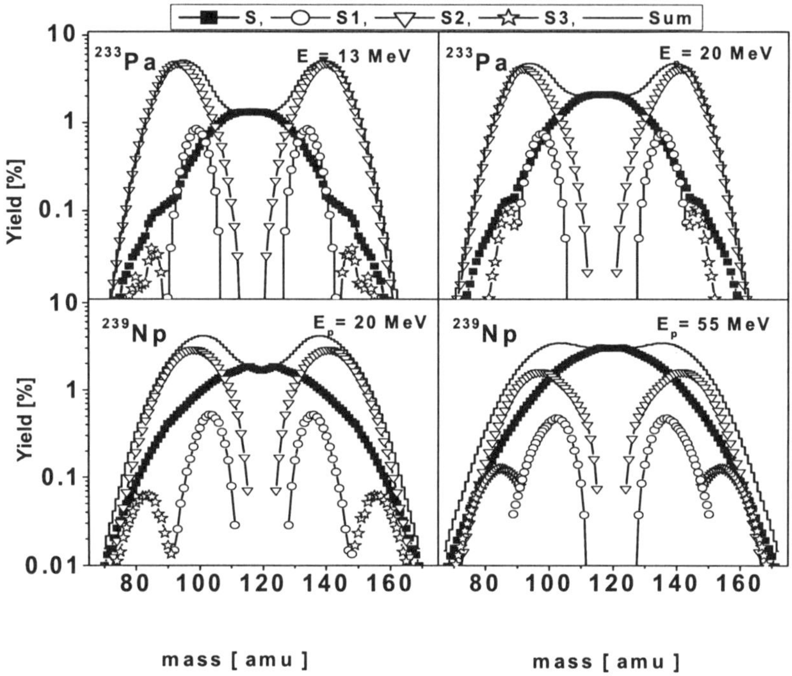

Fig.1 Results of decomposition of fragments mass distribution of ^{233}Pa and ^{239}Np. Solid line-experiment, solid squares - S, open circles - S1, open triangles - S2, open asterisks –S.

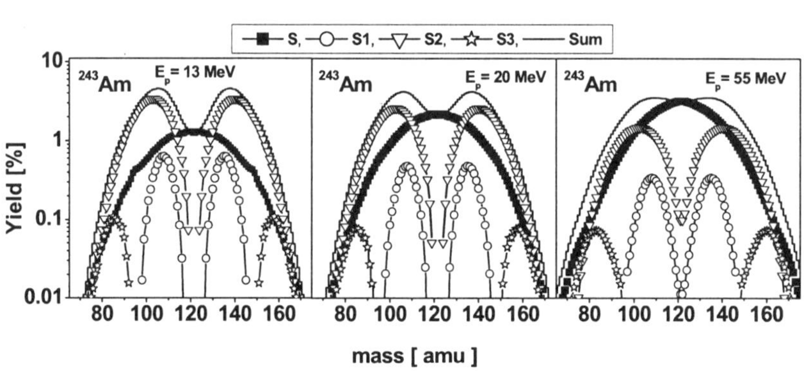

Fig.2 The same results as in Fig. 1 performed for ^{243}Am. Designation is the same as in Fig. 1.

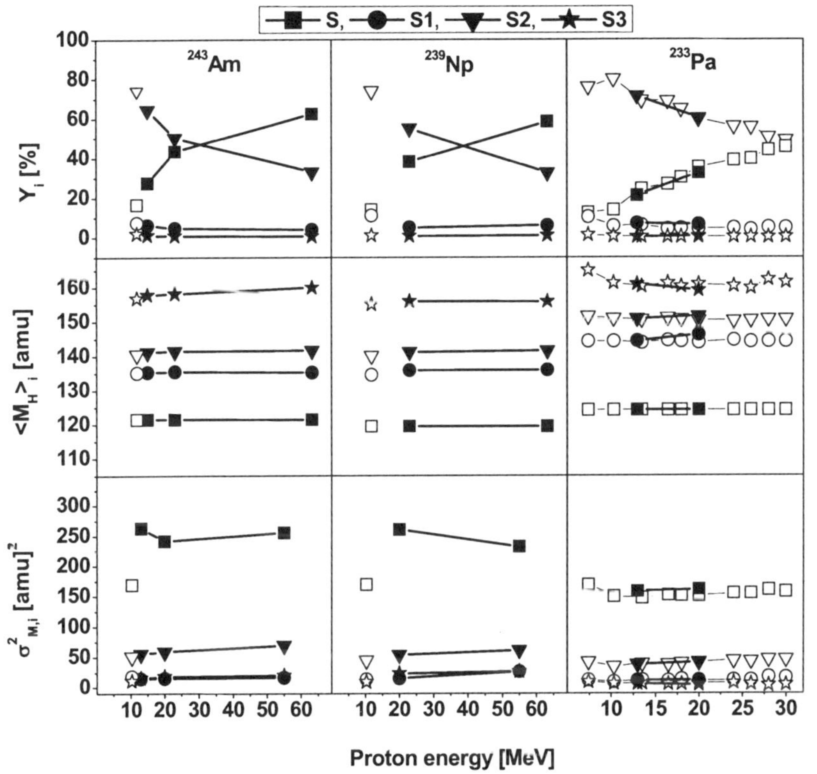

Fig. 3 Mass yields Y_i, average masses of heavy fragment $\langle M_H \rangle_i$ and mass yield dispersions $\sigma^2_{M,i}$ for distinct modes as a function of proton energy. Solid symbols - our results, open symbols - results taken from Ref [1]; squares - S mode, circles - S1 mode, triangles - S2 mode, asterisks - S3 mode.

shapes of nascent nuclei. S1 mode features can be attributed to the spherical shell closure in heavy fission fragment with the mass ranging from 132 to 134. S2 mode is affected by the deformed neutron shell N=88 formed in the heavy fragment with average mass around 140. Properties of S3 mode are due to the spherical neutron shell closure N=50 in light fragment whose partner heavy nucleus is from the region of lanthanoides.

As a further characteristic the average γ–ray multiplicity $\langle M_\gamma \rangle$ per event as a function of the total kinetic energy for each fission mode was studied. The

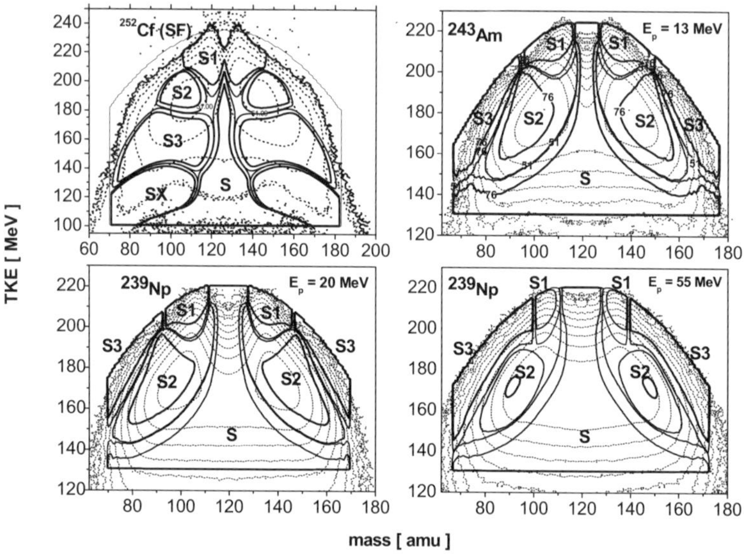

Fig.4 Contour plot of fragment mass versus TKE for fission fragments of $^{252}Cf(SF)$, ^{243}Am at E_p = 13 MeV and ^{239}Np at E_p = 20 and 55 MeV.

dependencies are plotted in Fig. 7 for the compound nuclei ^{243}Am, ^{239}Np and ^{233}Pa at proton energies 13, 20, 40 and 55 MeV. It is obvious that average γ–ray multiplicity $<M_\gamma>$ per event decreases with the raise in the total kinetic energy predominantly for S and S2 modes. The decrease in γ-ray multiplicity $<M_\gamma>$ is less steep for other modes. This behaviour again demonstrates the main characteristics of distinct fission modes.

Table I. Average γ-ray multiplicity per fission fragment for fission modes of ^{252}Cf.

Mode	Light Fragment	Heavy Fragment
S1	4.77	1.05
S2	4.95	3.22
S3	5.14	3.61
SX	2.33	5.60
S	3.32	3.32

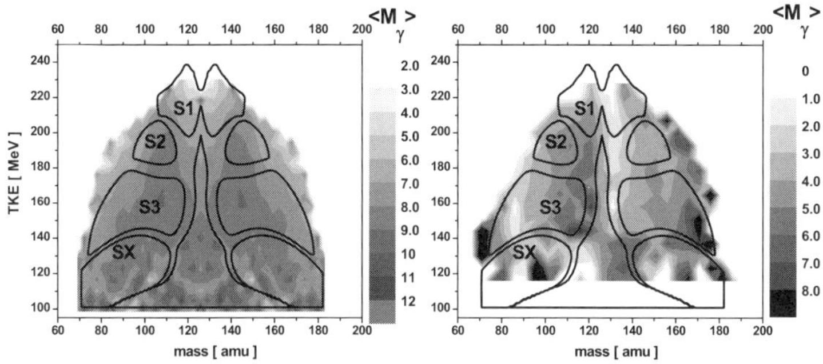

Fig.5. Left panel: Average γ-ray multiplicity for ^{252}Cf(SF) per fission event. Right panel: Average γ-ray multiplicity for ^{252}Cf(SF) per fission fragment.

Table II. Average γ-ray multiplicity per event for fission modes of ^{243}Am.

Mode	E_p = 13 MeV	E_p = 20 MeV	E_p = 55 MeV
S1	5.6	5.8	6.0
S2	8.1	8.3	8.5
S3	6.0	7.0	8.0
S	9.2	10.0	10.6

In the case of symmetric S mode energy available for the deformation and excitation of the fragments lowers with the increase in the TKE, hence the decrease in the γ-ray multiplicity $<M_\gamma>$ arises. The same feature is found for the S2 mode. Since S1 mode occurs at the high kinetic energies, fission fragments are considerably cold, so energy available for the neutron and γ-ray emission is low. Therefore the γ-ray multiplicity $<M_\gamma>$ declines only a little with the TKE increase. What concerns S3 mode, heavy fragment is strongly deformed whereas light fragment is spherical and its γ-ray multiplicity $<M_\gamma>$ is compared to the heavy fragment multiplicity rather low, almost negligible. Thus average $<M_\gamma>$ in fact reflects the spin of strongly deformed heavy fragment. Decrease in $<M_\gamma>$ with TKE growth shows the fact that heavy fragment becomes less deformed and more compact in shape.

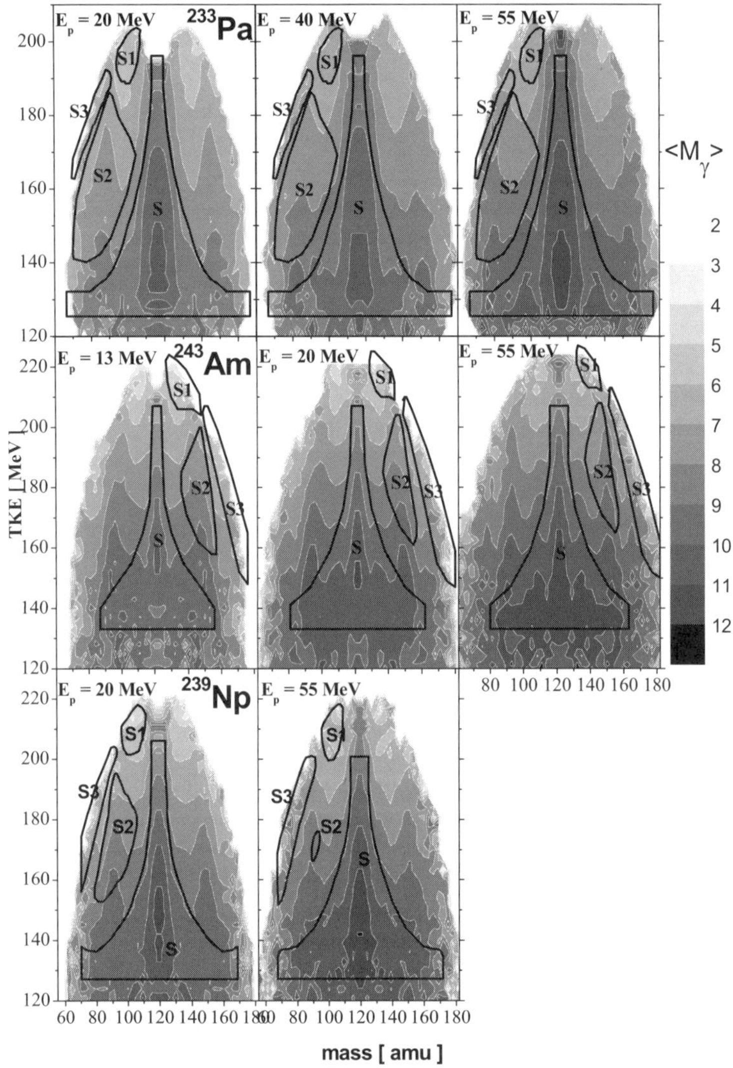

Fig. 6 Average γ-ray multiplicity $<M_\gamma>$ per event for reaction p+^{232}Th ->^{233}Pa (upper part), p+^{242}Pu ->^{243}Am (middle part) and p+^{238}U ->^{293}Np (lower part).

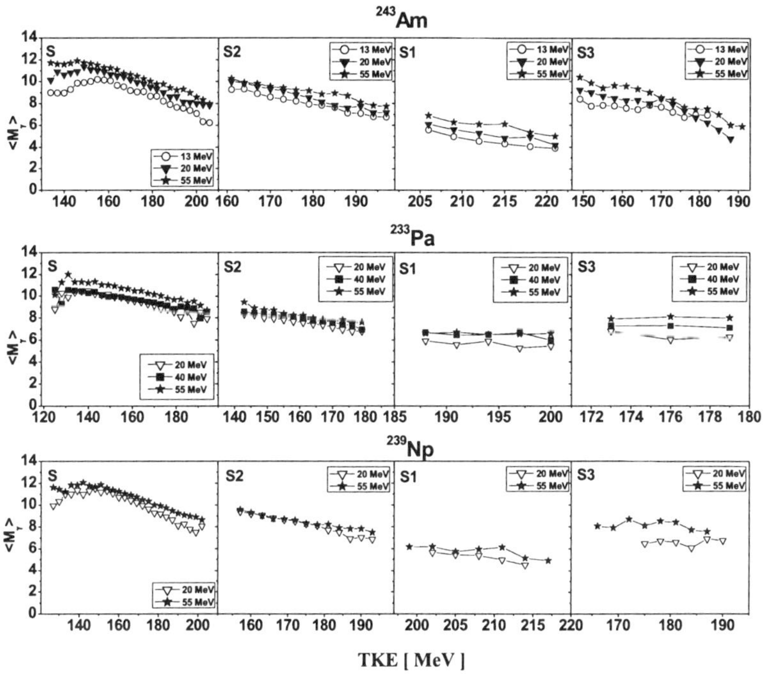

Fig. 7 Average γ-ray multiplicity $<M_\gamma>$ per event as a function of TKE for each fission mode for ^{243}Am (upper panel), ^{233}Pa (middle panel) and ^{239}Np (lower panel).

5 Discussion

Mass and energy distributions (MEDs) of fragments in the proton-induced fission of compound nuclei ^{233}Pa, ^{239}Np and ^{243}Am at the proton energies E_p=13, 20, 40 and 55 MeV have been analyzed in terms of the multimodal fission. The multimodal analysis method [1] was used to obtain the separate MEDs of the independent modes S1, S2, S3 and S from the experimental MEDs.

The most important characteristics of fission modes – mass yields Y_i, average heavy fragment masses $<M_H>_i$ and dispersions of mass yields $\sigma^2_{M,i}$ were investigated in dependence on the excitation energy for above-mentioned compound nuclei. For all the studied reactions increase in excitation energy results in redistribution of S and S2 mode yields. Fission mode characteristics are in good accordance with those of Ref [1].

The γ–ray multiplicities were presented as a function of fragment mass and TKE for ^{233}Pa, ^{239}Np and ^{243}Am. It can be inferred that with increasing proton energy substantial changes of γ–ray multiplicity as a function of fragment mass and TKE arise only in the symmetric mode. The dependencies of γ–ray multiplicity on the TKE for given excitation energies show decrease in $\langle M_\gamma \rangle$ with TKE raise for all of the fission modes.

References

1. D.M. Gorodisskiy, et al., In Proc. of the 5th International Conference on Dynamical Aspects of Nuclear Fission. Casta-Papiernicka, 23-27 October 2001, World Scientific, Singapore 2002.
2. L. Krupa, et al., In International Symposium on Exotic Nuclei. Peterhof, Russia, 5-7 July 2004, World Scientific, Singapore 2005.

YIELDS OF CORRELATED FRAGMENT PAIRS AND AVERAGE GAMMA-RAY MULTIPLICITIES AND ENERGIES IN ^{208}PB(^{18}O,F)

A.BOGACHEV[1], O.DORVAUX[2], E.KOZULIN[1], L.KRUPA[1], M.ITKIS[1], A.ASTIER[3], M.-G.PORQUET[3], L.STUTTGE[2]

[1]*JINR, Joliot-Curie 6, 141980, Dubna, Moscow region, Russia*

[2]*IReS, IN2P3-CNRS and Universite Louis Pasteur, 67037 Strasbourg, France*

[3]*CSNSM IN2P3-CNRS and Universite Paris-Sud, 91405 Orsay, France*

Measurements of fission fragment yields, neutron and gamma-ray multiplicities and average energies emitted by gamma rays have been carried out for the six fragment pairs (Z_L/Z_H = 44/46, 42/48, 40/50, 38/52, 36/54 and 34/56) in ^{208}Pb(^{18}O,f) at beam energy E=85 MeV, using the γ–ray spectroscopy technique to analyze γ–γ–γ coincidence data.

1. Introduction

Detailed studies of fission fragment charge, mass and energy distributions for a large variety of fissile systems are an important source of information about the mechanism of this process. The total kinetic energy (TKE) of fission fragments is mainly defined by the Coulomb potential as it appears just after scission. Therefore, TKE values, together with fragment mass asymmetry, provide the information on some essential characteristics of scission configurations. Prompt fission neutrons and γ-rays carry information about the excitation energy of the fission fragments and also are an important source of information about low-energy fission.

Access to new information, in addition to that usually obtained in previous experiments [1], could improve our ability to understand the fission process in more detail. One possible source of such information is spectroscopic study of prompt γ-rays, emitted by fission fragments. It was shown [2] and confirmed in other works [3,4] that the total intensities of the lowest $2^+\rightarrow 0^+$ rotation band transitions observed in the de-excitation of even-even fission products reflect, to a high degree of accuracy (<5%), the total independent yields of these isotopes. It was shown in [3,4] that the values of relative yields of pairs of fission fragments, as they appear after neutron evaporation, can be extracted from the analysis of γ–γ and γ–γ–γ coincidences, detected for prompt fission γ-rays with new larger detector arrays (GAMMASPHERE, EUROBALL IV). This led to the first measurements of the independent yields of six pairs, namely Ru/Pd, Mo/Cd, Zr/Sn, Sr/Te, Kr/Xe and Se/Ba and the prompt neutron multiplicity distribution for this pairs in the reaction with heavy ions.

Other information that one can extract from our experiment includes angular momenta of the fission fragments for specific fragment pairs. Such data are desirable since the fragment angular momentum gives the information about coupling between different collective degrees of freedom at the descent from saddle point to scission [5]. Results and analysis of the relative intensities of γ-ray transitions between the levels of the ground-state rotational bands of some light and its corresponding heavy partner-fragments are discussed. Mean angular momentum values of fission fragments obtained after neutron evaporation with γ–γ–γ coincidence method are also discussed here. In summarizing of our results, we also propose some prospects for future experiments.

This work was also partly devoted to the investigation of multimodal manifestation in fusion-fission process in the reaction $^{18}O + ^{208}Pb \rightarrow ^{226}Th$. Earlier it was found that fission modality exhibits itself not only in the properties of mass and total kinetic energy (TKE) distributions, but also in the fission fragment angular distributions [6], post-fission neutron multiplicities ν_{post} and their distributions [7,8]. In the present work we tried to make any conclusions concerning the multimodal fission based on the analysis of γ–rays emitted from secondary fission fragments originated from neutron-deficient Th isotopes.

2. Experimental setup

The measurements were carried out using the γ-ray spectrometer EUROBALL IV (HPGe-detectors) coupled with the Inner Ball (BGO-detectors). The spectrometer contained 71 Compton-suppressed Ge detectors (15 cluster germanium detectors placed in the backward hemisphere with respect to the beam, 26 Clover germanium detectors located around 90° perpendicular to beam axis, 30 tapered single-crystal germanium detectors located at forward angles) and an inner ball of 210 BGO crystals. Each Cluster and Clover detector included seven and four, respectively, smaller Ge crystals [9,10].

We used thick (100 mg/cm^2) enriched ^{208}Pb as a target. In the case of such thick target Doppler shift correction was not needed because fission fragments were completely stopped in ^{208}Pb. In order to suppress the Compton scattering events, anti-Compton shield was used. Add-back effect was also taken into account. The beam was provided by the VIVITRON accelerator of IReS (Strasbourg).

The data were recorded in an event-by-event mode with the requirement that the minimal number of Ge detectors (not suppressed with anti-Compton shield), fired in prompt coincidence, is equal to three. In this data set, prompt HPGe γ-ray events were selected by requiring a gate of 60 nsec in the TDC spectra. A set of 4×10^9 three- and higher-fold events were available for a subsequent analysis. The offline analysis was performed as an analysis of multi-gated spectra. The Radware [11] and [12] packages were used for this task.

3. Data analysis method.

The triple γ-γ-γ coincidence data were analyzed to extract the relative production cross-section for a given pair of final fragments. A double gate on two coincident γ-rays in a given nucleus (A_1, Z_1) (the $2^+ \rightarrow 0^+$ and $4^+ \rightarrow 2^+$ transitions) was applied to obtain a double-gated γ-ray spectrum where the γ-ray transitions of the partner fragments (several isotopes with different neutron numbers) were clearly identified. The intensities of the γ-ray transitions in the partners (the $2^+ \rightarrow 0^+$ lines) were corrected for the detection efficiency of the γ-rays involved in the selection and used to extract the relative yields of the isotopes.

For a fixed pair of fission products (A_1, Z_1) and (A_2, Z_2) with $Z_1+Z_2=90$, the number of emitted neutrons is uniquely defined as $\nu_n=226-(A_2+A_1)$. As a result, the isotope yields uniquely determine the distributions of the neutron multiplicities associated with the nucleus (A_1, Z_1). The total neutron multiplicity distribution for the fixed fission channel (e.g Pd-Ru pair) was then obtained by summing the yields for the same number of emitted neutrons associated with the different isotopes of fixed pair.

In this work, the results on production yields and neutron multiplicity distribution for the six fission channel pairs are reported. In order to find out the extent to which the results are influenced by the chosen gating transitions and by the available spectroscopic information, the data analysis was performed in two ways by double gating: first we put gates 2^+-0^+ on the light as well as on the heavy isotope and counted 4^+-2^+ transition events of both fragments, second we put gates on 4^+-2^+ and 2^+-0^+ transitions of one fragment, and counted 2^+-0^+ transition events in the complementary fragment. These counts were considered as yields of corresponding fragments.

This procedure was applied only for even-even isotopes, where the level schemes in most cases are well-known, and the ground-state rotational band is clearly identified. In case of odd-odd and even-odd nuclei the decay paths are usually much more complicated and the complete spectroscopic information is not available in some cases. Due to these reasons we refused the analysis of odd-odd and even-odd fission products.

Also, we did not take into account the side-feeding contribution to the yield. In other words, we took only isotopes which suffered the γ-ray transitions from the level at least 6^+ of the rotational band. It is expected that for even-even nuclei the side-feeding does not give very significant contribution to the yield (approximately 5-15%) in the case of double-gated spectra analysis. But in fact we performed triple-gated spectra analysis, of course, it should enlarge the errors of our yield estimations. Moreover, the efficiency correction and background subtraction procedure also contribute to the errors. Taking into account all of that, we estimate our errors to about 10-30% depending on the statistics.

4. Results and discussion

The example of double-gated spectrum is given in the Fig. 1. Two gates were set on ^{112}Pd (which correspond to transitions $2^+ \to 0^+$ and $4^+ \to 2^+$) to obtain this spectrum. The peaks corresponding to the complementary fragments (Ru isotopes) and to the higher transitions of ^{112}Pd (up to $12^+ \to 10^+$) are shown. After subtraction of the background and making corrections of the measured peak areas for the known detection efficiencies, we obtained the relative transition intensities.

The detailed experimental yield distributions of correlated fragments and distributions of numbers of emitted neutrons for the most intensive pair Mo/Cd are given in Fig. 2. The total relative production yield of the Mo and Cd isotopes was obtained by summing over single partition yields (for Mo isotope summing up the numbers in the rows and for the Cd summing up the numbers in the columns). The yield distributions for this pair are shown in Fig. 2. using different ways of γ-ray gating. As one can see, the differences between the two distributions usually reach 20-30%. For some pairs the errors are much higher. For example for mass 122 (Fig 2.) there is a great difference in value when gated on Cd compared to value when gated on Mo. This is due to fact that some of the transitions which should contribute to the mass yield were not seen. In this case the higher value was taken into account in the fission fragment isotopic distributions of studied pairs (see Fig. 3).

In the same way the yield distributions of other five pairs (Mo/Cd, Zr/Sn, Sr/Te, Kr/Xe and Se/Ba) were obtained. The results are shown in Fig. 3. In this case the average values of both data sets (first, the two gates are set on heavy fragment and one gate on the light one and then vice versa) are presented.

4.1 Fission fragment yields and neutron multiplicities

In the Fig. 3 the yields of different nuclides are given. As it was expected, these distributions are rather narrow. Just about 4-6 isotopes are visible with number of neutrons, emitted from both fission fragments up to ten. So, it is clear that we may restore the total neutron multiplicity for the products of the fission process (we discuss it below). Making the yield summation over masses we obtain the curve of mass distribution. It is shown on the same Fig. 3 as solid line. One can observe the “structure” of the curve: the number of peaks and valleys. Such behavior of the curve is simple to explain. In our analysis we considered only even-even isotopes. And the valleys are caused by the lack of odd-even and odd-odd isotope yields. We estimate the error bars of about 10-30% (see the text above).

In the Fig. 4 we compare our mass distribution deduced from the present data with other experiments. The first experiment was carried out at TAMU accelerator [13] (Chubarian et al.). The energy of ^{18}O ions was 78 MeV. Another experiment was conducted in Strasbourg at VIVITRON at ^{18}O energy of the same 85 MeV [1] (Pokrovsky et al.). In both experiments the measurements were carried out with TOF-TOF method. In fact, the TOF-TOF method gives the possibility to measure the fission fragment masses before scission (pre-fission masses). But the method that we applied (γ-γ-γ coincidence) gives fission fragment masses after scission (post-fission masses). Due to this reason we had to shift the mass distributions obtained with TOF-TOF method in the region of smaller masses on the number of pre-scission neutrons ν_{pre}. In the Fig 5 the number of pre-scission neutrons is given for different fragment pairs. The mean values calculated with these distributions show that the mean number of emitted neutrons (for all fragment pairs) is about 5.5 per fission. We shifted the mass distributions obtained in previous experiments on this number. Of course, this number of emitted neutrons differs a bit for different fission products, but we neglect this difference, because it is about 1 amu or maybe less. In our opinion, our distribution is in good agreement with the mass distributions obtained by TOF-TOF method. For the central part (symmetric fission) the maximums of the even-even isotopes are "sitting" on these curves within error bars. For the asymmetric fission region our distribution follows the distribution of Pokrovsky et al. [1] with very good agreement and has evident difference with Chubarian et al. distribution [13]. Chubarian et al. distribution is wider, but the excitation energy is lower than that for our experiment and in the experiment of Pokrovsky et al [1]. It could be explained as follows: in case of Chubarian et al. measurements [13] we have lower excitation energy. This fact gives the possibility to observe the superasymmetric fission mode which is allocated in the A<80 region, which is explained by the influence of the shell effects. But in the case of Pokrovsky et al. [1] experiment the excitation energy is higher and the shell effects disappear. At least it is seen that the manifestation of the superasymmetric mode for higher energy (Pokrovsky et al. experiment [1]) is strongly suppressed, whereas the symmetric part grows up very significantly. The width of symmetric fission mode with increasing excitation energy grows up not so drastically. That is the reason that the mass distribution for 85 MeV is narrower than that for 78 MeV.

Unfortunately, in the case of our experiment it is difficult to say something about the influence of the shell effects and its manifestation because we have mass distribution which is not complete due to the lack of even-odd and odd-odd isotope yields. But nevertheless, the width of our mass distribution is very close to that which was observed by Pokrovsky et al [1] at the energy of ^{18}O ions equal to 86 MeV.

One of the problems of the presented analysis was the fact that some of nuclei of interest could populate the metastable states with rather high probability. For example, almost all analyzed isotopes of Sn de-excite through these states. The life-times of these states of Sn isotopes is about 7 microseconds. But the time-gate of the data acquisition system was just 800 nanoseconds. It means that we should underestimate the yield of Sn isotopes and, respectively, the yield of Zr isotopes which are complementary to Sn. Due to this reason we assume that the yields of Sn and Zr isotopes should be higher than it is observed in the analysis.

On the basis of the yields of all six studied pairs we obtained the total neutron multiplicity distributions shown in Fig. 5. Also, the average total neutron multiplicities $<\nu_n>$ are given for each fragment pair. These neutron multiplicity distributions in first approximation could be described with Gaussian curves with average value ranging from 5 up to 6.2 neutrons.

4.2 Average spin and average energy per one gamma quantum distributions

The mechanism of spin generation in spontaneous fission products has been investigated in the past [14,15,16] and it was found to be associated mainly with orbital angular momentum transfer to the fission fragments and with the excitation of various angular momentum bearing modes, such as wriggling, bending, twisting and tilting in the nascent fragments [17,18]. The present study offers a unique possibility for a detailed analysis of the spin distribution of single fission fragments. In this work the average angular momenta for individual fission fragments were derived from the measured populations of different spin levels.

Such data analysis was performed for the most intense channels of the studied pairs (usually two or three correlated fragment pairs) employing the gates, used for obtaining the correlated fragment pair yields. First, we set two gates on heavy fragment γ-ray transitions ($2^+ \rightarrow 0^+$ and $4^+ \rightarrow 2^+$) and one gate on the light fragment γ-ray transition ($2^+ \rightarrow 0^+$). The average spin for both correlated fragments was then calculated using all transitions found in sorted triple-coincident spectrum with weighting factors and corrected for detection efficiency. The example of such a spectrum in the case of ^{108}Ru-^{112}Pd pair is presented in Fig. 6. Then we set two gates on light fragment ($2^+ \rightarrow 0^+$ and $4^+ \rightarrow 2^+$) and one gate on heavy one ($2^+ \rightarrow 0^+$) and procedure for average spin calculation was repeated. For example, in the case of mass A=108, the average spin was obtained as follows: the highest yields for this mass come from the fragment ^{108}Ru. All other contributions from ^{108}Pd and ^{108}Mo were neglected. Then we calculated the average spins for ^{108}Ru in coincidence with fission fragments ^{112}Pd and ^{114}Pd and weighted by their yields (The contribution of both fission fragment pair independent yields ^{108}Ru-^{112}Pd and ^{108}Ru-^{114}Pd into the total isotopic yield of ^{108}Ru is more than 90%). In the similar way we calculated the average energy emitted by γ-rays. The resulting average spins and average energies of secondary fission fragments as a function of mass are shown in Fig. 7.

The average spins calculated in such a way are of course smaller than real ones. We do not take into account side-feeding to levels 0^+ and 2^+ in the first fragment and 0^+ in second one. In addition there are some transitions we do not see because their yield is comparable to the background. The statistical dipole (E1) contribution of γ-rays to the average spin is equal to $\alpha < M_\gamma^{stat} >$, where a~0.3 and $< M_\gamma^{stat} >$ is average multiplicity of statistical γ-rays. In our case the $< M_{\gamma}^{stat} >$ is about 1-2 and so its contribution is quite small. We estimate that the overall systematic accuracy in calculation of average spins may be as large as 20-30 percent.

As one can see from Fig. 7 the average spin curve as a function of single fragment mass is characterized by a sawtooth behavior similar to that which is well known for spontaneous fission of ^{252}Cf [[19]] and for thermal-neutron induced reactions [[20]]. The similar behaviour was also recently observed in the case of proton induced reactions on actinides [[21]]. The curve has too local minima. The first one around the mass m=110, the second one in the range of masses m=120-135. The dip in the curve for masses around m=130 probably reflects the influence of shell effects on average spins of fission fragments. The average spin of fission fragment slightly increases with increasing of mass except the mass region where the shell effects manifest. On the other hand the average energy has opposite tendency. The highest values are reached for the lowest spins. These features was observed in previous experiments with spontaneous fission of ^{252}Cf and in some thermal-neutron and proton induced reactions [19,20,21].

The quite low value of fission fragment spin for masses around m=120 is probably connected with very low observed mass yields of fission fragment pair Zr/Sn. As it was already mentioned above, for isotopes ^{120}Sn, ^{122}Sn and ^{124}Sn there exists isomer states with spin 7^-. Their life times are more than 1 μs and γ-transitions from these states could not be detected in our experiment. If most of transitions go through this state then the obtained yields and average spins are much lower than that for neighboring fission fragment pairs. That also means that average spin of Sn isotopes have to be higher than 7 , otherwise the mass yields of fission fragment pair Zr/Sn can not be so low. The low yields and spins can also be explained if the initial spin distribution of secondary fragments is low (has small average value). In this case the contribution from side-feeding to levels 0^+ and 2^+ should be much higher and that results to low yields for pair Zr/Sn. Unfortunately for this moment we can not ambiguously conclude which of these two assumptions is true.

4.3 Fission modes

This reaction was studied intensively in many experiments and there exist data concerning not only the MED, but also the pre and post-scission neutron multiplicities as well as γ-ray multiplicities [1,13]. As it is well known from previous experiments, the mass distribution of fission fragments produced in $^{208}Pb(^{18}O,f)$ at the beam energy E_{lab}=78 MeV show four fission modes. These are symmetric mode (S) and three asymmetric modes, standard-1 (S1), standard-2 (S2), and standard-3 (S3). The multimodality of MED is very well seen also in our experiment where the beam energy was higher E_{lab}=85 MeV (see Fig. 4). In the Fig. 4 one can see two dips in mass distribution of secondary fission fragments. The first one, around the mass m=120, is probably concerned with lacks of events due to existence of isomeric states in Sn isotopes as it was discussed in previous sections. The second dip, around the mass m=136, is, in our opinion mainly, due to manifestation of asymmetric fission modes. In this mass region the mass distribution has a bump which is even more outstanding for E_{lab}=78 MeV. In addition, the average post-scission neutron multiplicity and average angular momentum of secondary fission fragments reach a local minimum. All of that is an indication of multimodal fission manifestation of fission fragments.

Earlier it was found that fission modality exhibits itself not only in properties of MED's, but also in the fission fragment angular distributions [22], postfission neutron multiplicities ν_{post} and their distributions [23,24]. In this paper we will show that the phenomenon of multimodal fission also manifests itself in the γ-rays emitted from secondary fission fragments of the neutron-deficient Th isotopes.

5. Summary

A method based on measurements of intensities of γ-transitions from correlated pairs of secondary fragments in a γ-γ-γ coincidence experiment was used for the first time to determine the detailed characteristics of the reaction ^{208}Pb(^{18}O,f) at the beam energy E_{lab} = 85 MeV. By applying this method, we measured directly the yields of six charge splits produced in the fission of Th. Summing up these yields, we obtained the mass, charge, and neutron multiplicity distributions of fission fragments. For a long time these data have been obtained only with use of integral methods in low and medium energy fission. The agreement of our data with those which were known previously proves the validity of the approach made in this work. In some aspects (independent yields of fission fragments, mean neutron multiplicity obtained for different charge splits) our results complement considerably the previously known data. In addition, an approach used in this work allowed us to obtain yields and multiplicity distributions of prompt neutrons emitted at various charge splits. These data are not accessible in previous experimental methods. The attractive feature of these new distributions is that they were obtained, practically, as a result of direct measurements of the triple γ-γ-γ coincidence peaks. That was made for the first time for the reaction with heavy ions. This makes a strong difference between these distributions and those which are derived from the neutron detection experiments involving sophisticated unfolding procedures applied to the raw data. In addition, average angular momentum as a function of secondary fission fragments was obtained directly from spectroscopic studies of discrete γ–ray lines. The average fragment spins show some structure as a function of secondary fission fragment mass.

The future investigations should combine fission fragments measurements (by TOF-TOF method or similar) in coincidence with γ-ray measurements (by the array of Ge-detectors with high energy resolution). The Total Kinetic Energy of the fission fragments will improve the procedure conditions. In this case the target should be rather thin in order to the fission fragments not to lose too much energy in it, because big energy losses will merge the mass distribution. But in the case of thin target one should include Doppler shift correction to the γ–ray procedure. Nevertheless, we suppose that the compromise could be found. Such comprehensive measurements will considerably improve our ability to recognize the scission configurations and draw more precise conclusions about the final energy partition at fission. New possibilities will arise for learning the level schemes of fission fragments and, therefore, they could shed light on the problem of the fragment angular momentum origin.

Acknowledgements

This work was performed with the partly support of the Russian Foundation for Basic Research under Grant No. 3-02-16779 and INTAS under Grant No. 03-51 6417.

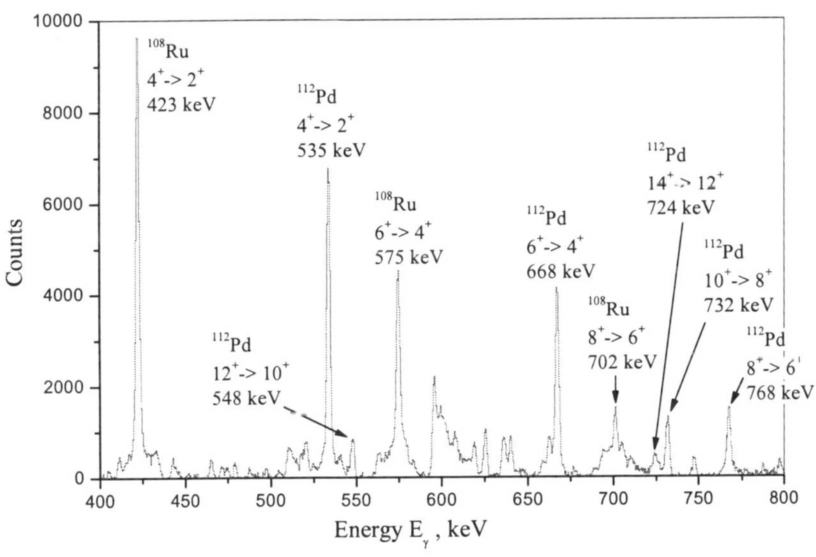

Fig. 1. Double gated spectrum of the ^{108}Ru-^{112}Pd isotope pair. Gates on the transitions $2^+ \rightarrow 0^+$ for both isotopes are set.

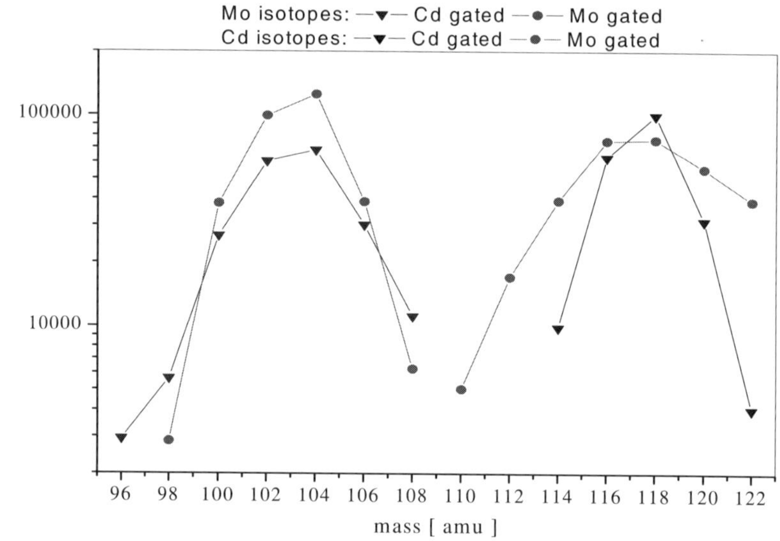

Fig. 2. Total relative yield of Mo and Cd isotopes. In the first case the $4^+ \rightarrow 2^+$ transition was set on the Cd fragments and in the second one the $4^+ \rightarrow 2^+$ transition was set on the Mo fragments.

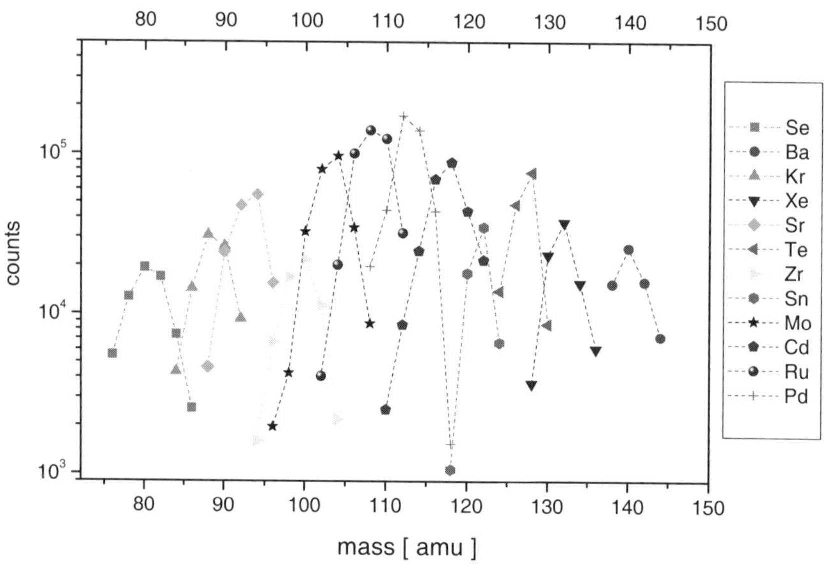

Fig. 3. Summary of fission fragment isotopic distributions (for fragment pair partitions Ru/Pd, Mo/Cd, Zr/Sn, Sr/Te, Kr/Xe and Se/Ba) deduced from the fragment pair independent yields is presented.

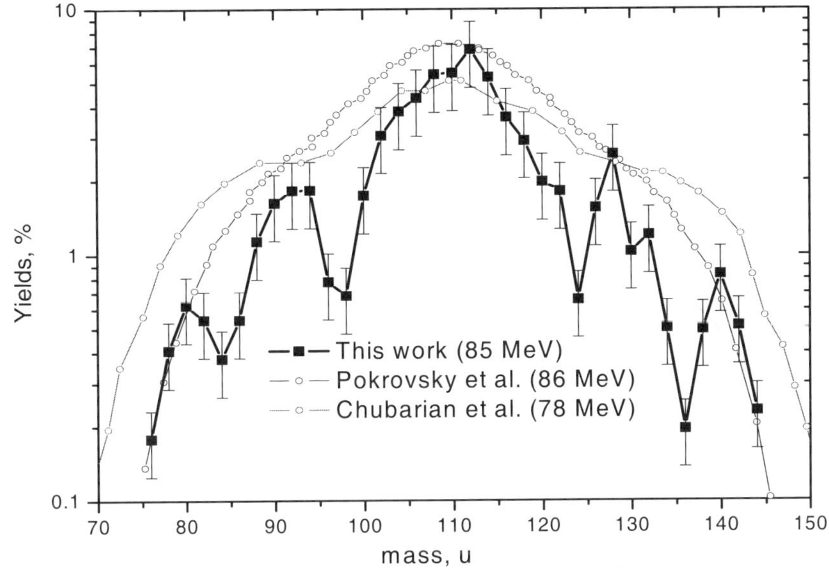

Fig.4. Mass distributions: green triangles - Our work (not normalized to other work); black squares-our work normalized to Chubarian et al.; Blue solid squares – Pokrovsky et al. (86 MeV, VIVITRON, 2003); Blue empty circles – Pokrovsky et al. – pre-fission neutrons subtracted; Red empty circles – Chubarian et al. – pre-fission neutrons subtracted.

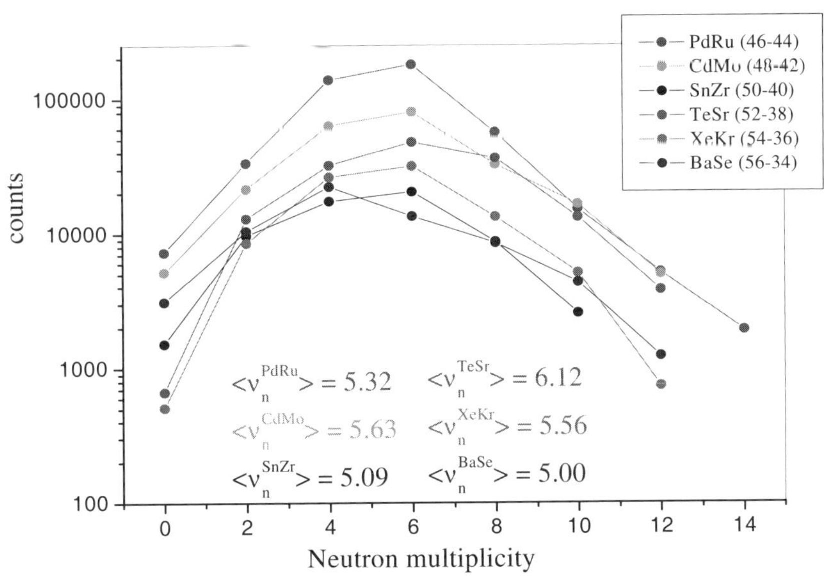

Fig. 5. Total neutron multiplicity distributions for six pair partitions from our experiment.

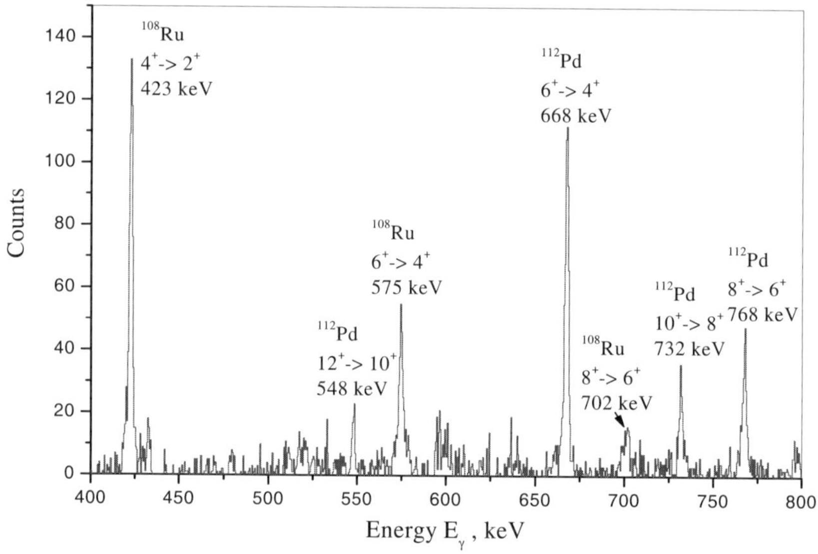

Fig. 6. Triple gated spectrum of the ^{108}Ru-^{112}Pd isotope pair. Two gates on the transitions $2^+ \rightarrow 0^+$ for both isotopes are set. The third gate is set on the transition $4^+ \rightarrow 2^+$ of ^{112}Pd.

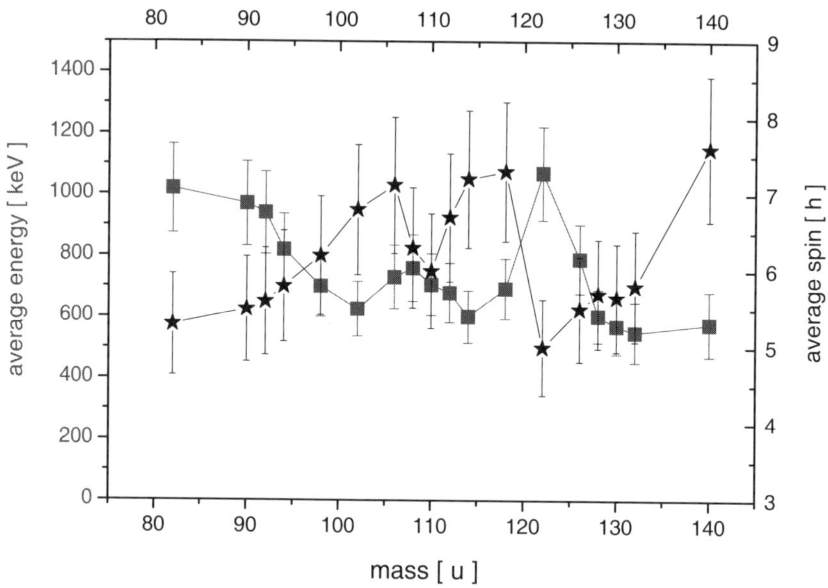

Fig. 7. Average angular momentum and energy emitted by γ-rays of secondary fission fragments as a function of mass.

References

[1] I.V. Pokrovsky et al., Phys. Rev. C 62 (2000) 014615.

[2] E. Cheifetz, J. B. Wilhelmy, R. C. Jared, and S. G. Thompson, Phys. Rev. C **4**, 1913 (1971).

[3] G. M. Ter-Akopian et al., Phys. Rev. C **55**, 1146 (1997).

[4] D.C.Biswas et al., Eur.Phys.J. **A7**, 189-195 (2000).

[5]

[6] F. Steiper *et al.,* Nucl. Phys. **A563**, 282 (1993); A. A. Goverdovski *et al.,* Phys. At. Nucl. **58**, 188 (1995).

[7] J. van Aarle, *et al.,* Nucl. Phys. **A578**, 77 (1994); in *Second International Workshop Nuclear Fission and Fission Product Spectroscopy, Seyssins, France, 1998,* edited by G. Fioni *et al.,* AIP Conf. Proc. No. 447 (AIP, Woodbury, New York, 1998), p. 283.

[8] J. F. Wild *et al.,* Phys. Rev. C **41**, 640 (1990); T. Ohsawa *et al.,* Nucl. Phys. **A653**, 17 (1999); **A665**, 3 (2000).

[9] H. Fann, J.P. Schiffer, U. Strohbusch, Phys. Lett. B 44, 19 (1973).

[10] J. Simpson, Z. Phys. A 358, 139 (1997).

[11] D. Radford, Nucl. Instrum. Methods A 361, 297; 306 (1995).

[12] M. Morhac et al, *Nucl. Instr. and Meth.* **A401** (1997) 385.

[13] G. Chubarian et al., Physical Review Letters **87,** 052701 (2001).

[14] J. B. Wilhelmy, E. Cheifetz, R. C. Jared, S. G. Thompson, H. R. Bowman, and J. O. Rasmussen, Phys. Rev. C **5**, 2041 (1972).

[15] H. Nifenecker *et. al.*, Nucl. Phys. **A189**, 285 (1972).

[16] P. Glassel *et. al.*, Nucl. Phys. **A502**, 315C (1989).

[17] L. G. Moretto and R. P. Schmitt, Phys. Rev. **C21**, 204 (1980).

[18] R. P. Schmitt and A. J. Pacheco, Nucl. Phys. **A379**, 313 (1982).

[19] H. Maier-Leibnitz, H.W. Schmitt, and P. Armbruster, in *Poceedings of the Symposium on the Physics and Chemistry of Fission, Salzburg, 1965 (International Atomic Energy, Vienna, Austria, 1965)* Vo. II, p. 143.

P. Armbruster, et al., Z. Naturfosch **26a**, (1971) 512.

S.A.E. Johansson, Nucl. Phys. **60** (1960) 378.

[20] F. Pleasenton, R.L. Ferguson, and H.W. Schmitt, *Phys. Rev.* **C6** (1972) 1023.

[21] L. Krupa et al., Proc. Int. Symp. On Exotic Nuclei, EXON 2004,July 5-12,2004, Peterhof, Russia, Ed.: Yu.E. Penionzhkevich and E.A. Cherepanov, World Scientific 2005, p.343.

[22] F. Steiper *et al.,* Nucl. Phys. **A563**, 282 (1993); A. A. Goverdovski *et al.,* Phys. At. Nucl. **58**, 188 (1995).

[23] J. van Aarle, *et al.,* Nucl. Phys. **A578**, 77 (1994); in *Second International Workshop Nuclear Fission and Fission Product Spectroscopy, Seyssins, France, 1998,* edited by G. Fioni *et al.,* AIP Conf. Proc. No. 447 (AIP, Woodbury, New York, 1998), p. 283.

[24] . F. Wild *et al.,* Phys. Rev. C **41**, 640 (1990); T. Ohsawa *et al.,* Nucl. Phys. **A653**, 17 (1999); **A665**, 3 (2000).

RECENT EXPERIMENTS AT GAMMASPHERE INTENDED TO THE STUDY OF ^{252}CF SPONTANEOUS FISSION

A. V. DANIEL, G. M. TER-AKOPIAN, A. S. FOMICHEV, YU. TS. OGANESSIAN, G. S. POPEKO and A. M. RODIN

Flerov Laboratory of Nuclear Reactions, JINR, Dubna, 141980, Russia

J. H. HAMILTOM, A. V. RAMAYYA, J. K. HWANG, D. FONG, C. GOODIN and K. LI

Department of Physics, Vanderbilt University, Nashville, TN 37235

J. O. RASMUSSEN, A. O. MACCHIAVELLI and I. Y. LEE

Lawrence Berkeley National Laboratory, Berkeley, CA 94720

D. SEWERYNIAK, M. CARPENTER, C. J. LISTER and SH. ZHU

Argonne National Laboratory, Chicago, IL 60439

J. KLIMAN and L. KRUPA

Institute of Physics, SAS, Bratislava 84511, Slovakia

J. D. COLE

Idaho National Engineering and Environment Laboratory, Idaho Falls, Idaho 83415

W.-C. MA

Mississippi State University, Mississippi State, MS 39762

S. J. ZHU

Tsinghua University, Beijing 100084, China

L. CHATURVEDI

Banaras Hindu University, Varanasi 221005, India

Recent experiments designed for the multi-parameter analysis of ^{252}Cf spontaneous fission are described. The technique of multiple γ-ray spectroscopy was supplemented with the measurements of kinematical characteristics of fission fragments and light charged particles emitted in ternary fission.

1. Introduction

A series of experiments[1] executed on Gammasphere using hermetically closed source of ^{252}Cf and a $\gamma-\gamma$ coincidence technique essentially extended the obtained experimental information and made deeper our insight of the fission process. In particular, the technique of double and triple gamma coincidences allowed us to identify the pairs of complementary fission fragments for the first time. As a result the yields of fission fragments pairs and the distributions of neutron multiplicities were obtained for various charge splits of a fissile nucleus. Using these data we could derive some conclusions about the excitation energy distributions of primary fragments.[1,2]

Future evolution of this work resulted in new experiments designed for the multi-parameter analysis of the ^{252}Cf spontaneous fission.[3–5] The traditional technique of multiple γ-ray spectroscopy was supplemented with the measurements of kinematical characteristics of fission fragments and light charged particles (LCPs). This implies the use of an open source of ^{252}Cf and the introduction of Doppler correction to the measured γ-ray energies.

2. Experiments

Two experiments have been carried out using Gammasphere at the Lawrence Berkeley National Laboratory and at the Argonne National Laboratory. Gammasphere was set to record γ rays with energy between ~80 keV and ~5.4 MeV. The γ-ray detection efficiency varied from a maximum value of ~17% to ~4.6% at the γ energy 3368 keV. The fission fragment and LCPs detectors were placed in a spherical chamber installed in the center of Gammasphere. The arrangements of these detectors are shown schematically in the Fig. 1 and Fig. 2 for the first and second experiments, respectively.

Details and sizes of the detector arrays are summarized in Table 1. The $\Delta E \times E$ telescopes were used to measure the LCPs emitted in the ternary fission. Double side silicon strip (DSS) detectors were used for measuring kinetic energy and flight directions of fission fragments. The sources were prepared from ^{252}Cf specimen deposited in a 5-mm spot on Ti foils (the foil thickness was 1.8μ and 2.0μ, respectively, in the first and second experiment). In the first experiment, the source was additionally covered by gold foils on both sides. These foils had the minimum thickness required for stopping all fission fragments.

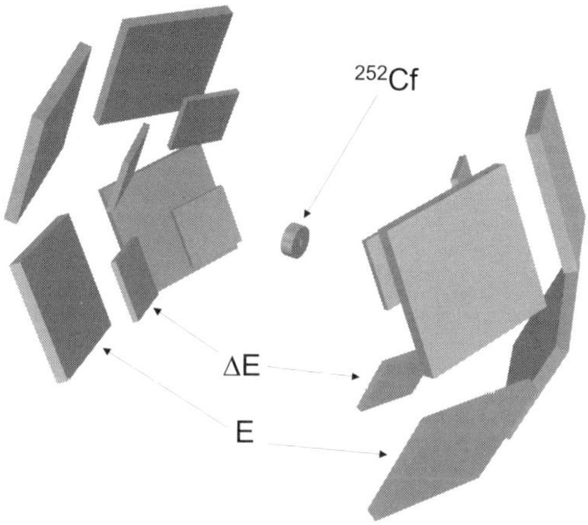

Fig. 1. Schematic diagram showing the detector array of the first experiment. Eight $\Delta E \times E$ telescopes intended for LCPs are placed around a ^{252}Cf source.

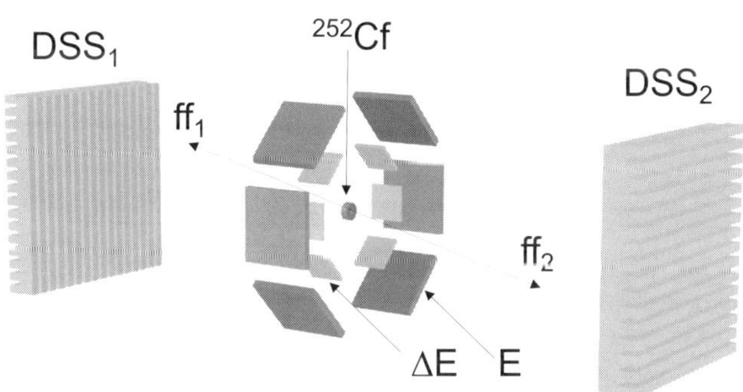

Fig. 2. Schematic diagram showing the detector array of the second experiment. The source of ^{252}Cf is in the center of the detector array. Two double side Si strip detectors, DSS1 and DSS2, are hit by fission fragments. Six $\Delta E \times E$ telescopes are used for the LCPs detection.

3. Results of the first experiment

During a two week experiment $\sim 1.6 \times 10^7$ events were recorded. Data acquisition was triggered by ΔE or E signals with amplitudes exceeding the

Table 1. Detector arrangement (D denotes the distance to the source)

	Experiment				
	I (made in LBNL)		II (made in ANL)		
Detector	ΔE	E	DSS	ΔE	E
Number	8	8	2	6	6
Area [mm^2]	10 × 10	20 × 20	60 × 60	10 × 10	20 × 20
Thickness [μ]	9.0-10.5	400	400	9.5-10	300
Strips	-	-	32	-	-
D [mm]	27	40	80	19	33

threshold values which were set to prevent the detection of twofold pileup events of α particles emitted in the radioactive decay of ^{252}Cf. Ternary fission events were stored at a condition that at least one γ ray was detected by Gammasphere within the time interval allocated for these events.

The resolution of the $\Delta E \times E$ telescopes allowed us to well identify helium, beryllium, boron and carbon nuclei when energy deposition in the E detector was greater then 5 MeV. It allowed us to refine data on the LCPs energy distributions using additional calibration measurements done with the open ^{252}Cf source.[6] Having these LCPs energy spectra we could estimated the portions of LCPs registered in the experiment. These data are summarized in Table 2.

Table 2. Detection conditions obtained for LCPs

LCP	E_th, MeV	P, %	Counts
He	9	93	4905767
Be	20	39	30960
B	26	26	1940
C	32	32	6445

The matrix of $\gamma - \gamma$ coincidences was built for the He ternary fission events. Using technique described in[1] we estimated the yields of fission fragment pairs shown in Tables 3 - 6. By summing the data of Tables 3-6 in the rows and columns one can obtained independent yields of fission fragments emitted in the He ternary fission of ^{252}Cf (see Fig. 3).

The numbers of recorded Be and C ternary fission events (see Table 2) were not high enough to build of the $\gamma - \gamma$ coincidences matrices. Instead, we created two linear γ spectra accordingly from these two data groups. Independent yields of 38 and 35 fragments, respectively, were obtained for the first time for the Be and C ternary fission of ^{252}Cf. The results are presented in Fig. 4 and Fig. 5.

Table 3. Independent yields of fission fragment pairs for the Ce-Sr charge split of ^{252}Cf (He ternary fission)

Ce - Sr	95	96	97	98
146	—	—	—	0.06(2)
148	0.08(2)	0.29(5)	0.17(5)	0.19(6)
150	0.19(4)	0.31(3)	0.12(3)	—
152	—	0.10(3)	—	—

Table 4. Independent yields of fission fragment pairs for the Ba-Zr charge split of ^{252}Cf (He ternary fission)

Ba - Zr	98	100	101	102	103
141	—	—	—	—	0.20(8)
142	—	0.14(6)	0.23(11)	0.51(9)	0.30(10)
143	—	—	0.86(40)	0.94(22)	0.30(17)
144	—	0.96(6)	1.33(45)	1.17(22)	0.18(7)
145	0.13(8)	0.70(17)	0.81(40)	0.65(17)	—
146	0.22(4)	0.60(5)	0.43(20)	—	—
147	0.16(8)	0.30(9)	—	—	—
148	0.07(5)	—	—	—	—

Table 5. Independent yields of fission fragment pairs for the Xe-Mo charge split of ^{252}Cf (He ternary fission)

Xe - Mo	104	105	106	107	108
136	—	—	—	—	0.05(2)
137	—		0.20(3)	0.08(3)	0.29(3)
138	—	0.63(30)	1.00(7)	0.31(9)	0.76(9)
139	—	0.66(23)	0.75(6)	0.31(5)	0.37(4)
140	0.77(5)	1.50(60)	1.24(7)	0.07(7)	—
141	0.18(5)	0.45(18)	0.21(4)	—	—
142	0.39(4)	0.20(10)	—	—	—

These results allowed us to present, for the first time, charge distributions obtained for fission fragments appearing in the ternary fission of ^{252}Cf in coincidence with helium, beryllium and carbon LCPs. These charge distributions are presented in Fig. 6. For comparison we show in Fig. 6 the fragment charge distribution known for the binary fission of ^{252}Cf.

From comparison made for the two charge distributions, one obtained

Table 6. Independent yields of fission fragment pairs for the Te-Ru charge split of ^{252}Cf (He ternary fission)

Te - Ru	109	110	111	112
134	0.11(4)	0.43(4)	0.14(2)	0.12(2)
135	—	0.13(2)	—	—
136	0.21(8)	0.68(11)	—	—

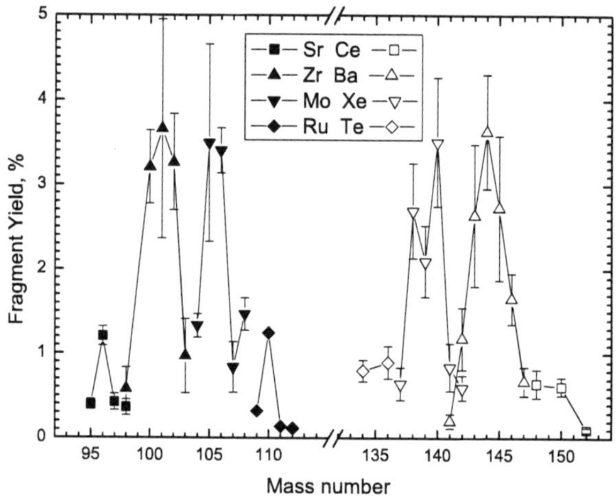

Fig. 3. Independent yields of fission fragments in the He ternary fission of ^{252}Cf.

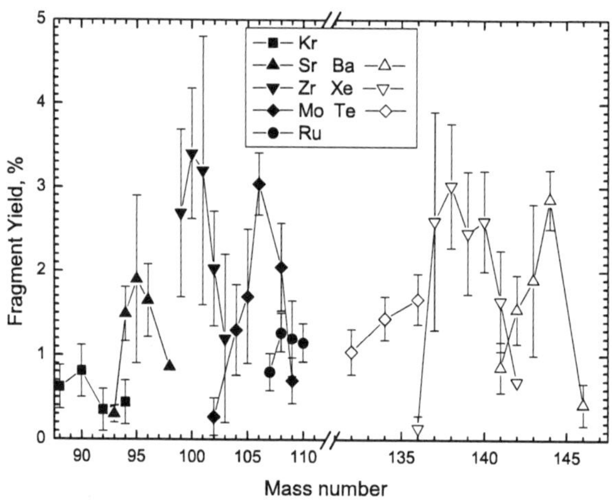

Fig. 4. Independent yields of fission fragments in the Be ternary fission of ^{252}Cf.

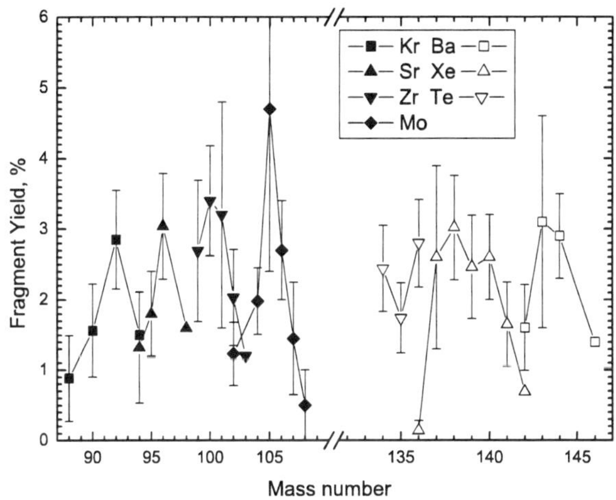

Fig. 5. Independent yields of fission fragments in the C ternary fission of ^{252}Cf.

for the He ternary fission and another one known for the binary fission of ^{252}Cf, we see that the two protons entering the He LCPs come from the light fragments, otherwise obtained as those emitted in the binary fission. Similar considerations show that Be nuclei take, on average, ~2.7 protons from the light fragment with the rest of charge coming from the heavy fragment. Finally,we see that both fission fragments contribute about the same proton number in the formation of carbon LCPs. The average proton numbers removed from the light and heavy binary fission fragments by He, Be and C LCPs are presented in Fig. 7.

4. Results of the second experiment

Energy calibration was done in accordance with the well-known method described in.[7] A general form for the energy calibration of the solid state detector may be written in the following form for a fission fragment:

$$E = (a + a')x + b + b'M, \tag{1}$$

where a, a', b and b' are constants for a particular detector; E and M are the kinetic energy and mass of fragment; and x is a pulse height. Usually, one can calculate parameters a, a', b and b' using four constants a_0, a'_0, b_0,

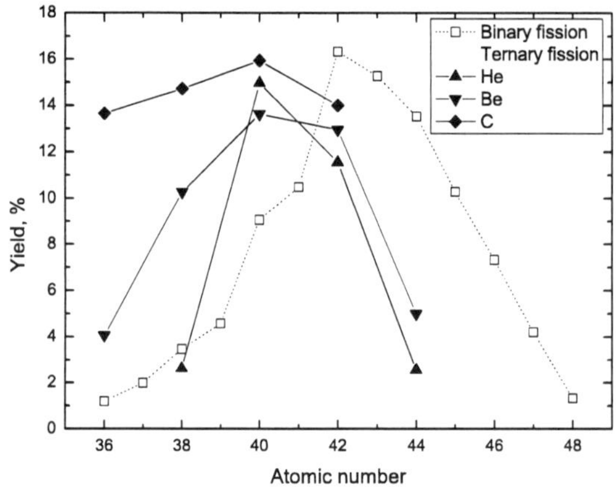

Fig. 6. Charge distributions of light fragments emitted in the He, Be and C ternary fission of ^{252}Cf. The dotted line shows the charge distribution known for the binary fission of ^{252}Cf.

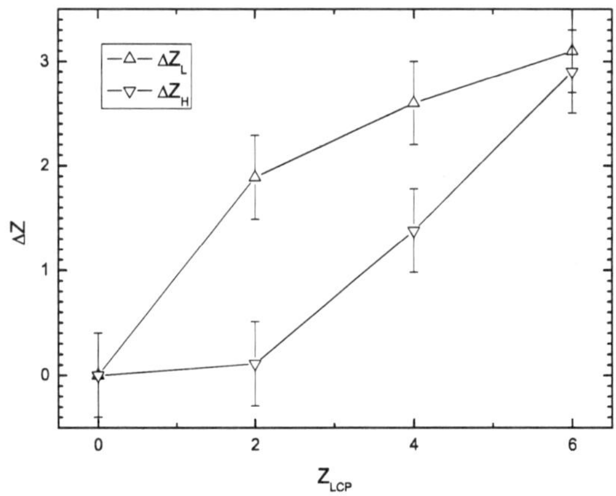

Fig. 7. Up and down triangles show the number of protons removed respectively from the light and heavy ternary fission fragments, which otherwise could be obtained as those emitted in binary fission.

b'_0 presented in[7] and the positions of the two peaks P_L and P_H corresponding to the light and heavy fission fragments in the pulse-height spectrum measured for the ^{252}Cf spontaneous fission:

$$\begin{aligned} a &= a_0/(P_L - P_H), \\ a' &= a'_0/(P_L - P_H), \\ b &= b_0 - a_0 \times P_L, \\ b' &= b'_0 - a'_0 \times P_L. \end{aligned} \tag{2}$$

The Constants a_0, a'_0, b_0 and b'_0 allow one to take account of the ionization defect in silicon and are universal for many types of silicon detectors. Taking into consideration the energy loss of fission fragments occurring in our experiment, we rewrote Eqs.1 and (2) in the following manner:

$$\begin{aligned} a_{0n}a'_0 &= a'_{0n}a_0, \\ a_{0n}[(b'_{0n} - b'_0)\Delta P - a'_{0n}P_L] &= a'_{0n}[(b_{0n} - b_0)\Delta P - a_{0n}P_L], \\ E_L\Delta P &= (a_{0n} + a'_{0n}m_L)P_L + (b_{0n} - a_{0n}P_L) \\ &\quad +(b'_{0n}\Delta P - a'_{0n}P_L)m_L, \\ E_H\Delta P &= (a_{0n} + a'_{0n}m_H)P_L + (b_{0n} - a_{0n}P_H) \\ &\quad +(b'_{0n}\Delta P - a'_{0n}P_H)m_H, \end{aligned} \tag{3}$$

where a_0, a'_0, b_0 and b'_0 are original coefficients from;[7] a_{0n}, a'_{0n}, b_{0n} and b'_{0n} are the new coefficients calculated for our case; E_L, E_H, m_L and m_H are fragment energies and masses associated with the two peaks in the experimental pulse-height spectrum; ΔP and P_L are respectively the distance between two peaks and the peak position of the light fission fragments. It was shown that the solution to system 4 relative to a_{0n}, a'_{0n}, b_{0n} and b'_{0n} does not depend on ΔP and P_L and can be written in the following form:

$$\begin{aligned} a_{0n} &= P_A + P_{AL}E_L + P_{AH}E_H, \\ a'_{0n} &= P'_A + P'_{AL}E_L + P'_{AH}E_H, \\ b_{0n} &= P_B + P_{AL}E_L, \\ b'_{0n} &= P'_{0n} + P'_{AL}E_L, \end{aligned} \tag{4}$$

where coefficients P_A, P'_A, P_B, P'_B, P_{AL}, P_{AH}, P'_{AL}, P'_{AH} depend on a_0, a'_0, b_0, b'_0, m_L and m_H only. It was shown[5] that for the fission fragments energy loss typical for our case it is possible to assign to m_L and m_H, respectively, the mean mass values of the light and heavy fragments known for the ^{252}Cf spontaneous fission. Also, the values of E_L and E_H could be taken as the result of subtracting the energy losses taken in the passive layers (2μ Ti foil and 1.5μ "dead" layers presented in our DSS detectors) by the light and heavy fragments having the mean mass values m_L and m_H, respectively, from their mean kinetic energies.

Having this energy calibration, we could calculate the loci corresponding to different fission fragment pairs in the two-dimensional plot X_A vs. X_B. Of course, different fragment pairs could not be separated totally using only the data coming from the DSS detectors. But the contributions of other fission fragment pairs are reduced in the $\gamma - \gamma$ coincidence matrices created for the selected pair.

Two variants of implementing Doppler correction were used for creating the $\gamma - \gamma$ coincidences matrices. When only γ transitions in the heavy or light fragment were of interest the Doppler correction was made with the assumption that all detected γ rays came either from the light or from the heavy fragment. Being interested in the $\gamma - \gamma$ events associated with the complementary fragments we made the Doppler correction two times for each γ ray. As a result, the number of γ rays was doubled. At first, one-half of the total number of γ rays was corrected assuming that they were emitted by the light fragment, whereas the other half was corrected assuming that these γ rays were emitted by the heavy fragment. Only coincidences between the γ rays of these two groups were placed in the $\gamma - \gamma$ energy matrix in such a way that the corrected energy values of γ rays from the two groups were placed on the two different axes of the matrix. The result of this procedure is demonstrated in Fig. 8. The two γ ray spectra shown in Fig. 8 correspond to the ^{104}Mo ^{146}Ba fission fragment pair. These spectra were created using the same gate $2^+ \rightarrow 0^+$ ^{146}Ba on the $\gamma - \gamma$ coincidence matrices built without (Fig. 8a) and with (Fig. 8b) Doppler correction. One can see clear peaks of the γ transitions of ^{104}Mo in spectrum (Fig. 8b), which are smeared in spectrum (Fig. 8a).

The locus corresponding to the one fission fragments pair can be divided into small loci by TKE. Then one may build a number of $\gamma - \gamma$ coincidence matrices corresponding to these small loci and estimate the yields of fission fragment pair in dependence of TKE. The preliminary results of this approach are demonstrated in Fig. 9 for the fission fragment pairs ^{106}Mo ^{140}Ba, ^{106}Mo ^{142}Ba, and ^{106}Mo ^{146}Ba for the first time.

5. Conclusion

The extraordinary capability of Gammasphere in the γ ray spectroscopy, combined with the LCPs and fission fragments detectors, significantly expanded our possibilities in the study of fission. Particulary, we for the first

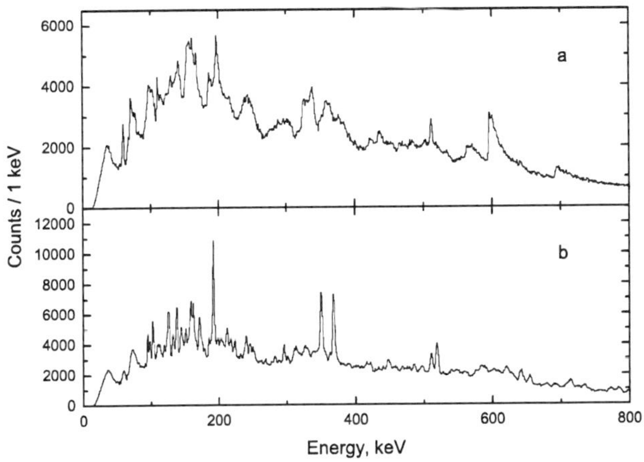

Fig. 8. Gamma spectra of (a) ^{146}Ba and ^{104}Mo fission fragments without Doppler correction and (b) predominantly ^{104}Mo with Doppler correction. Gates were opened at the 181-keV γ line of ^{146}Ba.

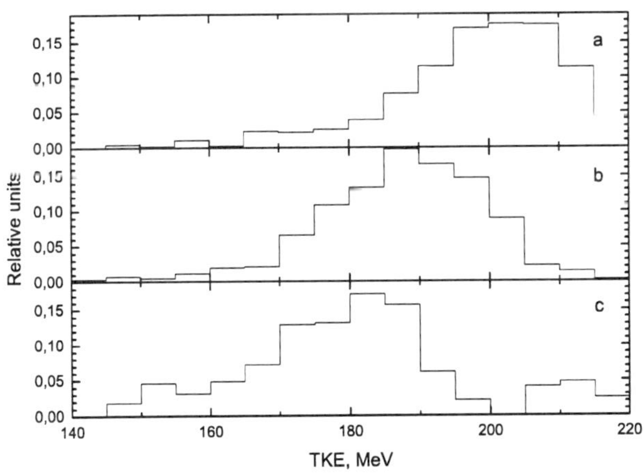

Fig. 9. Yields of fission fragment pairs are shown in function of TKE: (a) for ^{106}Mo ^{144}Ba, (b) for ^{106}Mo ^{142}Ba and (c) for ^{106}Mo ^{146}Ba.

time obtained data on the yields of fission fragment pairs emitted in the He ternary fission of ^{252}Cf, the yields of fission fragments formed in the ^{252}Cf ternary fission accompanied by the emission of He, Be and C LCPs, the TKE dependence of the yield of specific fragment pairs emitted on the binary fission of ^{252}Cf.

References

1. G. M. Ter-Akopian, J. H. Hamilton, Y. T. Oganessian, A. V. Daniel, J. Kormicki, A. V. Ramayya, G. S. Popeko, B. R. S. Babu, Q. Lu, K. Butler-Moore, W. C. Ma, E. F. Jones, J. K. Deng, D. Shi, J. Kliman, V. Polhorsky, M. Morhac, J. D. Cole, R. Aryaeinejad, N. R. Johnson, I. Y. Lee and F. K. MacGowan, *Physical Review C* **55**, 1146 (1997).
2. G. M. Ter-Akopian, J. H. Hamilton, Y. T. Oganessian, A. V. Daniel, J. Kormicki, A. V. Ramayya, G. S. Popeko, B. R. S. Babu, Q. Lu, K. Butler-Moore, W. C. Ma, S. Cwiok, W. Nazarevich, J. K. Deng, D. Shi, J. Kliman, M. Morhac, J. D. Cole, R. Aryaeinejad, N. R. Johnson, I. Y. Lee, F. K. MacGowan and J. X. Saladin, *Physical Review Letters* **77**, 32 (1996).
3. G. M. Ter-Akopian, A. V. Daniel, A. S. Fomichev, G. S. Popeko, A. M. Rodin, Y. T. Oganessian, J. H. Hamilton, A. V. Ramayya, J. Kormicki, J. K. Hwang, D. Fong, P. Gore, J. D. Cole, M. Jandel, J. Kliman, L. Krupa, J. O. Rasmussen, I. Y. Lee, A. O. Macchiavelli, P. Fallon, M. A. Stoyer, R. Donangelo, S. C. Wu and W. Greiner, *Physics of Atomic Nuclei* **67**, 1860 (2004).
4. A. V. Daniel, G. M. Ter-Akopian, J. H. Hamilton, A. V. Ramayya, J. Kormicki, G. S. Popeko, A. S. Fomichev, A. M. Rodin, Y. T. Oganessian, J. D. Cole, J. K. Hwang, Y. X. Luo, D. Fong, P. Gore, M. Jandel, J. Kliman, L. Krupa, J. O. Rasmussen, S. C. Wu, I. Y. Lee, M. A. Stoyer, R. Donangelo and W. Greiner, *Physical Review C* **69**, 041305(R) (2004).
5. A. V. Daniel, J. H. Hamilton, A. V. Ramayya, A. S. Fomichev, Y. T. Oganessian, G. S. Popeko, A. M. Rodin, G. M. Ter-Akopian, J. K. Hwang, D. Fong, C. Goodin, K. Li, J. O. Rasmussen, D. Seweryniak, M. P. Carpenter, C. J. Lister, S. H. Zhu, R. V. F. Janssens, J. Batchelder, J. Kliman, L. Krupa, W. C. Ma, S. J. Zhu, L. Chaturvedi and J. D. Cole, *Physics of Atomic Nuclei* **69**, 1405 (2006).
6. G. M. Ter-Akopian, J. H. Hamilton, A. V. Ramayya, A. V. Daniel, G. S. Popeko, A. S. Fomichev, A. M. Rodin, Y. T. Oganessian, J. D. Cole, J. Kormicki, J. K. Hwang, D. Fong, P. Gore, J. O. Rasmussen, A. O. Macchiavelli, I. Y. Lee, M. A. Stoyer, W. Greiner, R. Donangelo, M. Jandel, L. Krupa and J. Kliman, Spontaneous fission of 252cf in the light of prompt gamma rays, in *3 International Conference: Fission and Properties of Neutron-Rich Nuclei*, eds. J. H. Hamilton, A. V. Ramayya and H. K. Carter (World Scientific, River Edge New Jersey, 2003).
7. H. W. Schmitt, W. E. Kiker and W. W. Williams, *Physical Review* **137**, 837 (1965).

Nuclear Structure studies of Microsecond Isomers near A = 100

J. GENEVEY, J.A. PINSTON, G. SIMPSON

Laboratoire de Physique Subatomique et de Cosmologie,
IN2P3-CNRS/Grenoble Universités, F-38026 Grenoble Cedex, France
** E-mail: genevey@lpsc.in2p3.fr*

W. URBAN

Faculty of Physics, Warsaw University,
ul. Hoża 69, 00-681 Warsaw, Poland

A large variety of shapes may be observed in Sr and Zr nuclei of the $A = 100$ region when the number of neutrons increases from $N = 58$ to $N = 64$. The lighter isotopes are rather spherical. It is also well established that three shapes co-exist in the transitional odd-A, $N = 59$, Sr and Zr nuclei. For $N > 59$, strongly deformed axially symmetric bands are observed. Recently, a new isomer of half-life 1.4(2) μs was observed in ^{95}Kr, the odd-odd ^{96}Rb has been reinvestigated and a new high spin isomer observed in the even-even ^{98}Zr. Beyond $N = 60$ nuclei, the neutron-rich Mo isotopes represent well deformed nuclei, but at the same time, the triaxial degree of freedom plays an important role. We have re-investigated the odd ^{105}Mo and ^{107}Mo and found that odd and even Mo in the neutron range $N = 62$-66 have comparable quadrupole and triaxial deformation. These nuclei were studied by means of prompt γ-ray spectroscopy of the spontaneous fission of ^{248}Cm using the EUROGAM 2 Ge array and/or measurements of μs isomers produced by fission of ^{239,241}Pu with thermal neutrons at the ILL (Grenoble).

Keywords: Exotic nuclei, Shape coexistence

1. INTRODUCTION

The region of neutron-rich nuclei near $A = 100$ is distinctive for its sudden change in the ground state properties of nuclei.[1] In particular, for the even $_{38}$Sr and $_{40}$Zr isotopes a sudden onset of strong deformation is observed at $N = 60$, whereas the lighter isotopes up to $N = 58$ are rather spherical. The isotones with $N = 59$ neutrons are of special interest because they are just at the border of the two regions.

Previous experiments have shown that their ground and low-lying states

are rather spherical,[2,3] while deformed bands with $\beta_2 \sim 0.3$ are present at about 600 keV excitation energy[1] and the maximum deformation of the region, $\beta_2 \sim 0.4$, is reached for the $9/2^+[404]$ band recently observed at 829.8 and 1038.8 keV, in ^{97}Sr and ^{99}Zr respectively.[4–6] The large β_2 value found for this band is observed for several even and odd Sr and Zr nuclei above $N = 60$. A simple explanation of the shape-coexistence mechanism has been proposed. It is based upon the Nilsson diagram and stresses the fundamental importance of the unique parity states.[7,8] Beyond $N = 60$ nuclei, the neutron-rich Mo isotopes represent well deformed nuclei, but at the same time, the triaxial degree of freedom plays an important role.
This paper summarizes the results obtained in this field by our group. After the description of the experimental techniques in section 1, we will present our recents results in the $N = 58$-60 region (Section 2) and in the $N = 62$-66 region (Section 3).

2. Experimental techniques

The Lohengrin mass spectrometer was used to select nuclei, according to their mass-to-ionic charge ratios (A/q), recoiling from a thin $^{239 or 241}$Pu target which was undergoing thermal-neutron-induced fission. The flight time of the nuclei through the spectrometer was around 1.6 μs. The fission fragments (FFs) were detected in an ionization chamber filled with isobutane gas. Two different setups have been installed at the focal plane of the spectrometer.
In the first setup, the FFs were detected in a gas detector of 13 cm length, and subsequently stopped in a 12 μm thin Mylar foil. Behind the foil, two cooled adjacent Si(Li) detectors covering an area 2×6 cm^2 were placed to detect the conversion electrons and X-rays, while the γ-rays were detected by two Ge of 60 % placed perpendicular to the beam. This setup allows conversion electrons to be detected down to low energy (15 keV) and allows γ-electron coincidences to be obtained. Details on this experimental setup can be found in.[9,10]
In the second setup, the FFs were detected in an ionization chamber filled with isobutane at a pressure of 47 mb. This ionization chamber has good nuclear charge (Z) identification. It consists of two regions of gas, $\Delta E1$=9 cm and $\Delta E2$=6 cm, separated by a grid. This system is able to identify the nuclear charge in the Z$\sim$40 region, with a resolution $(FWHM)$ of about two units. The γ rays deexciting the isomeric states were detected by a Clover Ge detector and three single Ge crystals of the Miniball array[11] assembled in the same cryostat. All these detectors were placed perpendicular to the

ion beam. They were packed in a very close geometry, thanks to the small thickness (6 cm) of the ionization chamber. The total efficiency for the γ detection is 20 % and 4 % for photons of 100 keV and 1 MeV respectively. Any γ rays detected in the germanium detectors up to 40 μs after the arrival of an ion were recorded on the disk of the data acquisition system. A time window of 250 ns was used for γ-γ coincidences in the data-analysis software.

3. The N = 58-60 isotopes

3.1. ^{95}Kr

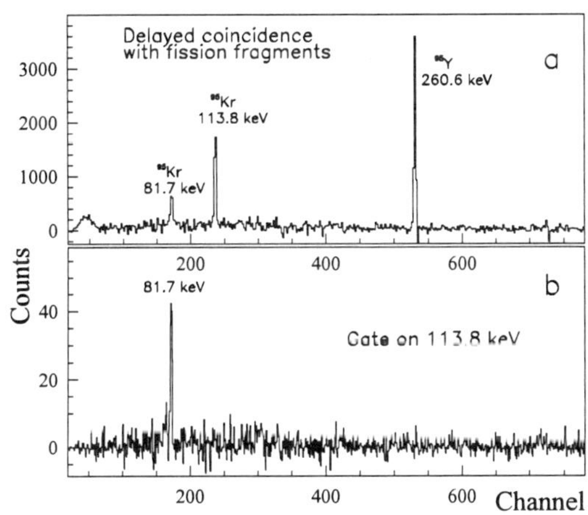

Fig. 1. (a) γ decay spectrum of the ^{95}Kr and ^{95}Y isomers. (b) coincidence spectrum gated on the 113.8 keV γ ray.

Two μs isomers have been observed in the $A = 95$ mass chain. The strong γ line of 260.6 keV in Fig. 1a deexcites an isomeric level of 56.2 μs half life, already assigned to ^{95}Y.[12] The two weaker γ-rays of 81.7(2) and 113.8(2) keV energies belong to a new isomer.[13] As seen in Fig. 1b, the two transitions are in coincidence with each other. This new isomer has a half life of 1.4(2) μs. The method used for the nuclear charge (Z) identification of this new isomer of mass 95, is shown in Fig. 2. The ^{95}Y isomer was already known and its $\Delta E1$ is shown in Fig. 2a. The $\Delta E1$ spectrum for the whole mass chain

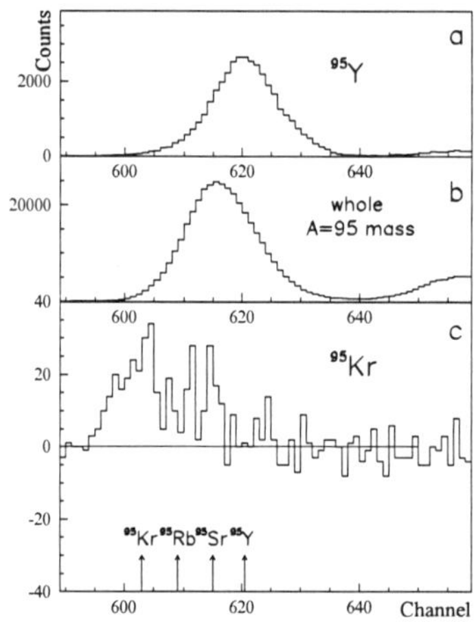

Fig. 2. Energy lost in the first step of gas Δ E1, for (a) the ^{95}Y isomer, (b) the whole $A = 95$ mass chain, and (c), the ^{95}Kr isomer.

95 is shown in Fig. 2b. The $\Delta E1$ spectrum of the new isomer is shown in Fig. 2c. It was obtained by coincidences between the incoming ions and the γ-rays of the isomer. The position of the centroid of the $\Delta E1$ distribution allows the nuclear charge $Z = 38$ to be assigned to this new isomer unambiguously. More details on the method can be found in Ref.[13]

In Fig. 3, the level scheme of the new isomer found in ^{95}Kr is shown and compared with the previously known isomers in the isotones of ^{97}Sr and ^{99}Zr.[2–5,14] In these three nuclei, the low-lying isomer decays by an $E2$ transition. The measured $B(E2)$ values, 1.33(5), 1.75(10) and 1.47(27) $Wu.$, for ^{99}Zr, ^{97}Sr and ^{95}Kr, respectively, are comparable, which suggests that the three transitions have an analogous nature and that the three isotones have the same spins. The ground and the two first excited states of these three isotones are very likely spherical, as suggested by the measured $B(E2)$ values, and their dominant configurations are the $\nu s_{1/2}$, $\nu d_{3/2}$ and $\nu g_{7/2}$ shell-model states, respectively. One may notice that their energies change very little between ^{99}Zr, which is quite close to the line of stability and ^{95}Kr, which is very far from it.

In contrast, we have not observed the $9/2^+[404]$ strongly deformed isomer, present in the two other isotones.[4] The non observation of this isomer means that either this level does not exist in ^{95}Kr, it is too weakly fed by fission, or its half life is shorter than about 0.5 μs, because the flight time throw the Lohengrin spectrometer is about 1.7 μs.

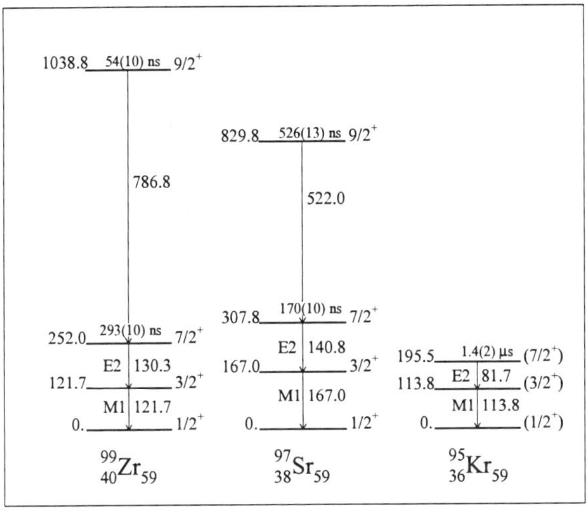

Fig. 3. Decay schemes of the ^{99}Zr, ^{97}Sr, and ^{95}Kr isomers.

3.2. ^{96}Rb

The N=59 odd-odd very neutron-rich ^{96}Rb nucleus has been reinvestigated.[15] It was previously measured by Genevey *et al.*[16] with the Lohengrin spectrometer at the ILL reactor in Grenoble. The γ-counting rates obtained in this new measurement are about ten times higher than in the previous one. Examples of $\gamma-\gamma$ coincidences are reported in Fig. 4. The level scheme based on $\gamma-\gamma$ and $e-\gamma$ coincidences is shown in Fig. 5.
This scheme is very similar to the one observed in ^{98}Y.[17] The low-lying levels are rather spherical as well as well as the 10^- isomeric state. The two rotational bands are fed by μs isomers close in excitation energy, 1181.5 keV in ^{98}Y and 1135 keV in ^{96}Rb, and the isomeric transitions have comparable $B(E2)$ values. All these features, strongly suggest that the two isomers have the same $(\pi(g_{9/2})\nu(h_{11/2}))_{10^-}$ configuration. A strongly attractive $n-p$ interaction explains the presence of these isomers at a relatively low energy. Consequently, the strong n-p interaction may induce a competition between high-spin, fully-aligned spherical configurations and the levels of rotational bands in this transitional region. Moreover, it is interesting to note that the neutron and proton orbitals present in the configuration of the spherical iso-

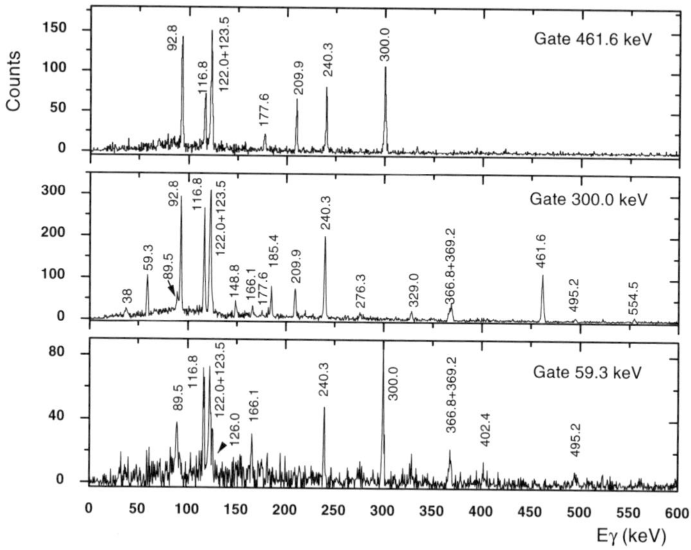

Fig. 4. Examples of $\gamma - \gamma$ coincidences in ^{96}Rb.

mer and in the deformed band of these odd-odd nuclei originate together from the same spherical unique-parity states $\pi(g_{9/2})$ and $\nu(h_{11/2})$.

3.3. ^{98}Zr

A new (17^-) μs isomeric state at 6603.3 keV has been observed for the first time in ^{98}Zr.[19] Mass and isotopic identification of the isomer were performed by examining coincidences between the mass-separated ions, detected in the ionization chamber, and the isomer-delayed γ rays. Much of the decay scheme below the isomer has already been assigned to ^{98}Zr. This nucleus has previously been studied by prompt γ-ray spectroscopy of secondary fission fragments populated by light-ion induced[6] and spontaneous fission.[1,20] The proposed level scheme is presented in figure 6.

This new 1.9 (2) μs isomer at 6603.3 keV observed in ^{98}Zr, with a proposed configuration of $\pi(g^2_{9/2})\nu(g^1_{7/2}h^1_{11/2})$ and a single particle nature, decays by a pure, or almost pure, $E2$ transition into a 15^- state, which then decays into two collective bands, one of positive parity, the other negative, the latter of which is observed for the first time. The existence of a spherical, single-particle state at such a high energy (6603.3 keV) and spin (17^-)

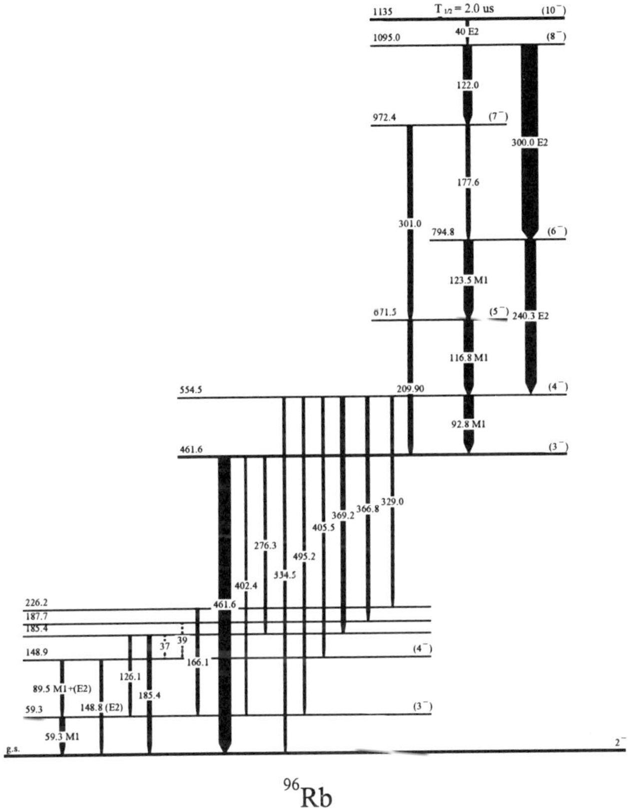

Fig. 5. Decay scheme of the 2.0 μs isomer in ^{96}Rb obtained in the present work. The low-lying levels and the isomer at 1135 keV have rather spherical configurations, while a rotational band develops above 460 keV.

is quite unusual, in fact both these values are the highest known for a μs isomer in this region. These high-spin shape coexisting states again demonstrate the richness of nuclear structure phenomena in this region.

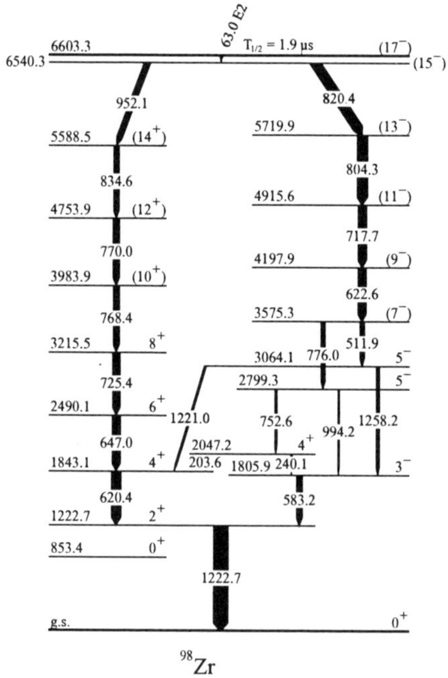

Fig. 6. Decay scheme of the 1.9 (2) μs isomer observed in the present work. The 853.4 keV 0^+ bandhead[1] is also included.

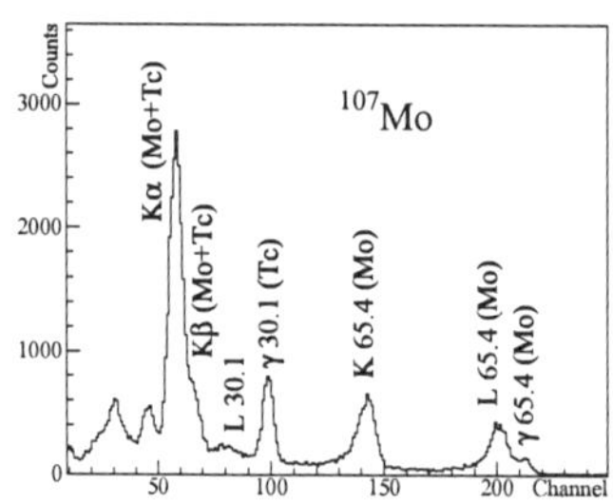

Fig. 7. Si(Li) spectrum of the A=107 isomers observed in the present experiment. Narrow peaks correspond to X- and γ-rays, while the broader peaks are due to conversion electrons.

4. The N=62-66 region : ^{105}Mo and ^{107}Mo

In the even-even $^{104-108}$Mo a new situation occurs.[21–23] These nuclei are strongly deformed ($\beta_2 \approx 0.37$) but at the same time, the levels of the

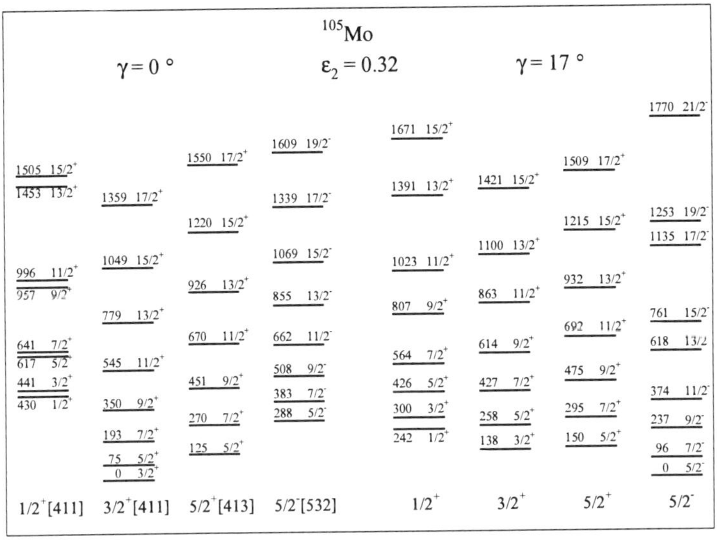

Fig. 8. Calculated levels for the four bands in ^{105}Mo performed for γ deformations 0° and 17°, respectively. It is worth noting that the staggering in the $1/2^+$ and $5/2^-$ differs strongly for these two deformations.

$K^\pi = 2^+$ γ-band go down in energy with the increase of the neutron number, suggesting that the triaxial degree of freedom plays an important role in these isotopes. The ^{105}Mo nucleus was populated in spontaneous fission of ^{248}Cm and prompt γ-rays following fission were measured using the EUROGAM2 array. Four well developed bands were observed in ^{105}Mo.[24] In ^{107}Mo, three well developed bands were previously reported from γ-ray measurements of the spontaneous fission of ^{248}Cm.[25] To complete the level scheme, this nucleus has been produced through thermal-neutron induced fission reaction of a ^{241}Pu target, at the ILL reactor in Grenoble. Figure 7 shows the Si(Li) spectrum of the A=107 isomers observed in this experiment. A new isomer of 420 ns half live has been observed in this work. It has been tentatively assigned as a $1/2^+$ state deexciting by an $E2$ transition to the $5/2^+$ ground state. This wealth of information in ^{105}Mo and ^{107}Mo, which is unique in the neutron-rich odd Mo isotopes, makes a comparison in the frame of the rotational model meaningful.[24]
The excited states of the bands in ^{105,107}Mo and their γ-ray decay patterns were calculated using the code ASYRMO.[26] Figure 8 gives the results of the calculation for the lowest four bands in ^{105}Mo and Figure 9 for ^{107}Mo. A satisfactory fit to the experimental data has been obtained using the simple

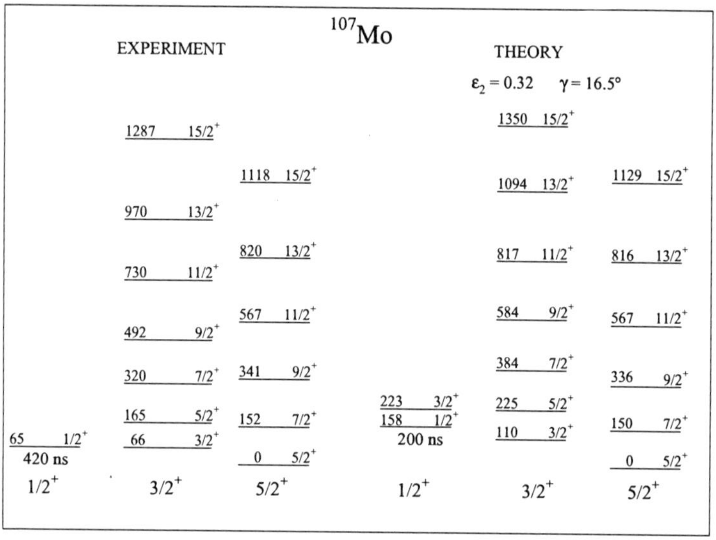

Fig. 9. Comparison of the experimental levels of the three positive parity bands in ^{107}Mo with theory.

particle-rotor calculations for both ^{105,107}Mo assuming that these nuclei are asymmetric rotors with remarkably similar deformations, $\epsilon_2 = 0.32$ and $\gamma \approx 17°$. It is worth noting that four bands of the same origin observed in these two nuclei are well reproduced by these parameters. For the even-even $^{104-108}$Mo, the parameters of γ and quadrupole deformation can be deduced from the experimental data. These values are comparable to the ones observed in the odd Mo. Moreover, a comparable triaxial deformation ($\epsilon_2 \approx 0.32, \gamma \approx 22°$) was also reported for ^{107}Tc, with $N = 64$ neutrons.[27] All these data strongly suggest that the cores have similar shapes in the heavy even-even and odd-neutron Mo nuclei, as well as in the odd-proton Tc nuclei, and that the odd-A nuclei are not strongly affected by the unpaired particle.

5. Conclusion

In Table 1 are reported the nuclear shapes observed in the $A \sim 100$ region. A great wealth of information was recently gained in the odd-mass and odd-odd N=59 isotones. It is now well established that three shapes coexist in ^{97}Sr and ^{99}Zr, while two different shapes were seen in ^{96}Rb. The two

Table 1. Nuclear shapes in the A ~ 100 region

Nuclei	A	Z	N	β_2	γ
Kr	95	36	59	<0.20 ; ?	0
Rb	96	37	59	<0.20; 0.28-0.39	0
Sr	97	38	59	<0.20; 0.32; 0.41	0
Zr	99	40	59	<0.15; 0.30; 0.41	0
Sr	98	38	60	0.40	0
Zr	98	40	60	0.40	0
Mo	104-108	42	62-66	0.38	~20°
Mo	105-107	42	63-65	0.38	~17°
Tc	107	43	64	0.38	~22°

unique-parity states $\pi(g_{9/2})$ and $\nu(h_{11/2})$ play a considerable role in all these nuclei. The relative occupation of these orbitals is able to change drastically the shape of the nucleus.

For $N = 60$ Sr and Zr nuclei the β_2 deformation is at the maximum value of the region. For even and odd Mo isotopes, the triaxial degree of freedom becomes important. A satisfactory fit to the experimental data has been obtained for both ^{105}Mo and ^{107}Mo in simple particle-rotor calculations, assuming that these nuclei are asymmetric rotors. The parameters of the model are very similar for both nuclei. It is important to note, that in both nuclei all four bands are well reproduced using the same γ parameters. Moreover, in the neutron range N=62-66, the γ deformation of the odd Mo are very close to the values of the even-even Mo, suggesting that all these nuclei have comparable core deformations.

References

1. W. Urban, J.L. Durell, A.G. Smith, W.R. Phillips, M.A. Jones, B.J. Varley, T. Rząca-Urban, I. Ahmad, L.R. Morss, M. Bentaleb, N. Schulz Nucl. Phys. **A689**, 605 (2001)
2. K.L. Kratz *et al.*, Z. Phys. A, **312**, 43 (1983).
3. L.K. Peker, Nucl. Data Sheets **73**, 1 (1994).
4. W. Urban, J.A. Pinston, J. Genevey, T. Rząca-Urban, A. Złomaniec G. Simpson, J.L. Durell, W.R. Phillips, A.G. Smith, B.J. Varley, I. Ahmad, and N. Schulz, Eur. Phys. J. A**22**, 241 (2004).
5. J.K. Hwang *et al.*, Phys. Rev. C **67**, 054304 (2003).
6. C.Y. Wu, H. Hua, D. Cline, A.B. Hayes, R. Teng, R.M. Clark, P. Fallon, A. Goergen, A.O. Macchiavelli, K. Vetter. Phys. Rev. C **70**, 064312 (2004).
7. J.A. Pinston, J. Genevey, G. Simpson, W. Urban in *3rd Internatinal Workshop on Nuclear Fission and Fission-product Spectoscopy* (AIP conference proceedings 798 p.149, 2005)
8. J.A. Pinston, Science and Technology Journal, Bulgarian Nuclear Society

ISSN1310-8727 P. 48, 2005.
9. J.A. Pinston, and J. Genevey, J. Phys. J. G **30**, R57 (2004).
10. J. Genevey, J.A. Pinston, H. Faust, C. Foin, S. Oberstedt, and M. Rejmund, Eur. Phys. J. A **9**, 191 (2000).
11. J. Eberth *et al.*, Rep. of Progr. in Nucl. Phys. 46, 389 (2001).
12. B. Pfeiffer, E. Monnand, J. A. Pinston, F. Schussler, G. Jung, J. Munzel and H. Wolnik, Proc. Int. Conf. Far from Stability, Helsingor, Denmark 7-13 June 1981, CERN report 81-09, p. 423.
13. J. Genevey, R. Guglielmini, R. Orlandi, J.A. Pinston, A. Scherillo, G. Simpson, I. Tsekhanovich, N. Warr, J. Jolie Phys. Rev. C **73**, 037308 (2006).
14. A. Zlomamaniec, H. Faust, J. Genevey, J.A. Pinston, T. Rząca-Urban, G.S. Simpson, I. Tsekhanovich and W. Urban, Phys. Rev. C **72**, 067302 (2005).
15. J. A. Pinston, J. Genevey, R. Orlandi, A. Scherillo, G.S. Simpson, I.Tsekhanovich, W.Urban, H.Faust, N.Warr. Phys. Rev. C **71**, 064327 (2005).
16. J. Genevey, F. Ibrahim, J.A. Pinston, H. Faust, T. Friedrichs, M. Gross, and S. Oberstedt, Phys. Rev. C **59**, 82 (1999).
17. S. Brant, G. Lhersonneau, and K. Sistemich, Phys. Rev. C **69**, 034327 (2003).
18. C. Thibault *et al.*, Phys. Rev. C **23**, 2720 (1980).
19. G.S. Simpson, J.A. Pinston, D. Balabanski, J. Genevey, G. Georgiev, J. Jolie, D.S. TJudson, R. Orlandi, A. Scherillo, I.Tsekhanovich, W. Urban, N. Warr. Phys. Rev. C . To be published
20. J.H. Hamilton, A.V. Ramaya, S.J. Zhu, G.M. Ter-Akopian, Y.T.Oganessian, J.D. Cole, J.O. Rasmussen, M.A. Stoyer Prog.Part.Nucl.Phys. **35**, 635 (2001).
21. A. G. Smith, J.L. Durell, W.R. Phillips, M.A. Jones, M. Leddy, W. Urban, B.J. Varley, I. Ahmad, L.R. Morss, M. Bentaleb, A. Guessous, E. Lubkiewicz, N. Schulz, and R. Wyss, Phys. Rev. Lett. **77**, 1711 (1996).
22. A. Guessous, N. Schulz, M. Bentaleb, E. Lubkiewicz, J.L. Durell, C.J. Pearson, W.R. Phillips, J.A. Shannon, W. Urban, B.J. Varley, I. Ahmad, C.J. Lister, Morss, K.L. Nash, C.W. Williams, and S. Khazrouni, Phys. Rev. C **53**, 1191 (1996).
23. H. Hua, C. Y. Wu, D. Cline, A. B. Hayes, R. Teng, R. M. Clark, P. Fallon, A. Goergen, A. O. Machiavelli, and K. Vetter, Phys. Rev. C **69**, 014317 (2004).
24. J.A. Pinston, W. Urban, Ch. Droste, J. Genevey, T. Rzaca-Urban, G. Simpson, J.L. Durell, A. G. Smith, B.J. Varley, I. Ahmad. Phys. Rev. C . To be published
25. W. Urban, T. Rzaca-Urban, J. A. Pinston, J. L. Durell, W. R. Phillips, A. G. Smith, B. J. Varley, I. Ahmad, and N. Schultz, Phys. Rev. C **72**, 027302 (2005).
26. P. Semmes and I. Ragnarsson, *The Particle plus Triaxial Model: a User's Guide, distributed at the Hands-on Nuclear Physics Workshop, Oak Ridge, 5-16 August 1991 (unpublished).*
27. Y. X. Luo *et al.*, Phys. Rev. C**70**, 044310 (2004).

Covariant Density Functional Theory: Isospin properties of nuclei far from stability

G. A. Lalazisis

Department of Theoretical Physics, Aristotle University of Thessaloniki Thessaloniki, GR-54124 Greece
**E-mail: glalazis@auth.gr*

The Relativistic Hartree-Bogoliubov (RHB) density functional with density dependent coupling constants is applied in the description of exotic nuclei. This approach provides an improved description of the isovector properties of finite nuclei at and away from stability line.

1. Introduction

Experiments with radioactive nuclear beams have disclosed a wealth of structure phenomena in exotic nuclei with extreme isospin values, and the next generation of radioactive-beam facilities will present new exciting opportunities for the study of the nuclear many-body laboratory. In the neutron-rich side, the exotic phenomena include the weak binding of the outermost neutrons, regions of nuclei with very diffuse neutron densities, formation of the neutron skin and halo structures, the disappearance of spherical magic numbers, and the onset of deformation and shape coexistence. Isovector quadrupole deformations could develop at the neutron drip-lines, and experimental evidence for the occurrence of low-energy "pygmy" excitations has been reported.

Extremely proton-rich nuclei are important both for nuclear structure studies and in astrophysical applications. These systems are characterized by exotic ground-state decay modes such as direct emission of charged particles and β-decays with large Q-values. The phenomenon of proton emission from the ground-state has been extensively investigated in medium-heavy and heavy, spherical and deformed nuclei. The properties of many proton-rich nuclei play an important role in the process of nucleosynthesis by rapid-proton capture.

Some of the most interesting examples of nuclear systems with large isospin values have been found in recent experimental studies of the synthesis and stability of the heaviest elements. The periodic system has been extended with elements that are found beyond the macroscopic limit of nuclear stability and are stabilized only by quantal shell effects.

Covariant density functional (CDF) theory[1] is a powerful tool for the description of nuclear structure properties at and away from beta-stability line. The density functional of the CDF theory can be expressed by means of the Relativistic Hartree-Bogoliubov (RHB) model.[2] This model manifests the relativistic extension of Hartree-Fock-Bogoliubov theory. RHB theory, with a limited number of phenomenological parameters, is able to provide a unified description of ph-pp correlations. The RHB model has been succesfully employed in the analysis of a variety of nuclear structure phenomena, not only in nuclei along the β-stability line , but also in exotic nuclei with extreme isospin values and close to the particle drip-line. The Lorentz structure of the model allows the inclusion of an elegant method for nuclear saturation and the description of the spin-orbit properties in a systematic fashion without the need of additional parameters. The theory is able to describe not only ground-state properties, but also essential features of collective excitations such as rotations and vibrations.

The parameters of the model are the meson masses and the meson-nucleon coupling constants. They are adjusted to reproduce the nuclear matter equation of state and a set of global properties of spherical closed-shell nuclei. The density dependence of the effective nuclear force in the RHB model is essential for the correct description of the nuclear properties. Usually, in the standard version of the RHB model, the density dependence enters through the non-linear σ self interaction. This variation of the model is known as the non-linear RHB model. This model, particularly with the NL3[3] effective interaction for the Lagrangian of the Relativistic Mean Field (RMF) Theory, has been widely used in many nuclear structure studies. It turned out that NL3 parametrization is probably the best effective interaction for stable and exotic nuclei. However, the non-linear RMF model has certain limitations: It systematically overestimates the value of $r_n - r_p$. It predicts an equation of state for neutron matter which is very different from the standard microscopic many-body neutron equation of state of Friedman and Pandharipande.[4] Moreover, the model is not able to provide predictions for the masses that match the standards of nuclear astrophysics. It is, therefore, essential to improve the mass predictions particularly for those nuclei that take part in the r-process or rp-process of nucleosynthesis.

The RMF framework has recently been extended to include effective

Lagrangians with density-dependent meson-nucleon vertex functions. The functional form of the meson-nucleon vertices can be deduced from in-medium Dirac-Brueckner interactions, obtained from realistic free-space NN interactions, or a phenomenological approach can be adopted, with the density dependence for the σ, ω and ρ meson-nucleon couplings adjusted to properties of nuclear matter and a set of spherical nuclei. The latter was employed in Ref.,[5] where the relativistic Hartree-Bogoliubov (RHB) model was extended to include medium-dependent vertex functions.

Table 1. The parameter set DD-ME2. The masses are in Mev.

$M = 939.000$	$m_\omega = 783.000$	$m_\rho = 763.000$	$m_\sigma = 550.124$
$g_\sigma(\rho_{sat}) = 10.5396$	$g_\omega(\rho_{sat}) = 13.0189$	$g_\rho(\rho_{sat}) = 3.6836$	$a_\rho = 0.5008$
$a_\sigma = 1.3854$	$a_\omega = 1.3879$	$b_\sigma = 0.9781$	$b_\omega = 0.8525$
$c_\sigma = 1.5342$	$c_\omega = 1.3566$	$d_\sigma = 0.4661$	$d_\omega = 0.4957$

Very recently, a new relativistic mean-field effective interaction with explicit density dependence of the meson-nucleon couplings has been proposed.[6] This interaction denoted by DD-ME2 provides a significant improvement in the description of nuclear structure properties. In section 2 the RHB model with density dependent coupling constants is discussed. Section 3 contains several applications of the model with the recently proposed force DD-ME2. Finally, section 4 summarizes our main conclusions.

2. The RHB model with density dependent coupling constants

Refs.[7–9] contain a very detailed discussion of the density-dependent nuclear hadron field theory. The relativistic Hartree-Bogoliubov (RHB) model and the random phase approximation (RPA) based on effective interactions with density dependent meson-nucleon couplings are described in Refs.[5] and,[10] respectively. For completeness, we include the essential features of the relativistic Lagrangian density with medium-dependent vertices

$$\begin{aligned}\mathcal{L} = &\bar{\psi}\left(i\gamma\cdot\partial - m\right)\psi + \frac{1}{2}(\partial\sigma)^2 - \frac{1}{2}m_\sigma\sigma^2 \\ &- \frac{1}{4}\Omega_{\mu\nu}\Omega^{\mu\nu} + \frac{1}{2}m_\omega^2\omega^2 - \frac{1}{4}\vec{\mathrm{R}}_{\mu\nu}\vec{\mathrm{R}}^{\mu\nu} + \frac{1}{2}m_\rho^2\vec{\rho}^{\,2} - \frac{1}{4}\mathrm{F}_{\mu\nu}\mathrm{F}^{\mu\nu} \\ &- g_\sigma\bar{\psi}\sigma\psi - g_\omega\bar{\psi}\gamma\cdot\omega\psi - g_\rho\bar{\psi}\gamma\cdot\vec{\rho}\vec{\tau}\psi - e\bar{\psi}\gamma\cdot A\frac{(1-\tau_3)}{2}\psi\,. \quad (1)\end{aligned}$$

Vectors in isospin space are denoted by arrows, and bold-faced symbols will indicate vectors in ordinary three-dimensional space. The Dirac spinor ψ

denotes the nucleon with mass m. m_σ, m_ω, and m_ρ are the masses of the σ-meson, the ω-meson, and the ρ-meson. g_σ, g_ω, and g_ρ are the corresponding coupling constants for the mesons to the nucleon. $e^2/4\pi = 1/137.036$. The coupling constants and unknown meson masses are parameters, adjusted to reproduce nuclear matter properties and ground-state properties of finite nuclei. $\Omega^{\mu\nu}$, $\vec{R}^{\mu\nu}$, and $F^{\mu\nu}$ are the field tensors of the vector fields ω, ρ, and of the photon: The coupling constants and unknown meson masses are parameters, adjusted to reproduce nuclear matter properties and ground-state properties of finite nuclei.

$$\Omega^{\mu\nu} = \partial^\mu \omega^\nu - \partial^\nu \omega^\mu \tag{2}$$

$$\vec{R}^{\mu\nu} = \partial^\mu \vec{\rho}^{\,\nu} - \partial^\nu \vec{\rho}^{\,\mu} \tag{3}$$

$$F^{\mu\nu} = \partial^\mu A^\nu - \partial^\nu A^\mu \,. \tag{4}$$

g_σ, g_ω, and g_ρ are assumed to be vertex functions of Lorentz-scalar bilinear forms of the nucleon operators. In practical applications of the density-dependent hadron field theory the meson-nucleon couplings are assumed to be functions of the baryon density $\psi^\dagger\psi$. In a relativistic framework the couplings can also depend on the scalar density $\bar{\psi}\psi$. Nevertheless, expanding in $\psi^\dagger\psi$ is the natural choice, because the baryon density is connected to the conserved baryon number, unlike the scalar density for which no conservation law exists. The scalar density is a dynamical quantity, to be determined self-consistently by the equations of motion, and expandable in powers of the Fermi momentum. For the meson-exchange models it has been shown that the dependence on baryon density alone provides a more direct relation between the self-energies of the density-dependent hadron field theory and the Dirac-Brueckner microscopic self-energies.[9] The explicit dependence of the vertex functions on the baryon density produces rearrangement contributions to the vector nucleon self-energy. The rearrangement terms result from the variation of the vertex functionals with respect to the baryon fields in the density operator (which coincides with the baryon density in the nuclear matter rest-frame). For a model with density dependent couplings, the inclusion of the rearrangement self-energies is essential for energy-momentum conservation and thermodynamical consistency (i.e. for the pressure equation derived from the thermodynamic definition and from the energy-momentum tensor).[7,8]

The meson-nucleon vertex functions are determined either by mapping the nuclear matter Dirac-Brueckner nucleon self-energies in the local density approximation ,[7,9,11] or the parameters of an assumed phenomenological density dependence of the meson-nucleon couplings are adjusted to

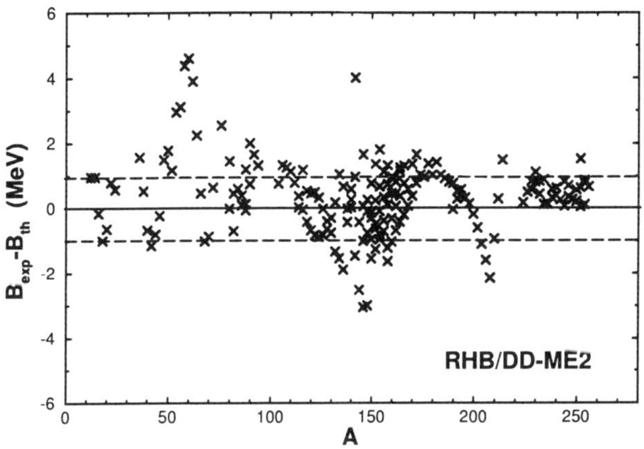

Fig. 1. Absolute deviations of the binding energies calculated with the DD-ME2 interaction from the experimental values.[15]

reproduce properties of symmetric and asymmetric nuclear matter and finite nuclei.[5,8] In the phenomenological approach of Refs.[5,8,9] the coupling of the σ-meson and ω-meson to the nucleon field reads

$$g_i(\rho) = g_i(\rho_{\text{sat}}) f_i(x) \quad \text{for} \quad i = \sigma, \omega \;, \tag{5}$$

where

$$f_i(x) = a_i \frac{1 + b_i(x + d_i)^2}{1 + c_i(x + d_i)^2} \tag{6}$$

is a function of $x = \rho/\rho_{\text{sat}}$, and ρ_{sat} denotes the baryon density at saturation in symmetric nuclear matter. The eight real parameters in (6) are not independent. The five constraints $f_i(1) = 1$, $f''_\sigma(1) = f''_\omega(1)$, and $f''_i(0) = 0$, reduce the number of independent parameters to three. Three additional parameters in the isoscalar channel are: $g_\sigma(\rho_{\text{sat}})$, $g_\omega(\rho_{\text{sat}})$, and m_σ - the mass of the phenomenological sigma-meson. For the ρ-meson coupling the functional form of the density dependence is suggested by Dirac-Brueckner calculations of asymmetric nuclear matter[11]

$$g_\rho(\rho) = g_\rho(\rho_{\text{sat}}) \exp\left[-a_\rho(x-1)\right] \;. \tag{7}$$

The isovector channel is parameterized by $g_\rho(\rho_{\text{sat}})$ and a_ρ. Usually the free values are used for the masses of the ω and ρ mesons: m_ω = 783 MeV

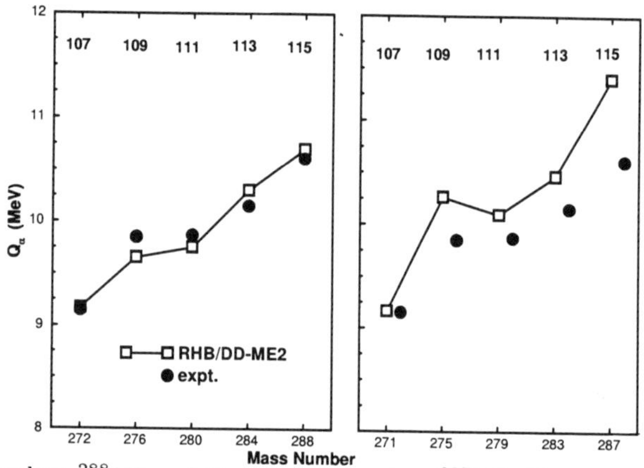

Fig. 2. Odd nucleus $^{288}115$ and the odd-even nucleus $^{287}115$. The experimental data are from Ref.,[20] and the calculated values correspond to transitions between the ground-states calculated in the RHB model with the DD-ME2 interaction plus Gogny D1S pairing.

and $m_\rho = 763$ MeV. In principle, one could also consider the density dependence of the meson masses. However, since the effective meson-nucleon coupling in nuclear matter is determined by the ratio g/m, the choice of a phenomenological density dependence of the couplings makes an explicit density dependence of the masses redundant.

The eight independent parameters: seven coupling parameters and the mass of the σ-meson, are adjusted to reproduce the properties of symmetric and asymmetric nuclear matter, binding energies, charge radii and neutron radii of spherical nuclei .[6]

3. Numerical results

Ground-state properties have been calculated in the RHB model with the DD-ME2 effective interaction in the particle-hole channel, and with the Gogny interaction[12] in the pairing channel

$$V^{pp}(1,2) = \sum_{i=1,2} e^{-((\mathbf{r}_1-\mathbf{r}_2)/\mu_i)^2} (W_i + B_i P^\sigma - H_i P^\tau - M_i P^\sigma P^\tau), \quad (8)$$

with the set D1S for the parameters μ_i, W_i, B_i, H_i, and M_i $(i = 1, 2)$.

The theoretical binding energies of approximately 200 nuclei calculated in the RHB model with the DD-ME2 plus Gogny D1S interactions, are compared with experimental values in Fig. 1. Except for a few Ni isotopes with $N \approx Z$ that are notoriously difficult to describe in a pure mean-field approach, and several transitional medium-heavy nuclei, the calculated binding energies are generally in very good agreement with experimental data. Although this illustrative calculation cannot be compared with microscopic mass tables that include more than 9000 nuclei,[13,14] we emphasize that the rms error including all the masses shown in Fig. 1 is less than 900 keV. This is a significant improvement compared with the previous RMF calculations with non linear-forces, where the rms error was around 2.5 MeV.[19] Moreover, since a finite-range pairing interaction is used, the results are not sensitive to unphysical parameters like, for instance, the momentum cut-off in the pairing channel. When compared with data on absolute charge radii and charge isotope shifts from Ref.,[16] the calculated charge radii exhibit an rms error of only 0.017 fm.

An important field of applications of self-consistent mean-field models includes the structure and decay properties of superheavy nuclei. The relativistic mean-field framework has recently been very successfully employed in calculations of chains of superheavy isotopes . Since generally relativistic density-dependent effective interactions provide a very realistic description of asymmetric nuclear matter, neutron matter and nuclei far from stability, one can also expect a good description of the structure of superheavy nuclei. In Table 2 we have shown that the interaction DD-ME2 reproduces ground-state properties of superheavy nuclei with high accuracy. Of course it is also interesting to analyze predictions for decay chains. In a very recent work[20] evidence has been reported for the synthesis of element $Z = 115$. In Fig. 2 we compare the calculated and experimental Q_α values for two α-decay chains starting from the odd-odd nucleus $^{288}115$ and the odd-even nucleus $^{287}115$. The two superheavy nuclides with $N = 173$ and $N = 172$ were produced in the $3n$- and $4n$-evaporation channels following the reaction ^{243}Am+^{48}Ca.[20] The theoretical Q_α values correspond to transitions between the ground-states calculated in the RHB model with the DD-ME2 effective interaction and with the Gogny interaction D1S in the pairing channel. The Dirac-Hartree-Bogoliubov equations and the equations for the meson fields are solved by expanding the nucleon spinors and the meson fields in terms of the eigenfunctions of a deformed axially symmetric oscillator potential. A simple blocking procedure is used in the calculation of odd-proton and/or odd-neutron systems. The blocking calculations are performed without breaking the time-reversal symmetry. We notice that

Table 2. RHB model (DD-ME2 plus Gogny D1S pairing) results for the binding energies, radii of charge and neutron density distributions, quadrupole and hexadecupole moments of heavy and superheavy nuclei, in comparison with experimental data.[15–18]

Nucleus	B.E (MeV)	r_c (fm)	r_n (fm)	Q_p (b)	H_p (b^2)
^{224}Ra	1720.47 (1720.31)	5.71	5.85	4.93 (6.33)	0.45
^{226}Ra	1731.13 (1731.61)	5.74	5.88	6.22 (7.19)	0.65
^{228}Ra	1741.67 (1742.49)	5.76	5.92	7.44 (7.76)	0.79
^{230}Ra	1751.94 (1753.05)	5.79	5.95	8.39	0.86
^{228}Th	1743.04 (1742.49)	5.78	5.90	7.64 (8.42)	0.88
^{230}Th	1751.94 (1753.05)	5.80	5.93	8.57 (8.99)	0.97 (1.09)
^{232}Th	1766.10 (1766.92)	5.82	5.96	9.28 (9.66)	1.00 (1.22)
^{234}Th	1776.80 (1777.68)	5.84	5.99	9.78 (8.96)	0.96
^{232} U	1766.39 (1765.97)	5.83	5.94	9.57 (10.00)	1.10
^{234} U	1778.66 (1778.57)	5.85	5.97	10.10 (10.35)	1.10 (1.40)
^{236} U	1790.29 (1790.42)	5.87	6.00	10.46 (10.80)	1.03 (1.30)
^{238} U	1801.38 (1801.69)	5.88	6.02	10.74 (11.02)	0.94 (0.83)
^{240} U	1811.82 (1812.44)	5.90	6.05	11.03	0.86
^{238}Pu	1801.85 (1801.27)	5.89	6.01	11.09 (11.26)	1.00 (1.38)
^{240}Pu	1813.84 (1813.46)	5.91	6.03	11.32 (11.44)	1.00 (1.15)
^{242}Pu	1825.26 (1825.01)	5.92	6.05	11.55 (11.61)	0.90
^{244}Pu	1836.00 (1836.06)	5.94	6.08	11.61 (11.73)	0.79
^{246}Pu	1845.97 (1846.66)	5.95	6.10	11.52 (11.52)	0.66
^{244}Cm	1836.67 (1835.85)	5.95	6.06	12.03 (12.14)	0.91
^{246}Cm	1848.17 (1847.83)	5.96	6.08	12.08 (12.26)	0.80
^{248}Cm	1858.94 (1859.20)	5.97	6.11	12.01 (12.28)	0.67
^{250}Cm	1869.20 (1869.75)	5.98	6.13	11.81	0.54
^{250}Cf	1870.20 (1870.00)	6.00	6.11	12.41 (12.70)	0.62
^{252}Cf	1881.31 (1881.28)	6.01	6.13	12.22 (12.95)	0.49
^{254}Cf	1892.02 (1892.12)	6.02	6.15	11.97	0.36
^{252}Fm	1879.55 (1878.93)	6.02	6.12	12.86	0.57
^{254}Fm	1891.85 (1890.99)	6.03	6.14	12.58	0.41
^{256}Fm	1903.21 (1902.55)	6.04	6.16	12.45	0.31
^{252}No	1872.83 (1871.31)	6.03	6.10	13.23	0.56
^{254}No	1886.39 (1885.61)	6.04	6.12	13.22	0.45
^{256}Rf	1892.38 (1890.67)	6.07	6.13	13.57	0.34
^{260}Sg	1910.95 (1909.05)	6.10	6.16	13.70	0.15
^{264}Hs	1929.96 (1926.75)	6.13	6.18	13.42	-0.05

for both α-decay chains the trend of experimental transition energies is accurately reproduced by our calculations. For the odd-odd nucleus $^{288}115$, in particular, the theoretical Q_α values are in excellent agreement with the experimental data. For completeness, in Fig. 3 we also include the ground-state quadrupole deformation parameters β_2 of the superheavy nuclei that belong to the two α-decay chains.

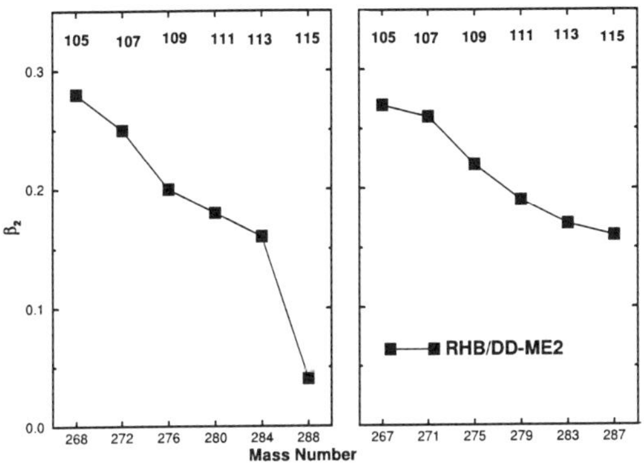

Fig. 3. Calculated ground-state quadrupole deformation parameters β_2 of the superheavy nuclei that belong to the two α-decay chains shown in Fig. 2.

The fully self-consistent RRPA[10] and RQRPA[21] have been used to calculate excitation energies of giant resonances in doubly-closed and open-shell nuclei, respectively. The RQRPA is formulated in the canonical basis of the RHB model and, both in the *ph* and *pp* channels, the same interactions are used in the RHB equations that determine the canonical quasiparticle basis, and in the matrix equations of the RQRPA.

For ^{208}Pb the RRPA results for the monopole and isovector dipole response are displayed in Fig. 4. For the multipole operator $\hat{Q}_{\lambda\mu}$ the response function $R(E)$ is defined

$$R(E) = \sum_i B(\lambda_i \to 0_f) \frac{\Gamma/2\pi}{(E - E_i)^2 + \Gamma^2/4}, \tag{9}$$

where Γ is the width of the Lorentzian distribution, and

$$B(\lambda_i \to 0_f) = \frac{1}{2J+1} |\langle 0_f || \hat{Q}_\lambda || \lambda_i \rangle|^2. \tag{10}$$

In the examples considered here the continuous strength distributions are obtained by folding the discrete spectrum of R(Q)RPA states with the Lorentzian (see eq. (9)) with constant width $\Gamma = 1$ MeV. The calculated peak energies of the ISGMR: 13.9 MeV, and IVGDR: 13.5 MeV should be compared with the experimental excitation energies: $E = 14.1 \pm 0.3$ MeV[22] for the monopole resonance, and $E = 13.3 \pm 0.1$ MeV[23] for the dipole resonance, respectively.

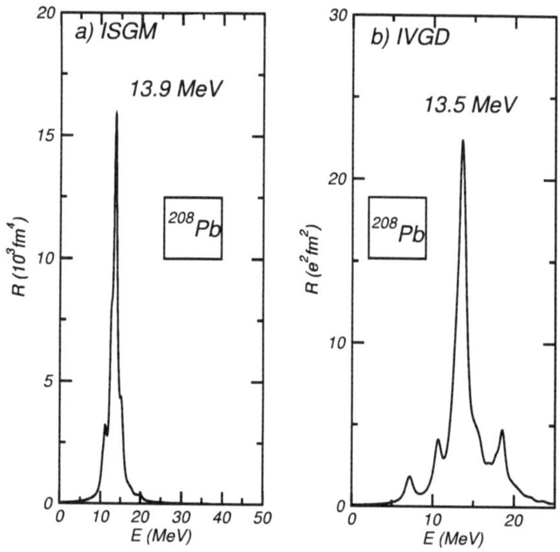

Fig. 4. The isoscalar monopole (a), and the isovector dipole (b) strength distributions in ^{208}Pb calculated with the effective interaction DD-ME2. The experimental excitation energies are: 14.1 ± 0.3 MeV[22] for the monopole resonance, and 13.3 ± 0.1 MeV[23] for the dipole resonance, respectively.

In Fig. 5 we compare the RQRPA results for the Sn isotopes with experimental data on IVGDR excitation energies.[24] In contrast to the case of ^{208}Pb, the strength distributions in the region of giant resonances exhibit fragmentation and the energy of the resonance E_{GDR} is defined as the centroid energy $\bar{E} = m_1/m_0$, calculated in the same energy window as the one used in the experimental analysis (13–18 MeV). The RHB+RQRPA calculation with the DD-ME2 interaction reproduces in detail the experimental excitation energies and the isotopic dependence of the IVGDR.

4. Conclusions

The relativistic density functional is extended to include density-dependent meson-nucleon coupling constants. Special attention is paid for the density dependence of the isovector channel. The new model provides a singificantly improved description of nuclear structure properties. Numerical calculations with the RHB model using the DD-ME2 force in the particle-hole channel and the Gongy interaction D1S in the pairing channel have been carried out for ground state properties and excitation energies of giant resonances. It turned out that the new model is able to provide very accurate predictions

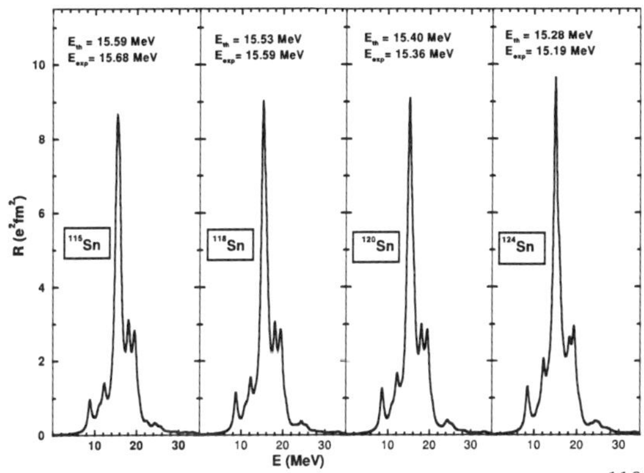

Fig. 5. The RHB+RQRPA isovector dipole strength distributions in ^{116,118,120,124}Sn. The experimental IVGDR excitation energies for the Sn isotopes are compared with the RHB+RQRPA results calculated with the DD-ME2 effective interaction.

for the nuclear masses and an excellent description of superheavy nuclei. The fully self-consistent relativistic (Q)RPA has been used to calculate excitation energies of giant resonances of spherical nuclei. The calculations are in excellent agreement with experiment.

5. Acknowledgements

This work has been partly supported by the Programe Pythagoras II of the Greek MoE and RA and the European Union under project 80661.

References

1. G. A. Lalazissis, P. Ring, and D. Vretenar (Eds.), *Extended Density Functionals in Nuclear Structure Physics*, Lecture Notes in Physics 641, (Springer, Berlin Heidelberg 2004).
2. D. Vretenar, A.V. Afanasjev, G.A. Lalazissis, P. Ring *Phys. Rep.* 409 (2005) 101 .
3. G.A. Lalazissis, J. König, and P. Ring, *Phys. Rev.* C 55 (1997) 540 .
4. B. Friedman and V.R. Pandharipande, *Nucl. Phys.* A 361 (1981) 502 .
5. T. Nikšić, D. Vretenar, P. Finelli, P. Ring, *Phys. Rev.* C 66 (2002) 024306 .
6. G. A. Lalazissis, T. Nikšić, D. Vretenar, P. Ring, *Phys. Rev.* C 71 (2005) 024312 .
7. C. Fuchs, H. Lenske, and H.H. Wolter, Phys. Rev. C 52, 3043 (1995).
8. S. Typel and H. H. Wolter, Nucl. Phys. A 656, 331 (1999).
9. F. Hofmann, C. M. Keil, and H. Lenske, Phys. Rev. C 64, 034314 (2001).

10. T. Nikšić, D. Vretenar, P. Ring, *Phys. Rev.* C 66 (2002) 064302 .
11. F. de Jong and H. Lenske, *Phys. Rev.* C 57 (1998) 3099 .
12. J. F. Berger, M. Girod, and D. Gogny, *Nucl. Phys.* A 428 (1984) 23 .
13. M. Samyn, S. Goriely, and J. M. Pearson, *Nucl. Phys.* A 725 (2003) 69 .
14. S. Goriely, M. Samyn, M. Bender, and J. M. Pearson, *Phys. Rev.* C 68 (2003) 054325 .
15. G. Audi, A. H. Wapstra, and C. Thibault, *Nucl. Phys.* A 729 (2003) 337 .
16. E.G. Nadjakov, K.P. Marinova, Yu.P. Gangrsky, At. Data Nucl. Data Tables 56, 133 (1994).
17. A. Krasznahorkay et al., Phys. Rev. Lett. 82, 3216 (1999).
18. S. Raman, C. W. Nestor Jr., P. Tikkanen, At. Data Nucl. Data Tables 78, 1 (2001).
19. G.A. Lalazissis, S. Raman and P. Ring, *Atom. Data Nucl. Data Tables* 71 (1999) 1.
20. Yu. Ts. Oganessian et al., *Phys. Rev.* C 69 (2004) 021601 .
21. N. Paar, P. Ring, T. Nikšić, D. Vretenar, *Phys. Rev.* C 67 (2003) 034312 .
22. D.H. Youngblood, H.L. Clark,and Y.W. Lui, *Phys. Rev. Lett.* 82 (1999) 691
23. J. Ritman et al., *Phys. Rev. Lett.* 70 (1993) 533 .
24. B. L. Berman and S. C. Fultz, *Rev. Mod. Phys.* 47 (1975) 713 .
Phys. Rev. C 68 (2003) 024310 .

RELATIVISTIC MEAN-FIELD DESCRIPTION OF LIGHT NUCLEI

J. LEJA

Faculty of Mechanical Engineering, Slovak University of Technology
Bratislava, Slovakia
E-mail: jozef.leja@stuba.sk

Š. GMUCA

Institute of Physics, Slovak Academy of Sciences
Bratislava, Slovakia
E-mail: stefan.gmuca@savba.sk

We have calculated binding energies of light even-even atomic nuclei. The calculations have been performed in the framework of the relativistic mean-field theory. The comparision with experimental data is not satisfactory because two systematical deviations have been discovered. The possibility to improve model predictions incorporating phenomenological Wigner term has been studied and new parameters of Wigner term have been obtained.

Keywords: Relativistic mean-field theory, Wigner term

1. Introduction

In recent years the relativistic mean-field theory has been frequently used for calculations of bulk and single-particle properties of atomic nuclei. The calculations have been successfully performed from light to superheavy nuclei, but still exist nuclei where the predictions of this model are in poor agreement with experimental results.

2. The relativistic mean-field theory

The relativistic mean-field theory is a relativistic quantum field theory describing atomic nucleus as a system of relativistic particles interacting through effective meson exchange. The model[1] starts from a Lagrangian density including nucleon field (ψ_i), isoscalar-scalar meson field (σ), isoscalar-vector meson field (ω^μ), isovector-vector meson field ($\vec{\rho}^\mu$)

and electromagnetic field (A^μ):

$$\mathcal{L} = \bar{\psi}_i \left[i\gamma_\mu \partial^\mu - M + g_\sigma \sigma - g_\omega \gamma_\mu \omega^\mu - g_\rho \gamma_\mu \vec{\rho}^\mu \cdot \vec{\tau} - e\gamma_\mu \frac{(1-\tau_3)}{2} A^\mu \right] \psi_i + \\ + \frac{1}{2}\partial_\mu \sigma \partial^\mu \sigma - m_\sigma^2 \sigma^2 - \frac{1}{3} b_\sigma \sigma^3 - \frac{1}{4} c_\sigma \sigma^4 - \frac{1}{4} O^{\mu\nu} O_{\mu\nu} + \frac{1}{2} m_\omega^2 \omega_\mu \omega^\mu + \\ - \frac{1}{4} \vec{R}_{\mu\nu} \vec{R}^{\mu\nu} + \frac{1}{2} m_\rho^2 \vec{\rho}_\mu \vec{\rho}^\mu - \frac{1}{4} F_{\mu\nu} F^{\mu\nu}. \quad (1)$$

The M, m_σ, m_ω, m_ρ denote nucleon mass and meson masses, g_σ, g_ω, g_ρ, are nucleon-meson coupling constants and the strengths of the selfinteractions are given by constants b_σ and c_σ.

The field equations follow from the Euler-Lagrange equations in a standard way. Two approximations are necessary for solution of field equations:

- The mean-field approximation introduced by replacing the field operators for meson fields and electromagnetic field by their expectation values.
- The no-sea approximation realized by exclusion of the filled Dirac sea of negative energy states.

The nucleon spinor can by written in form:

$$\psi_i(r_\perp, z, \varphi) = \frac{1}{\sqrt{2\pi}} \begin{pmatrix} f_i^+(r_\perp, z) \exp^{i(\Omega_i - \frac{1}{2})\varphi} \\ f_i^-(r_\perp, z) \exp^{i(\Omega_i + \frac{1}{2})\varphi} \\ i g_i^+(r_\perp, z) \exp^{i(\Omega_i - \frac{1}{2})\varphi} \\ i g_i^-(r_\perp, z) \exp^{i(\Omega_i + \frac{1}{2})\varphi} \end{pmatrix} \zeta_{t_i}. \quad (2)$$

The components of nucleon spinor obey the set of Dirac equations:

$$(M + S + V_0) f_i^+ + \partial_z g_i^+ + \left[\partial_{r_\perp} + \frac{(\Omega_i + \frac{1}{2})}{r_\perp} \right] g_i^- = \varepsilon_i f_i^+, \quad (3)$$

$$(M + S + V_0) f_i^- - \partial_z g_i^- + \left[\partial_{r_\perp} - \frac{(\Omega_i - \frac{1}{2})}{r_\perp} \right] g_i^+ = \varepsilon_i f_i^-, \quad (4)$$

$$(M + S - V_0) g_i^+ + \partial_z f_i^+ + \left[\partial_{r_\perp} + \frac{(\Omega_i + \frac{1}{2})}{r_\perp} \right] f_i^- = -\varepsilon_i g_i^+, \quad (5)$$

$$(M + S - V_0) g_i^- - \partial_z f_i^- + \left[\partial_{r_\perp} - \frac{(\Omega_i - \frac{1}{2})}{r_\perp} \right] f_i^+ = -\varepsilon_i g_i^-. \quad (6)$$

The S is the scalar potential:

$$S = g_\sigma \sigma. \tag{7}$$

The V_0 represents the time-like component of vector potential:

$$V_0 = g_\omega \omega_0 + g_\rho \tau_3 \rho_0^{(3)} + e\frac{(1-\tau_3)}{2} A_0. \tag{8}$$

Meson and electromagnetic fields obey the set of Klein-Gordon equations:

$$\left(-\frac{1}{r_\perp}\partial_{r_\perp} r_\perp \partial_{r_\perp} - \partial_z^2 + m_\sigma^2\right)\sigma = -g_\sigma \rho_S - b_\sigma \sigma^2 - c_\sigma \sigma^3, \tag{9}$$

$$\left(-\frac{1}{r_\perp}\partial_{r_\perp} r_\perp \partial_{r_\perp} - \partial_z^2 + m_\omega^2\right)\omega_0 = g_\omega \rho_V, \tag{10}$$

$$\left(-\frac{1}{r_\perp}\partial_{r_\perp} r_\perp \partial_{r_\perp} - \partial_z^2 + m_\rho^2\right)\rho_0^{(3)} = g_\rho \rho_I, \tag{11}$$

$$\left(-\frac{1}{r_\perp}\partial_{r_\perp} r_\perp \partial_{r_\perp} - \partial_z^2\right) A_0 = e\rho_C. \tag{12}$$

The nucleon densities can be calculated through expressions:

$$\rho_S = 2\sum_{i>0}^{occ} \left[\left(|f_i^+|^2 + |f_i^-|^2\right) - \left(|g_i^+|^2 + |g_i^-|^2\right)\right], \tag{13}$$

$$\rho_V = 2\sum_{i>0}^{occ} \left[\left(|f_i^+|^2 + |f_i^-|^2\right) + \left(|g_i^+|^2 + |g_i^-|^2\right)\right], \tag{14}$$

$$\rho_I = 2\sum_{i>0}^{occ} \tau_3 \left[\left(|f_i^+|^2 + |f_i^-|^2\right) + \left(|g_i^+|^2 + |g_i^-|^2\right)\right], \tag{15}$$

$$\rho_C = 2\sum_{i>0}^{occ} \frac{(1-\tau_3)}{2} \left[\left(|f_i^+|^2 + |f_i^-|^2\right) + \left(|g_i^+|^2 + |g_i^-|^2\right)\right]. \tag{16}$$

The total energy is given by formula:

$$E = \sum_{i}^{occ} \varepsilon_i - \frac{1}{2} \int \left(g_\sigma \sigma_0 \rho_S + g_\omega \omega_0 \rho_V + g_\rho \rho_0^{(3)} \rho_I + e A_0 \rho_C \right) d^3x +$$

$$- \int \left(\frac{1}{3} b_\sigma \sigma_0^3 + \frac{1}{4} c_\sigma \sigma_0^4 \right) d^3x. \qquad (17)$$

Proton and neutron pairing correlations have been included using the BCS theory.[2]

3. Results and discussion

We have calculated binding energies of even-even Ne, Mg and Si isotopes. The calculations have been performed with NL-BA[3] parameter set which proved the best agreement with experimental data in our previous study of O isotopes.[7] The comparision of our results with experimental data[4] is not satisfactory because two systematical deviations are observed:

- The model predicts low binding energies in N=Z region. The deviation has a form of symmetric peak around N=Z nucleus.
- The model predicts high binding energies for neutron rich nuclei. The deviation is proportional to the neutron excess.

We have compared also the relativistic mean-field calculations by Lalazissis[5] performed with NL3[6] parameter set with experimental data but they exhibit the same deviations.

The discrepancy in N=Z region was observed also in other nuclear models, but their modern versions like finite-range droplet model by Myers and Swiatecki[8] or Hartree-Fock mass formula by Goriely[9] introduce various phenomenological terms suppressing this discrepancy.

The term convenient for supressing deviations observed in our calculations is a phenomenological Wigner term:

$$E_W = V_W \exp^{-\lambda\left(\frac{N-Z}{A}\right)^2} + V_W' |N - Z| \exp^{-\left(\frac{A}{A_0}\right)^2} \qquad (18)$$

introduced in Hartree-Fock mass formula. It consits of two parts:

- The first part has a form of Gaussian function. It is attractive and it represents the additional energy in N=Z region. The unique aspect of these nuclei is that neutrons and protons occupy very close orbitals. The large spatial overlaps between neutron and proton wave functions enhance neutron-proton correlations, especially the neutron-proton pairing. This part is interpreted as a consequence of the T=0 neutron-proton pairing.[10,11]

- The second repulsive part is based on the Wigner's supermultiplet theory of approximative symmetry of nuclear Hamiltonian under a combined spin-isospin symmetry leading to the term in binding energy proportional to $|N - Z|$.[12,13]

The incorporation of Wigner term using original parameters brought only small inprovement of results, therefore we have fitted new parameters suitable for relativistic mean-field calculations. The impact of Wigner term on binding energy differences for Mg isotopes we can see in Fig. 1.

Original and new parameters of Wigner term are in Table 1. Reason for radical discrepancy in parameter λ is not clear, but we must take into consideration that original values were fitted to over 2000 nuclei covering all periodic table and new parameters were fitted to only even-even nuclei from very small region.

Table 1. Parameters of Wigner term.

	original parameters[9]	new parameters
V_W (MeV)	-2.05	-5.193
λ	485.0	6.052
V'_W (MeV)	0.697	1.060
A_0	28	50.598

4. Conclusion

Two systematical deviations were found in calculated binding energies of even-even Ne, Mg and Si isotopes. These deviations can be suppressed by incorporation of phenomenological Wigner term used in non-relativistic Hartree-Fock mass formula.

5. Acknowledgments

This work was supported by the Slovak Grant Agency for Science VEGA under grant No. 2/4098/04 .

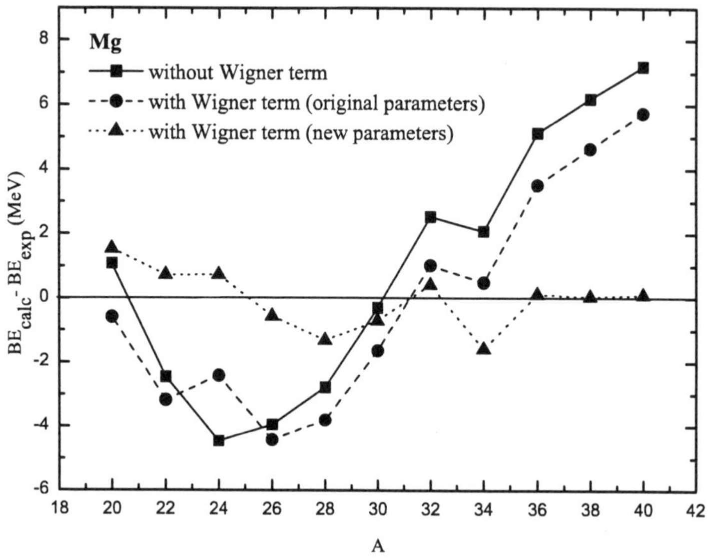

Fig. 1. Binding energy differences for even-even Mg isotopes.

References

1. B. D. Serot and J. D. Walecka, *Adv. Nucl. Phys.* **16**, 1 (1986).
2. P. Ring and P. Schuck, *The Nuclear Many-Body Problem*, (Springer-Verlag, New York, 1980).
3. Š. Gmuca, in *Proc.* 2nd *Int. Conf. Fission and Properties of Neutron Rich nuclei*, (World Scientific, Singapore, 2000).
4. G. Audi, A. H. Wapstra and C. Thibault, *Nucl. Phys. A* **729**, 129 (2003).
5. G. A. Lalazissis, S. Raman and P. Ring, *At. Data Nucl. Data Tabes* **71**, 1 (1999).
6. G. A. Lalazissis, J. König and P. Ring, *Phys. Rev. C* **55**, 540 (1997).
7. J. Leja and Š. Gmuca, *Acta Phys. Slovaca* **51**, 201 (2001).
8. W. D. Myers and W. J. Swiatecki, *Nucl. Phys. A* **612**, 249 (1997).
9. S. Goriely et al., *Phys. Rev. C* **66**, 024326 (2002).
10. N. Zeldes, *Phys. Lett. B* **429**, 20 (1998).
11. W. Satula and R. Wyss, *Nucl. Phys. A* **676**, 120 (2000).
12. E. Wigner, *Phys. Rev.* **51**, 106 (1937).
13. P. Van Isacker, D. D. Warner and D. S. Brenner, *Phys. Rev. Lett.*, **74** 4607 (1995).

ENERGY NUCLEON SPECTRA FROM REACTIONS AT INTERMEDIATE ENERGIES

OLEG GRUDZEVICH

State Technical University, Studgorodok, 1, Obninsk, Kaluga rgn., 249020 Russia

SERGEY YAVSHITS

Khlopin Radium Institute, 2nd Murinsky av. 28, St. Petersburg, 194021, Russia

YULIA MARTIROSYAN

State Technical University, Studgorodok, 1, Obninsk, Kaluga rgn., 249020 Russia

New exciton model of preequilibrium decay (*M*onte *C*arlo *P*reequilibrium) to compute spectra of multiparticle emission during an establishment of statistical equilibrium in the composite system is proposed. MCP stage of calculation was included into the standard scheme of the nucleon spectra calculation between the intranuclear cascade stage and statistical model stage. Systematic comparison of calculation results with the experimental spectra of nucleons from (p,xn), (p,xp), (n,xn) and (n,xp) reactions in a wide projectile energy region from 10 up to 160 MeV for targets from ^{27}Al up to ^{209}Bi has been carried out. The short description of the MCP model and results obtained are presented in the given work.

1. Introduction

The development of modern nuclear technologies requires the large amount of nuclear data to supply needs in the working out of the conceptual and design solutions in different fields of applications first at all the technologies of the radioactive waste transmutations and power productions, radiotherapy, shielding problem and so on.

There are two ways of nuclear data supply for practical goals – to include nuclear data generator into the transport codes or produce nuclear data files outside. We guess the second way is more reliable due to possibility of modern and sufficiently complicated codes applications. The development of the nuclear data libraries as well as corresponding computer codes has to be done for nuclear reactions induced by proton and neutron beams in projectile energy region 20 MeV – 1 GeV.

The MCFx code [1] is based on the detailed description of all stages of nuclear reaction induced by the intermediate energy nucleons. It uses the well-checked and reliable models for the entrance channel simulation (coupled channel method), direct processes (intranuclear cascade model), pre-equilibrium particle emission (exciton model with multiple particle emission [2,3]), and compound nuclear decay (statistical model).

2. Monte Carlo simulation of multiparticle preequilibrium emission

The spectrum of emitted nucleons is determined mainly by two quantities [3] that are density of particle-hole states $\omega(p,h,E)$ with number of particles p, number of holes h and excitation energy of composite system E as well as matrix element of two-particle interaction. Thus the nucleon may be emitted at different stages of motion to equilibrium state, i.e. from states with different number of particles p and holes h. Time development of process is governed by the system of master equations:

$$\frac{dP(n,t)}{dt} = P(n-2,t)\lambda_+(n-2,E) + P(n+2,t)\lambda_-(n+2,E) - \\ - P(n,t)\left[\lambda_+(n,E) + \lambda_-(n,E) + \sum_v L_v(n,E)\right] \tag{1}$$

where λ_+, λ_- are transition probabilities of a nucleus to more complicated or more simple state, correspondingly, L is the probability of a particle emission, P is the population probability of configuration $n=p+h$ at the moment of time t, E is the excitation energy of composite system.

The basic idea of the model [2] is to simulate the particle emission at the stage of equilibration. At this stage when particle hole configuration is complicating by two body interaction the nucleon may be emitted. After the nucleon emission the daughter nucleus is putted into the equilibration processes. The steps of model calculations are as follows: 1) calculation of all necessary preequilibrium parameters of all possible nuclei and excitation energies for given initial composite system, 2) a random number is used to select one of the way of continuation: two particle interaction or neutron/proton emission, 3) after two particle interaction the system goes to more complicated configuration, 4) if particle emission took place a random number is used to select the particle energy, 5) after the particle emission we go to the stage 2 with new initial data. The Monte Carlo simulation cascade was used for steps 2-5.

The calculated final spectrum of escaped particles is the sum of all reaction mechanism as:

$$S(\varepsilon) = S_{INC}(\varepsilon) + \sum_{Z,A}\sum_{E} Y_{INC}(Z,A,E,ph)\cdot\sum_{i} S^{i}_{MCP}(Z,A,E,\varepsilon) + \\ + \sum_{Z,A}\sum_{E} Y_{MCP}(Z,A,E)\cdot\sum_{i} S^{i}_{HF}(Z,A,U,\varepsilon), \quad (2)$$

where $S_{INC}(\varepsilon)$, $Y_{INC}(Z,A,E,ph)$ are the spectrum and the yields of residual nuclei calculated in the intranuclear cascade model [1], $S_{MCP}(\varepsilon)$, $Y_{MCP}(Z,A,E,ph)$ are the same but for preequilibrium decay and $S_{HF}(\varepsilon)$ is the statistical spectrum of evaporated particles with energy ε.

3. Results

We tested the above described procedure of multiparticle preequilibrium emission by comparison with results of the model [3] for single particle emission. After testing we included the procedure into MCFx code system. The comparison of the calculated neutron spectra with the experimental data is shown in Figs.1-2. One can see that as projectile energy increases the contribution of the fastest stage of reaction (intranuclear cascade stage) increases too. On the other hand, the total calculated nonequilibrium spectra describe existing experimental data reasonably good. The examples of the spectra for two different targets are shown in Figs. 3 and 4. It is seen from the figures that the proton and neutron experimental spectra are described reasonably well. All calculated results were obtained with the same model parameters.

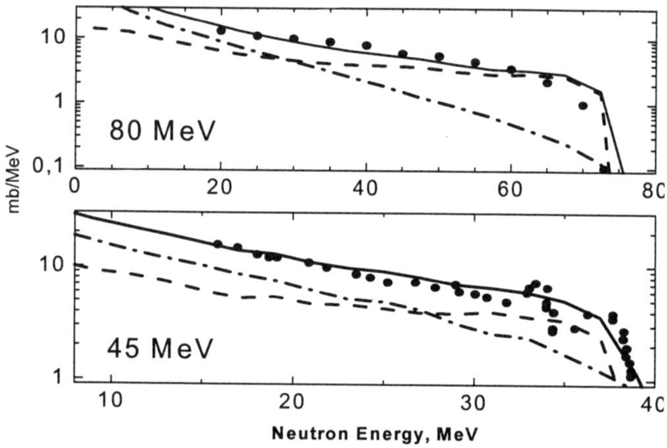

Fig.1. The comparison of calculated neutron spectra of $^{90}Zr(p,xn)$ reaction at different projectile energies (45 and 80 MeV) with experimental data (symbols, [4]), the curves are the calculation results: dash - intranuclear cascade calculation, dash dot- preequilibrium calculation, solid - sum of nonequlibrium spectra .

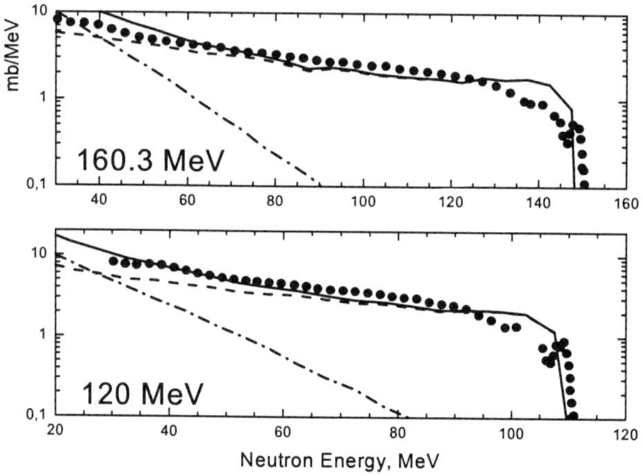

Fig.2. The same as in fig.1, but for 120 and 160 MeV proton energies.

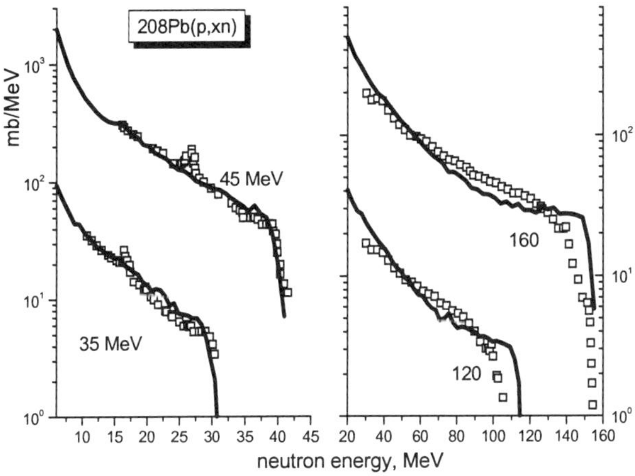

Fig.3. Comparison of calculated and experimental neutron spectra [4] for $^{208}Pb(p,xn)$ reaction at 35, 45 120 and 160 MeV proton energies.

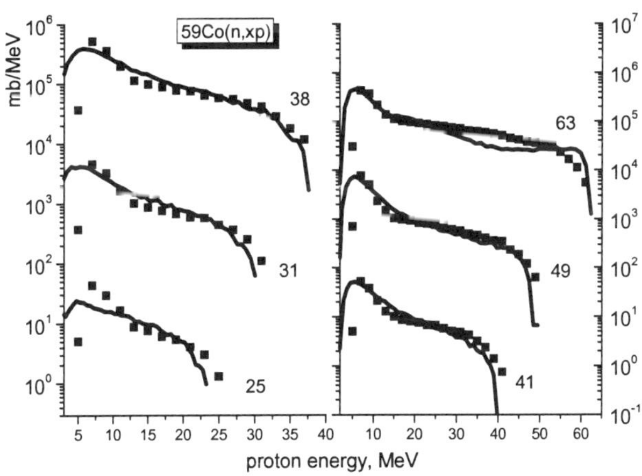

Fig.4. Comparison of calculated and experimental proton spectra [5] for $^{59}Co(n,xp)$ reactions at neutron energies 25, 31, 38, 41, 49, 63 MeV.

4. Conclusions

Multiparticle preequilibrium model was proposed and tested by comparison with experimental data on nucleon spectra for projectile energies up to 160 MeV. After the testing we included the procedure into MCFx code system [1]. Results of calculations of nonequilibrium spectra describe existing experimental data rather well.

MCFx code system was used to generate the first version of complete nuclear data file for proton-induced reactions on ^{208}Pb with energies up to 1 GeV. The file contains total cross-sections, double differential elastic cross-sections, fission cross-sections, double differential nucleon emission cross-sections, and fission fragment yields.

The work was performed under ISTC project # 2524.

References

1. Yavshits S.G., Ippolitov , Goverdovsky A.A., Grudzevich O.T., Theoretical approach and computer code system for nuclear data evaluation of 20-1000 MeV neutron induced reactions on heavy nuclei, Proc. of Int. Conf. on Nucl. Data for Sci. and Tech., Tsukubo, Japan, pp.104-107 (2001).
2. Akkermans J.M. and Gruppelaar H., Z. Phys , A300, p.345 (1981).
3. Griffin T.T., Statistical model of intermediate structure, Phys. Rev. Letters, v.17, p.478 (1966).
4. Blann M.,Doering P.R., Galonsky A., Patterson D.M., Serr F.E., Preequilibrium analysis of (p,n) spectra on various targets at proton energies of 25 to 45 MeV., Nucl. Phys., A257, p.15 (1976).
5. Nica N., Benck S., Raeymackers E., Slypen I., Meulders J.P., Corcalciuc V., Light charged particle emission induced by fast neutrons (25 to 65 MeV) on Co-59, Phys. Rev. C 51, p.1303 (1995).

ANALYSIS, PROCESSING AND VISUALIZATION OF MULTIDIMENSIONAL DATA USING DAQPROVIS SYSTEM

M. Morháč*[,1], V. Matoušek[1], I. Turzo[1] and J. Kliman[1,2]

[1] *Institute of Physics, Slovak Academy of Sciences, Dúbravská cesta 9, 845 11 Bratislava, Slovakia*

[2] *Flerov Laboratory of Nuclear Reactions, JINR Dubna, Russia*

* *E-mail: Miroslav.Morhac@savba.sk*

The multidimensional data acquisition, processing and visualization system for analysis of experimental data in nuclear physics is briefly described in the paper. The system includes a large number of sophisticated algorithms of the multidimensional nuclear spectra processing, including background elimination, deconvolution, peak searching and fitting.

Keywords: Data acquisition system, nuclear spectra analysis, storing and compression of histograms, background estimation, deconvolution, peak identification, fitting, visualization.

1. Introduction

In many nuclear physics laboratories a large number of home-made acquisition systems, ranging from small, through medium sized up to large ones, were designed. In the paper we describe a DaqProVis system developed at the Institute of Physics, Slovak Academy of Sciences in Bratislava. It integrates a large scale of routines dedicated for acquisition, sorting, storing, histogramming, analysis and presentation of multidimensional experimental data in nuclear physics [1]. The system is continuously being developed, improved and supplemented with new additional functions and capabilities.

2. Basic features and capabilities of the DaqProVis system

A data flow chart of the system is presented in Fig. 1. The raw events can be read either directly from experimental modules (CAMAC, VME) or from another DaqProVis system working in server mode or from list files collected in other experiments (e.g. Gammasphere). The basic element of the event is a variable (one value) read out from an address, which is called

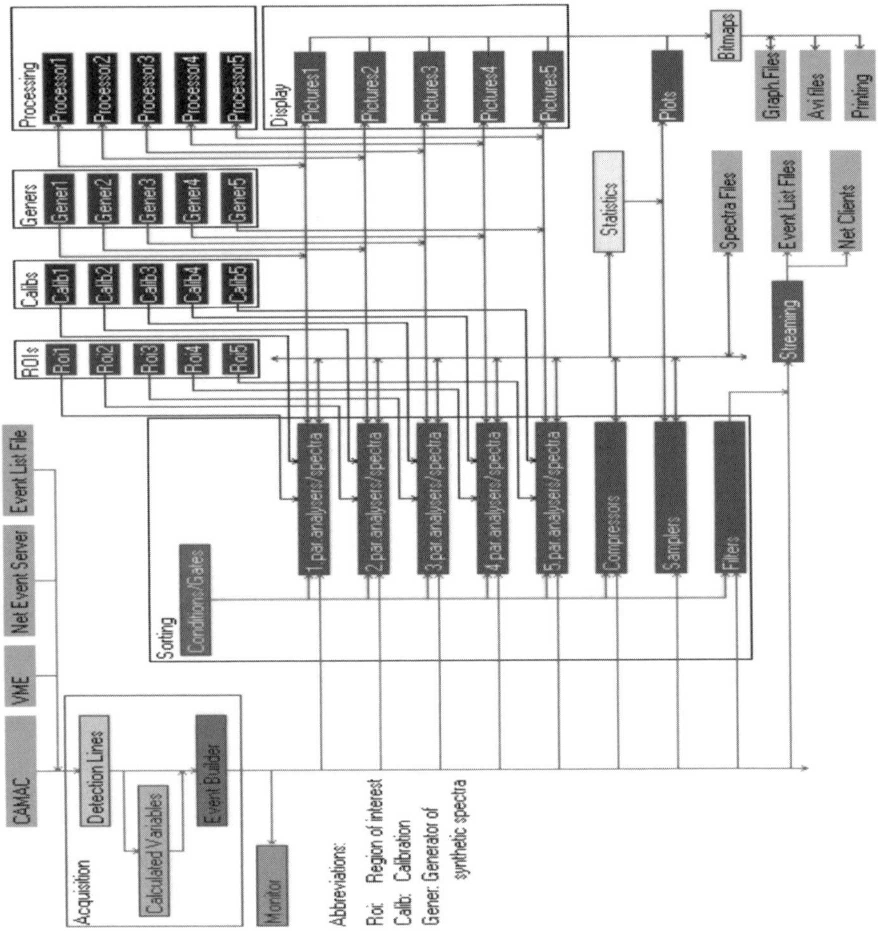

Fig. 1. Flow chart of data acquisition, processing and visualization system DaqProVis.

"detection line". It has its name and in hardware it is represented by an input register (ADC, QDC, TDC, counter etc.). If desired, events can be supplemented with variables calculated from read-out parameters. One can utilize a set of standard mathematical operators (+, -, *, /, , sqr, log, sin, cos, exp). The names of employed detection lines can stand for operands in the mathematical expressions.

The events can be written unchanged to an event list file, or/and to other DaqProVis systems (clients). They can be sorted according to predefined criteria (gates) and written to sorted streams as well. The event variables can be analyzed to create one-, two-, three-, four-, five-dimensional histograms – spectra, analyzed and compressed using on-line compression procedure, sampled using various sampler modes (sampling, multiscaling, or stability measurement of a chosen event variable). From acquired multidimensional spectra, one can make slices of lower dimensionality. Continuous scanning aimed at looking for and localizing interesting parts of multidimensional spectra, with automatic stop when the attached condition is fulfilled, is also possible. The condition is connected either with the contents of counts or with the maximum value in given region of interest.

Once collected the analyzed data can be further processed using both conventional and new developed sophisticated algorithms (Processor 1-5 blocks). One can also define regions of interests (ROI 1-5 blocks) and calibrations (Calib 1-5 blocks) for up to five-dimensional spectra. To facilitate the development of the processing algorithms we have implemented generators of synthetic spectra (blocks Gener 1-5).

The system allows one to display up to five-dimensional spectra using a great variety of conventional as well as sophisticated (shaded isosurface, volume rendering, projections of inserted subspaces, etc.) visualization techniques. If desired, all changes of individual pictures or entire screen can be recorded in an avi file. It proved to be very efficient tool mainly in the analysis of iterative processing methods.

3. Event sorting

After taking events from any of the above mentioned sources the first step of event processing is their selection or separation. The experimenter is interested only in the events satisfying the predetermined conditions or gates. Based on the gates the events can be broken up into different output streams written in the list mode either to files or sent to other clients. The gates can be used also for the decision about the acceptance of events for subsequent analysis in the analyzers or compressors (see Fig. 1).

The basic element of the data sorting is gate. To satisfy typical experimental needs in DaqProVis we have implemented the following types of gates:

- rectangular window
- polygon
- arithmetic function
- spherical gates
- composed gates.

Rectangular window specifies a set of event variables with lower and upper channels determining the region of event acceptance. This is the classical gating method commonly used.

The proper choice of gates can lead to an improvement in spectral quality, in particular the peak – to – background ratio, and to decrease the number of uncorrelated events in the projected spectrum. An efficient and simple way to choose the region of event acceptance in two-dimensional space of event variables is interactive setting of appropriate closed polygon. The advantage of this kind of gate is that one can design easily irregular shape. Its disadvantage is that it cannot be extended to higher dimensions and that it must be set manually.

The gate can be also represented by mathematical function of event variables (detection lines) x_1, x_2,...,x_n

$$f(x_1, x_2, ..., x_n) \leq 0. \tag{1}$$

The allowed operators are $+$, $-$, $*$, $/$, $\backslash$, sqr, log, exp, cos, sin. The built-in syntax analyzer is able to recognize the expressions written in Fortran-like style using names of event variables for operands, above given operators and parentheses. During the sorting for each event the value of the function (1) is calculated. If the value is less or equal to zero the event is accepted, i.e., the logical value of the condition is "true". By employing a suitable analytical function, one can specify more exactly the region of interesting parts in the spectrum.

When sorting events with Gaussian or quasi Gaussian distribution the gates with elliptic base are of special interest. The radii of ellipses are proportional to standard deviations σ_i or to the $FWHM_i = \sqrt{2\log 2}\sigma_i$ (full width at half maximum) of the photopeak distribution

$$r_i = \frac{1}{2}R \cdot FWHM_i. \tag{2}$$

Then for symmetrical n-dimensional spherical gates one can write

$$\sum_{i=1}^{n} \left(\frac{x_i - c_i}{0.5FWHM_i} \right)^2 - R^2 \leq 0. \tag{3}$$

However, due to various effects in detectors the peaks exhibit left-hand tailing. In [2] special gates reflecting the tails in spectrum peaks were proposed. The example of three-dimensional spherical gate is given in Fig. 2.

The result of application of any of the above defined gates (conditions) is either the value "true", i.e., the event is accepted for further processing or the value "false" (event is ignored). Every gate in DaqProVis has its own name. By applying logical operators (AND, OR, NOT) to operands (previously defined gate names) and using parentheses one can write very complex logical expressions defining the shape of the composed gate. The shape can be very complicated. One can define even the composition of disjoint subsets.

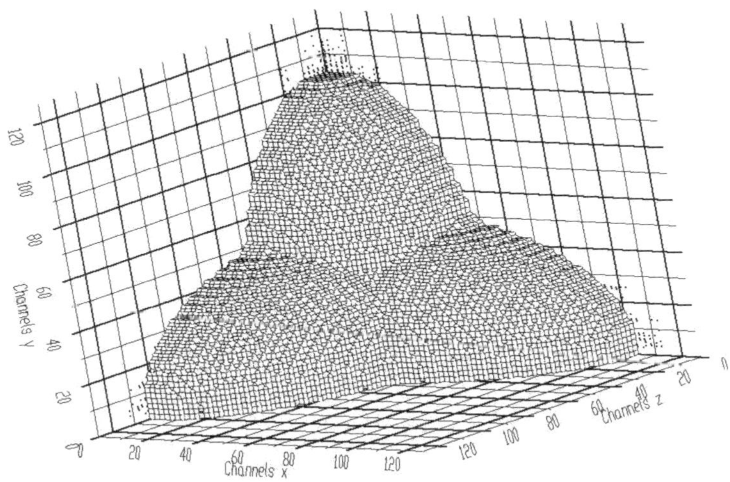

Fig. 2. Three-dimensional spherical gate

4. Storing and compression of multidimensional histograms

After eventual separating of interesting events from non-interesting ones the storing and possible compression, which is compelled by limited technical facilities, is the next element in the chain of processing of multidimensional experimental data arrays. It should be emphasized that because of practical reasons, e.g. interactive analysis, handling etc, the compression of large multidimensional arrays is in some situations unavoidable. The following methods of compression are implemented in DaqProVis system

- binning channels,
- utilizing the symmetry of multidimensional γ-ray spectra [3],
- classical orthogonal transform,
- adaptive orthogonal transforms [4], [5],
- randomizing transforms [6], [7].

5. Background estimation

The determination of the position and net areas of peaks due to γ-ray emissions requires the accurate estimation of the spectral background. A very efficient method of background estimation has been developed in [8]. The method is based on Statistics-sensitive Non-linear Iterative Peak-clipping algorithm. In [9] the algorithm has been extended to two-, and three-dimensional and subsequently generalized to n-dimensional case. The SNIP algorithm, together with its extensions and modifications tailored to special kinds of data, have been implemented in the DaqProVis system for up to five-dimensional spectra.

6. Deconvolution

The goal of the deconvolution operation is the improvement of the resolution in spectra. The principal results of the deconvolution operation were presented in [10]. Later we have optimized the Gold deconvolution algorithm that allowed to carry out the deconvolution operation much faster and to extend it to three-dimensional spectra. The results of the optimized Gold deconvolution for all one-, two-, and three-dimensional data are given in [11]. We have proposed improvements, modifications, extensions of existing deconvolution methods as well as new regularization techniques, e.g. boosted deconvolution, Tikhonov regularization with minimization of squares of negative values. All these methods are included in DaqProVis system.

7. Peak identification

The basic aim of one-dimensional peak searching procedure is to identify automatically the peaks in a spectrum with the presence of the continuous background and statistical fluctuations - noise. The essential peak searching algorithm is based on smoothed second differences (SSD) that are compared to its standard deviations [12]. We have extended the SSD based method of peak identification for two-dimensional and in general for multidimensional spectra [13]. In addition to the above given requirements the algorithm must be insensitive also to lower-fold coincidences peak-background (ridges) and their crossings.

However the resolution capability of the SSD based searching algorithms is quite limited. Therefore we have developed the high resolution peak searching algorithm based on the Gold deconvolution method. Let us illustrate its capabilities using the synthetic spectrum with several peaks located very close to each other. Detail of the spectrum with cluster of peaks is shown in Fig. 3. In the upper part of the figure one can see original data and in the bottom part the deconvolved data. The method finds also the peaks about existence of which it is impossible to guess from the original data.

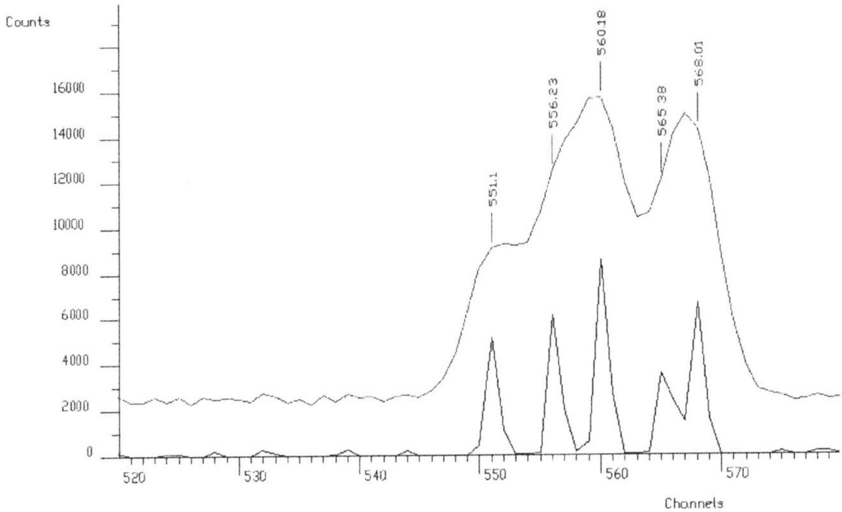

Fig. 3. Example of synthetic spectrum with cluster of peaks and its deconvolved spectrum.

8. Fitting

The final step and the key-stone of the nuclear spectra analysis consists is the fitting of the peak shape parameters of the identified peaks. The positions of peaks identified in the peak searching procedure are fed as initial estimates into the fitting function. In DaqProVis we have implemented several methods of fitting (Newton, conjugate gradients, Stiefel-Hestens, algorithm without matrix inversion [14], etc). Specific problem in the analysis of multidimensional γ-ray spectra is that connected with simultaneous fitting of large number peaks in large blocks of multidimensional γ-ray spectra and hence enormous number of fitted parameters. Therefore the fitting algorithms without matrix inversion, which allow a large number of parameters to be fitted, are of special attention. We have modified this algorithm and studied its properties in [15].

9. Visualization

The power of computers to collect, store and process multidimensional experimental data in nuclear physics has increased dramatically. Without visualization much of this increased power however, would be wasted because experiments are poor at gaining insight from data presented in numerical form.

We have developed several direct visualization algorithms to visualize two-, three-, and four-dimensional data. However, with increasing dimensionality of nuclear spectra the requirements in developing of multidimensional scalar visualization techniques becomes striking. The dimensionality of the direct visualization techniques is limited to four.

We have proposed and implemented the technique of inserted subspaces up to five-dimensional spectra. The goal is to allow one to localize and scan interesting parts (peaks) in multidimensional spectra. Moreover it permits to find correlations in the data, mainly among neighboring points, and thus to discover prevailing trends around multidimensional peaks.

The conventional as well as newly developed sophisticated visualization techniques and graphical models were described in [16]. The structure and complexity of the algorithms lend themselves for the implementation in on-line live mode during the data acquisition or processing. One can select various attributes of the display, e.g. color of the spectrum, the limits of the displayed part of the spectrum, window, marker, type of scale, and various display modes, slices, rotations of two-, or more-dimensional data.

10. Conclusions

The paper describes briefly the capabilities of the DaqProVis system. It integrates a large number of standard conventional methods as well as new developed algorithms of background elimination, deconvolution, peak searching, fitting etc. The modular structure of the system and the object oriented style make it possible to extend it continuously for new methods, algorithms and higher dimensions.

References

1. M. Morháč et al., *Nucl. Instr. and Meth.* **A502**, (2003) 728.
2. Ch. Theisen et al., *Nucl. Instr. and Meth.* **A432**, (1999) 249.
3. D.C. Radford, *Nucl. Instr. and Meth.* **A361**, (1995) 290.
4. M. Morháč et al., *Nucl. Instr. and Meth.* **A370**, (1996) 499.
5. V. Matoušek et al., *Nucl. Instr. and Meth.* **A502**, (2003) 725.
6. V. Bonačic et al., *Nucl. Instr. and Meth.* **66**, (1968) 213.

7. B. Souček et al., *Nucl. Instr. and Meth.* **66**, (1968) 202.
8. C.G. Ryan et al., *Nucl. Instr. and Meth.* **B34**, (1988) 396.
9. M. Morháč et al., *Nucl. Instr. and Meth.* **A401**, (1997) 113.
10. M. Morháč et al., *Nucl. Instr. and Meth.* **A401**, (1997) 385.
11. M. Morháč et al., *Digital Signal Processing* **13**, (2003) 144.
12. M.A. Mariscotti, *Nucl. Instr. and Meth.* **50**, (1967) 309.
13. M. Morháč et al., *Nucl. Instr. and Meth.* **A443**, (2000) 108.
14. I.A. Slavic, *Nucl. Instr. and Meth.* **A134**, (1976) 285.
15. M. Morháč et al., *Applied Spectroscopy* **57**, (2003) 753.
16. M. Morháč et al., *Acta Physica Slovaca* **54**, (2004) 385.

LIST OF PARTICIPANTS

Yoshihiro ARITOMO
Flerov Laboratory of Nuclear Reactions
Joint Institute for Nuclear Research
Joliot-Curie 6
141980 Dubna, Moscow region
Russia
aritomo24@muj.biglobe.ne.jp

Martina BEREŠOVÁ
Flerov Laboratory of Nuclear Reactions
Joint Institute for Nuclear Research
Joliot-Curie 6
141980 Dubna, Moscow region
Russia
fyzimabe@savba.sk

Alexey BOGACHEV
Flerov Laboratory of Nuclear Reactions
Joint Institute for Nuclear Research
Joliot-Curie 6
141980 Dubna, Moscow region
Russia
bogachev@nr.jinr.ru

Nicolae CARJAN
Bordeaux University - IN2P3
CENBG , BP 120 33175
Gradignan
France
carjan@in2p3.fr

Andrey DANIEL
Flerov Laboratory of Nuclear Reactions
Joint Institute for Nuclear Research
Joliot-Curie 6
141980 Dubna, Moscow region
Russia
daniel@jinr.ru

Herbert FAUST
Institut Laue-Langevin
6 rue Jules Horowitz
F-38000 Grenoble
France
faust@ill.fr

Janine GENEVEY
Laboratoire de Physique Subatomique et de Cosmologie
53, Avenue des Martyrs
38000 Grenoble
France
genevey@lpsc.in2p3.fr

Štefan GMUCA
Institute of Physics
Slovak Academy of Sciences
Dúbravská cesta 9
SK-842 28 Bratislava
Slovakia
gmuca@savba.sk

Chris GOODIN
Vanderbilt University
Department of Physics and Astronomy
1807 Station B
37235 Nashville
USA
christopher.t.goodin@vanderbilt.edu

Dmitry GORELOV
Flerov Laboratory of Nuclear Reactions
Joint Institute for Nuclear Research
Joliot-Curie 6
141980 Dubna, Moscow region
Russia
gorelovd@yandex.ru

Walter GREINER
Frankfurt Institute for Advanced Studies (FIAS)
J.W. Goethe Universität Frankfurt am Main
Max-von-Laue-Str. 1
60438 Frankfurt am Main
Germany
greiner@fias.uni-frankfurt.de

Mikhail ITKIS
Flerov Laboratory of Nuclear Reactions
Joint Institute for Nuclear Research
Joliot-Curie 6
141980 Dubna, Moscow region
Russia
itkis@flnr.jinr.ru

Dmitry KAMANIN
Flerov Laboratory of Nuclear Reactions
Joint Institute for Nuclear Research
Joliot-Curie 6
141980 Dubna, Moscow region
Russia
kamanin@fobos.jinr.ru

Ján KLIMAN
Flerov Laboratory of Nuclear Reactions
Joint Institute for Nuclear Research
Joliot-Curie 6
141980 Dubna, Moscow region
Russia
kliman@flnr.jinr.ru

Alexander KARPOV
Flerov Laboratory of Nuclear Reactions
Joint Institute for Nuclear Research
Joliot-Curie 6
141980 Dubna, Moscow region
Russia
karpov@jinr.ru

Galina KNYAZHEVA
Flerov Laboratory of Nuclear Reactions
Joint Institute for Nuclear Research
Joliot-Curie 6
141980 Dubna, Moscow region
Russia
galina.kniajeva@mail.ru

Yuri KOPATCH
Frank Laboratory of Neutron Physics
Joint Institute for Nuclear Research
Joliot-Curie 6
141980 Dubna, Moscow region
Russia
kopatch@nf.jinr.ru

Eduard KOZULIN
Flerov Laboratory of Nuclear Reactions
Joint Institute for Nuclear Research
Joliot-Curie 6
141980 Dubna, Moscow region
Russia
kozulin@jinr.ru

Nina KOZULINA
Flerov Laboratory of Nuclear Reactions
Joint Institute for Nuclear Research
Joliot-Curie 6
141980 Dubna, Moscow region
Russia
kozulin@jinr.ru

Ľuboš KRUPA
Flerov Laboratory of Nuclear Reactions
Joint Institute for Nuclear Research
Joliot-Curie 6
141980 Dubna, Moscow region
Russia
krupa@nrmail.jinr.ru

Georgios A. LALAZISSIS
Department of Theoretical Physics
Aristotle University of Thessaloniki
Gr-54006 Thessaloniki
Greece
glalazis@auth.gr

Jozef LEJA
Slovak University of Technology
Faculty of Mechanical Engineering
Department of Physics
Námestie Slobody 17
812 31 Bratislava
Slovakia
jozef.leja@post.sk

Taras LOKTEV
Joint Institute for Nuclear Research
Flerov Laboratory of Nuclear Reactions
Joliot-Curie 6
141980 Dubna, Moscow region
Russia
loktev@nrmail.jinr.ru

Yu-Gang MA
Shanghai Institute of Applied Physics
2019 Jia-Luo Road
Shanghai
China
ygma@sinap.ac.cn

Vladislav MATOUŠEK
Institute of Physics
Slovak Academy of Sciences
Dúbravská cesta 9
SK-84511 Bratislava 45
Slovakia
fyzimatu@savba.sk

Serban MISICU
National Institute for Nuclear Physics and Engineering
Horia Hulubei, Atomistilor, MG-6, Magurele
Romania
misicu@theory.nipne.ro

Miroslav MORHÁČ
Institute of Physics
Slovak Academy of Sciences
Dúbravská cesta 9
SK-84511 Bratislava 45
Slovakia
fyzimiro@savba.sk

Manfred MUTTERER
Institut Fur Kernphysik
Technische Universität Darmstadt
Schlossgartenstrasse 9
64289 Darmstadt
Germany
mutterer@email.de

Yuri PYATKOV
Moscow Engineering Physics Institute
Kashirskoe shosse 31
Moscow
Russia
yvp_nov@mail.ru

Karl-Heinz SCHMIDT
G S I
Planckstrasse 1
D-64291 Darmstadt
Germany
k.h.schmidt@gsi.de

Gavin SMITH
The University of Manchester
Oxford Road
M13 9PL, Manchester
UK
gavin.smith@man.ac.uk

Adam SOBICZEWSKI
Soltan Institute for Nuclear Studies
ul. Hoza 69
00-681 Warsaw
Poland
sobicz@fuw.edu.pl

Louise STUTTGÉ
IReS
Rue du Loess, BP 28
F-67037 Strasbourg
France
stuttge@in2p3.fr

Ivan TURZO
Institute of Physics
Slovak Academy of Sciences
Dúbravská cesta 9
SK-84511 Bratislava 45
Slovakia
fyziturz@savba.sk

Vladimir UTYONKOV
Flerov Laboratory of Nuclear Reactions
Joint Institute for Nuclear Research
Joliot-Curie 6
141980 Dubna, Moscow region
Russia
utyonkov@sungns.jinr.ru

Emanuele VARDACI
University of Naples "Federico II" - INFN
Complesso Universitario M.S. Angelo, via Cinthia, Edificio G
80126 Naples
Italy
vardaci@na.infn.it

Martin VESELSKÝ
Institute of Physics
Slovak Academy of Sciences
Dúbravská cesta 9
SK-84511 Bratislava 45
Slovakia
martin.veselsky@savba.sk

Sofie VERMOTE
University of Gent
Proeftuinstraat 86
B-9000 Gent
Belgium
sofie.vermote@ugent.be

Cyriel WAGEMANS
University of Gent
Proeftuinstraat 86
B-9000 Gent
Belgium
cyrillus.wagemans@ugent.be

Kun WANG
Shanghai Institute of Applied Physics
2019 Jia-Luo Road
Shanghai
China
ygma@sinap.ac.cn

Valery ZAGREBAEV
Flerov Laboratory of Nuclear
Reactions
Joint Institute for Nuclear Research
Joliot-Curie 6
141980 Dubna, Moscow region
Russia
zagreb@jinr.ru

AUTHOR INDEX

A

B

C

D

E

F

G

H

I

J

K

L

M

N

O

W

Y

Z